MOBILE AGENTS IN NETWORKING AND DISTRIBUTED COMPUTING

Wiley Series in Agent Technology

Series Editor: Michael Wooldridge, *University of Liverpool, UK*

The 'Wiley Series in Agent Technology' is a series of comprehensive practical guides and cutting-edge research titles on new developments in agent technologies. The series focuses on all aspects of developing agent-based applications, drawing from the Internet, telecommunications, and Artificial Intelligence communities with a strong applications/technologies focus.

The books will provide timely, accurate and reliable information about the state of the art to researchers and developers in the Telecommunications and Computing sectors.

Titles in the series:

Padgham/Winikoff: Developing Intelligent Agent Systems 0-470-86120-7 (June 2004)

Bellifemine/Caire/Greenwood: Developing Multi-Agent Systems with JADE 0-470-05747-5 (February 2007)

Bordini/Hübner/Wooldrige: Programming Multi-Agent Systems in Agent-Speak using Jason 0-470-02900-5 (October 2007)

Nishida: Conversational Informatics: An Engineering Approach 0-470-02699-5 (November 2007)

Jokinen: Constructive Dialogue Modelling: Speech Interaction and Rational Agents 0-470-06026-3 (April 2009)

Castelfranchi/Falcone: Trust Theory: A Socio-Cognitive and Computational Model 0-470-02875-0 (March 2010)

Cao/Das: Mobile Agents in Networking and Distributed Computing 0-471-7516-0 (January or July? 2012)

MOBILE AGENTS IN NETWORKING AND DISTRIBUTED COMPUTING

Edited by

Jiannong Cao

Sajal K. Das

A JOHN WILEY & SONS, INC., PUBLICATION

Library of Congress Cataloging-in-Publication Data:

Cao, Jiannong.
   Mobile agents in networking and distributed computing / Jiannong Cao, Sajal K. Das.
     p. cm—(Wiley series in agent technology ; 3)
   ISBN 978-0-471-75160-1 (hardback)
    1. Mobile agents (Computer software) 2. Electronic data processing—Distributed
processing. I. Das, Sajal K. II. Title.
QA76.76.I58C36 2012
006.3—dc23

2011017017

10 9 8 7 6 5 4 3 2 1

"To my wife, Miao Yan, for her tolerance, support and caring"
—Jiannong Cao

"To my professors—Late A. K. Choudhury (Calcutta University), Lalit M. Patnaik (Indian Institute of Science, Bangalore), and Narsingh Deo (University of Central Florida)—for their mentoring and showing the beauty of research."
—Sajal K. Das

CONTENTS

FOREWORD

I have worked in the area of artificial intelligence, and specifically on challenges in machine learning and data mining, for twenty years. Originally these challenges focused on theoretical and algorithmic issues. Eventually, I became interested in applying these ideas to complex, real-world problems. Applied AI and machine learning not only allows researchers like me to see tangible benefits of the work, but it also introduces new algorithmic and theoretical challenges that need to be tackled.

As AI algorithms scale, they no longer exist just in the virtual world but find use in the real world. The result is that intelligent agents not only need to focus on their own problems but need to interact with other agents. As this book discusses, these agents may be components of a single system. Alternatively, they may be independent agents that are cooperating in order to solve a larger problem or they may actually be competing for resources. The agents may be pieces of software or they could be physical beings such as humans or robots. An intelligent agent may automatically discover a clever way to negotiate with others and such an agent may even harness the capabilities of other agents to boost its own performance.

I met Sajal Das, one of the editors of this book, when we both worked at the University of Texas at Arlington. Sajal is an expert in mobile computing, wireless networks, pervasive and distributed systems and has written numerous books, conference and journal articles on this topic. Together, we decided to tackle one particularly ambitious application of our respective fields: designing a smart home. We designed our smart home to perceive the state of the residents and physical surroundings, to reason about the state and its relationship to the goal of the home, and to change the state of the home using actuators in a way that achieved the goal of the home. Such a smart home relies on many components at the physical and software levels that seamlessly share information and work together to meet the goals of the home. These components include sensors, controllers, interfaces, networks, databases, machine learning algorithms, and decision-theoretic reasoners.

As an AI researcher, I find that practical application of AI and machine learning techniques can at times be overshadowed by the hurdles we face in trying to facilitate interaction and cooperation of our agents with others. This is certainly true for smart homes. During the first year that we designed our MavHome smart home, the bulk of the effort went into designing middleware (based on agent technologies), communication methodologies, database support, and interfaces. Each of these components needed to be able to work in a distributed fashion and cooperate with the other components in a seamless manner. The next evolution of the smart home project, the CASAS smart home, made even more effective use of mobile agent technology as described in this book and so was able to be up and running with less design time and a smaller software and physical footprint.

The danger of designing a real-world application is that the infrastructure of the application can start to dominate the project. In the smart home example, the design of communication and cooperation strategies can take over the project and detract from our goal of designing a home with learning and reasoning, rather than support this goal. The ideas expressed and topics covered in this book are a valuable step in designing mobile agents. The emphases on agent cooperation and transparent cooperation facilitate the design of complex and multi-agent systems, while the discussion of routing, resource discovery, and distributed security offer potential enhancements to such systems.

I find the coverage of topics in this book timely and comprehensive. The twelve chapters of the book present state of the art research, design methodologies and applications of mobile agents in the areas of networking and distributed computing. These topics range from principles of applying mobile agents for distributed coordination and communication to advanced mobile agent models and algorithms to mobile agent security to important case studies with implementation and performance evaluation.

I believe that this book will be valuable for researchers and practitioners interested in intelligent agents and mobile computing. The book will provide descriptions of cutting-edge research in technology in mobile agents and distributed computing. It will also offer practical guidance for those who, like me, want to see their ideas span the gap from concept to real-world applications.

Diane J. Cook
Washington State University

Dr. Diane J. Cook is a Huie-Rogers Chair Professor in the School of Electrical Engineering and Computer Science at Washington State University. Dr. Cook received a B.S. degree in Math/Computer Science from Wheaton College in 1985, a M.S. degree in Computer Science from the University of Illinois in 1987, and a Ph.D. degree in Computer Science from the University of Illinois in 1990. Her research interests include artificial intelligence, machine learning, graph-based relational data mining, smart environments, and robotics. Dr. Cook is an IEEE Fellow.

PREFACE

A mobile agent is a specific form of mobile code and has the features of mobility, autonomy, adaptability, and collaboration. It provides a paradigm and a powerful tool for implementing various applications in a computer networking environment. Over the past decades, the mobile agent technology has attracted a lot of attention from researchers and practitioners, thus leading to the development of theories, algorithms, systems, and platforms. Mobile agents indeed provide a means to complement and enhance existing technology in various application areas, such as information retrieval, e-commerce, parallel/distributed processing, network management, distributed data mining, event detection, and data aggregation in wireless sensor networks, to name a few.

In this book we focus on networking and distributed computing applications, and investigate how mobile agents can be used to simplify their development and improve system performance. For example, a mobile agent can structure and coordinate applications running in a networking and distributed computing environment because the agent can reduce the number of times one site contacts another and also help filter out non-useful information, thus reducing the consumption of communication bandwidth. Taking advantage of being in a network site and interacting with the site locally, a mobile agent allows us to design algorithms that make use of up-to-date system state information for better decision making. Moreover, a group of cooperating mobile agents can work together for the purpose of exchanging information or engaging in cooperative task-oriented behaviors. Agents can also support mobile computing by carrying out tasks for a mobile user temporarily disconnected from the (wireless) network.

Criticisms about mobile agents in the past were mainly concerned with the performance and security issues. However, with the advent of computer networks, mobile devices, and system dependability over the last decade, it is promising now to revisit these challenges and develop sound solution methodologies. Recent development in emerging areas like cloud computing and social computing also provides new opportunities for exploring the mobile agent technology.

This book is intended as a reference for researchers and practitioners and industry professionals, as well as postgraduate and advanced undergraduate students studying distributed computing, wireless networking, and agent technologies. It provides a clear and concise presentation of major concepts, techniques, and results in designing and implementing mobile agents based on networking and distributed computing systems and applications. The book consists of 12 chapters divided into four parts: (i) introduction, (ii) principles of applying mobile agents, (iii) mobile agent based techniques and applications, and (iv) system design and evaluation.

We gratefully acknowledge all the authors for their excellent contributions. We also thank Wiley's editorial and production team – Diana Gialo, Simone Taylor, Christine Punzo, and particularly Shanmuga Priya – for their dedicated professional service. It has been a real pleasure to work with them. Finally, we thank our respective families for their tremendous support and cheerful tolerance of our many hours spent at work. We owe them this book.

Jiannong Cao, Hong Kong Polytechnic University
Sajal K. Das, The University of Texas at Arlington

CONTRIBUTORS

Nigel Bean, University of Adelaide

Paolo Bellavista, University of Bologna

Giacomo Cabri, University of Modena and Reggio Emilia

Jiannong Cao, Hong Kong Polytechnic University

Panos K. Chrysanthis, University of Pittsburgh

Antonio Corradi, University of Bologna

Andre Costa, University of Melbourne

Sajal K. Das, University of Texas at Arlington

Anurag Dasgupta, University of Iowa

Xinyu Feng, State Key Laboratory for Novel Software Technology at Nanjing University

Paolo Flocchini, University of Ottawa

Sukumar Ghosh, University of Iowa

Carlo Giannelli, University of Bologna

Jian Lu, State Key Laboratory for Novel Software Technology at Nanjing University

Evaggelia Pitoura, University of Ioannina

Raffaele Quitadamo, University of Modena and Reggio Emilia

George Samaras, University of Cyprus

Nicola Santoro, Carleton University

Luís Moura Silva, University of Coimbra

Yudong Sun, Oxford University

Xianbing Wang, National University of Singapore

Cheng-Zhong Xu, Wayne State University

Ping Yu, State Key Laboratory for Novel Software Technology at Nanjing University

PART I
Introduction

1 Mobile Agents and Applications in Networking and Distributed Computing

JIANNONG CAO

Department of Computing, Hong Kong Polytechnic University

SAJAL K. DAS

Department of Computer Science and Engineering, The University of Texas at Arlington, USA

1.1 INTRODUCTION

Agent technology has evolved from two research areas: artificial intelligence and distributed computing. The purpose of AI research is to use intelligent computing entities to simplify human operations. An agent is just a computer program targeting that purpose [1]. Distributed computing, on the other hand, allows a complex task to be better executed by cooperation of several distributed agents on interconnected computers. So, networking and distribution bring out the true flavor of software agent technology in terms of agent autonomy, coordination, reactivity, heterogeneity, brokerage, and mobility.

Mobile agents refer to self-contained and identifiable computer programs that can move over the network and act on behalf of the user or another entity [2]. They can execute at a host for a while before halting the execution and migrating to another host and resuming execution there. They are able to detect the environment and adapt dynamically to changes. Mobile agents are widely used for handling disconnected operations in distributed, mobile, and wireless networking environments [3–6]. Also, many applications, including network diagnostic, e-commerce, entertainment and broadcasting, intrusion detection, and home health care, are benefited from the use of mobile agents [7, 8].

Mobile Agents in Networking and Distributed Computing, First Edition.
Edited by Jiannong Cao and Sajal K. Das.
© 2012 John Wiley & Sons, Inc. Published 2012 by John Wiley & Sons, Inc.

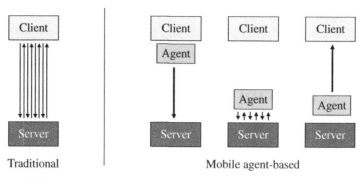

Traditional | Mobile agent-based

FIGURE 1.1   Mobile agent can reduce communication cost.

The term *mobile agent* contains two separate and distinct concepts: mobility and agency [9]. Some authors (e.g., [10]) classify a mobile agent as a special case of an agent, while others (e.g., [11]) separate the agency from mobility. Despite the differences in definition, most research on the mobile agent paradigm as reported in the literature has two general goals: reduction of network traffic and asynchronous interaction. Mobile agents can reduce the connecting time and bandwidth consumption by processing the data at the source and sending only the relevant results. By moving the agents to data-residing hosts, they can reduce communication costs. On the other hand, mobile agents support asynchronous interaction. They can continue computations even if the user that has started it, is no longer connected to the system. Mobile agents have been proposed as an alternative to the client–server paradigm which can be a more efficient and flexible mode of communication in certain application areas (Figure 1.1). It has been recognized that mobile agents provide a promising approach to dealing with dynamic, heterogeneous, and changing environments, which is tendency of modern Internet applications.

A mobile agent has the following properties or capabilities [12, 13]:

Mobility Transport itself from host to host within a network. This is the most distinguishing property from other kinds of agents. Note that a moving agent will carry its identity, execution state, and program code so that it can be authenticated and hence can resume its execution on the destination site after the move. Mobility refers to a wide range of new concepts. *Migration* is undoubtedly the most important of these concepts. Migration allows an agent to move from one location to another. The migration of a mobile agent requires the agent system to support execution stopping, state collection, data serialization and transfer, data deserialization, and execution resuming. From this point of view, mobile agents strongly rely on mobile code technology, which will be described in detail later in this chapter.

Intelligence Interact with and learn from the environment and make decisions. A most advanced agent should be able to decide its action based on its knowledge and the information it gets en route, and thus be able to generate new knowledge from its experience.

Autonomy Take control over its own actions. An agent should be able to execute, move, and settle down independently without supervision even in long-term running.

Recursion Create child agents for subtasks if necessary. An important concept is agent *cloning*: The agent can clone itself, that is, create a new mobile agent that is a copy of the parent. A pure cloning operation implies that the cloned agent has the same behavior (code) and the same knowledge (data) as the parent agent. A postcloning operation can initialize specific values in the cloned agent, which starts its life cycle in the same execution environment as the parent. Its location can however be different from the parent's.

Asynchrony In a distributed computing environment, perform computation concurrently and possibly on different sites. Also, performing computation on behalf of its user, an agent is responsible for the task assigned by a user and allows the user to offer and/or obtain resources and services in order to finish the task. All these can be done asynchronously with the user's action.

Collaboration Cooperate and negotiate with other agents. Complicated tasks can be carried out by collaboration of a group of agents.

1.2 MOBILE AGENT PLATFORMS

A mobile agent platform (MAP) is a software package for the development and management of mobile agents. It is a distributed abstraction layer that provides the concepts and mechanisms for mobility and communication on the one hand, and security of the underlying system on the other hand. The platform gives the user all the basic tools needed for creating some applications based on the use of agents. It enables us to create, run, suspend, resume, deactivate, or reactivate local agents, to stop their execution, to make them communicate with each other, and to migrate them.

Some agent standards enable interoperability between agent platforms so that software agents can communicate and achieve their objectives according to standardized specifications. The most popular agent standards are *FIPA* and *OMG-MASIF* as discussed below.

1.2.1 FIPA

The *Foundation for Intelligent Physical Agents* (FIPA) was formed in 1996 to produce software standards for heterogeneous interacting agents and agent-based systems. Currently, FIPA appears to be the dominant standards organization in the area of agent technology. Important efforts have been made to address the interoperability issues between the agent platforms. Figure 1.2

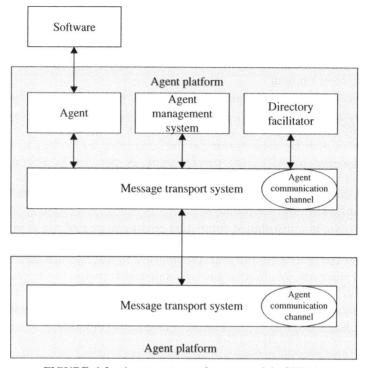

FIGURE 1.2   Agent system reference model of FIPA.

presents the overall architecture of an agent system as specified by FIPA. The message transport is the main underlying mechanism devoted to communication between agents based on Agent Communication Language (ACL); at this stage, mobile agents are not supported. The message transport itself relies on standard communication techniques used by distributed system frameworks, such as Common Object Request Broker Architecture (CORBA) or *Java* remote method invocation (RMI).

Both *Agent Management System* (AMS) and *Directory Facilitator* (DF) are FIPA agents: the AMS is responsible for the core management activities of the agent platform whereas the DF acts as a *yellow page* service. Agents are registered in the DF and can be localized from their types by other agents. In addition, agent communication is ensured through the *Message Transport System* (MTS), including the *Message Transport Protocol* (MTP) and the *Agent Communication Channel* (ACC), which directly provide agents with specific services for communication. The ACC may access information provided by the other agent platform services such as the AMS and DF to carry out its message transport tasks.

1.2.2   OMG-MASIF

In 1997, the Object Management Group (OMG) released a draft version of the *Mobile Agent System Interoperability Facilities* (MASIF) [14]. MASIF

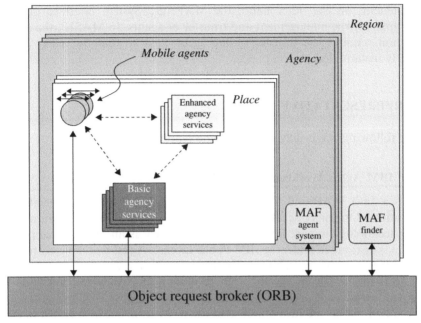

FIGURE 1.3 General architecture of OMG-MASIF mobile agent system.

proposes a specification of the communication infrastructure as well as interfaces defined in an interface definition language (IDL) to access mobility services in order to promote the interoperability and diversity of MAP. From the interoperability and heterogeneity perspectives, OMG follows the same objectives as FIPA. The objectives in terms of requirements and functionalities are clearly different, however. Whereas FIPA is concerned with a messagebased communication infrastructure, MASIF has to take into account the migration of the agent and must consequently focus on the way to dynamically *create the agent*, that is, to instantiate a new object at the right place and with the right class.

In Figure 1.3, the MASIF architecture appears to be a hierarchical organization of regions, agencies, and places [15]. The *place* is a context within an agent system in which an agent can execute its tasks and provide local access control to mobile agents. A place is associated with a location which consists of the place name and the address of the agent system within which the place resides. The *agency* represents the agent system itself or is the core part of the agent system. At a higher level, the *region* is a set of agent systems that have the same authority but are not necessarily of the same type.

Considering its origin, MASIF strongly relies on a CORBA architecture and therefore on the ORB. The services provided by the region, agency, and place are defined through IDL interfaces; the most important interfaces are MAFFinder and MAFAgentSystem: the MAFFinder supports the localization of agents, agent systems and places in the scope of a region or in the whole environment; on the other hand, the MAFAgentSystem interface provides operations for the management and transfer of agents. In MASIF, the agent's migration requires the transfer of the agent class so that the agent can be properly instantiated.

1.3   REPRESENTATIVE MAPs

In the following, we briefly describe some representative MAPs.

1.3.1   IBM Aglets Workbench (1997–2001)

This is a Java MAP and library that eases the development of agent-based applications. Originally developed at the IBM Tokyo Research Laboratory, the Aglets technology is now hosted at sourceforge.net as an open-source project, where it is distributed under the IBM Public License. Aglets is completely made in Java, granting a high portability of both the agents and the platform. The aglet represents the next leap forward in the evolution of executable content on the Internet, introducing program code that can be transported along with state information. Aglets are Java objects that can move from one host on the Internet to another. That is, an aglet that executes on one host can suddenly halt execution, dispatch itself to a remote host, and resume execution there. When the aglet moves, it takes along its program code as well as its data.

1.3.2   Agent Tcl (1994–2002, later known as D'Agents)

This does not formally specify a mobile agent model. Instead, a mobile agent is understood as a program that can be written in any language and that accesses features that support mobility via a common service package implemented as a server. This server provides mobile agent–specific services such as state capture, transfer facility, and group communication as well as more traditional services such as disk access, screen access, and CPU cycle. The philosophy was that all functionalities an agent ever wants are available in the server. Agent mobility then only concerns closure, which is the Tcl script (or scripts). There are no additional codes to load (i.e., no external references). In Agent Tcl, the state capture of an agent is handled automatically and transparently to the programmer. However, it is unclear what this state capture includes. Since Tcl is a script language, a frequent example given is that the executing script resumes after the instruction for mobility has been executed. There is also a plan to introduce process migration–like behavior such that the states of the agent would continue to evolve as it moves from place to place. However, this trend could have adverse effects in areas such as the complexity of the transfer mechanism and cost, adverse effects that are still being dealt with in the more traditional process migration.

1.3.3 Grasshopper (1998)

The agent development platform launched by IKV++ in August 1998, enables the user to create a wealth of applications based on agent technology. This platform is completely implemented in Java, a programming language that has become widely known among programmers, giving them the opportunity to work with Grasshopper without intensive further training. Companies with an urgent need for true distributed systems can therefore benefit almost immediately from the advantages of Java as well as from Grasshopper's unique suitability for such systems. Grasshopper is also the first mobile agent environment that is compliant to the industry standard supporting agent mobility and management (OMG-MASIF). This compliance ensures compatibility with other agent environments or applications based on the same standard, thus avoiding costly and time-consuming integration procedures. From Grasshopper version 1.2 released in 1999, it is also compliant with the specifications of the FIPA standards. Grasshopper can be used in many different application contexts, telecommunications being one of the most prominent application areas.

1.3.4 Concordia (1997)

This is another mobile agent framework built on Java. In Concordia an agent is regarded as a collection of Java objects. A Concordia agent is modeled as a Java program that uses services provided by a collection of server components that would take care of mobility, persistence, security, communication, administration, and resources. These server components would communicate among themselves and can run in one or several Java virtual machines; the collection of these components forms the agent execution environment (AEE) at a given network node. Once arriving at a node, the Concordia agent accesses regular services available to all Java-based programs such as database access, file system, and graphics, as in Aglet. A Concordia agent is considered to have internal states as well as external task states. The internal states are values of the objects' variables, while the external task states are the states of an itinerary object that would be kept external to the agent's code. This itinerary object encapsulates the destination addresses of each Concordia agent and the method that each would have to execute when arriving there. The designers of Concordia claim that this approach allows greater flexibility by offering multiple points of entry to agent execution, as compared to always executing an "after-move" method as in Agent Tcl, or Aglet. This concept of an externally located itinerary is similarly supported in Odyssey via task object. However, the infrastructure for management of these itinerary objects is not clear from the publicly available literature on Concordia which has support for transactional multiagent applications and knowledge discovery for collaborating agents.

1.3.5   In Mole (1997)

The agent is modeled as a cluster of Java objects, a closure without external references except with the host system. The agent is thus a transitive closure over all the objects to which the main agent object contains a reference. This island concept was chosen by the designers of Mole to allow simple transfer of agents without worrying about dangling references. Each Mole agent has a unique name provided by the agent system which is used to identify the agent. Also, a Mole agent can only communicate with other agents via defined communication mechanisms which offer the ability to use different agent programming languages to convert the information transparently when needed. A Mole agent can only exist in a host environment call *location* that serves as the intermediate layer between the agent and the operating system. Mole also supports the concept of *abstract location* to represent the collection of distributed physical machines. One machine can contain several locations, and locations may be moved among machines. Mole limits the abstract location to denote a configuration that would minimize cost due to communication. Thus, a collection of machines in a subnet is an acceptable abstract location, whereas a collection of machines that spans cities is not. Mole proposed the concept of a system agent which has full access to the host facilities. It is through interacting with these system agents that a given Mole agent (mobile) achieves tasks. A Mole mobile agent can only communicate with other agents (systems and mobile agents) and has no direct access to resources. The uniqueness of this agent model is its requirement for closure of objects, whereas other facilities such as static agent and communication are conceptually similar to other systems. What is unclear is how the Mole system enforces the closure requirement and whether there are mechanisms to handle closure management automatically. The concept of closure is technically convenient, but without helping tools it can be error prone and thus limiting.

1.3.6   The Odyssey

Project shares (or rather inherits) many features from a previous General Magic product: Telescript. However, the amount of open documentation on the Odyssey system is rather terse; therefore, its description is limited. The Odyssey mobile agent model also centers on a collection of Java objects, more similar in concept to Aglet than to Concordia or Mole. The top-level classes of the Odyssey system are *Agent, Worker,* and *Place.* Worker is a subclass of Agent and represents an example of what a developer can do with the Agent class. A Place class is an abstraction of where an Odyssey agent exists and performs work. A special facility such as directory service is associated with Place. Odyssey agents communicate using simple method calls, and do not support high-level communication. However, Odyssey agents can form and destroy meeting places to exchange messages. There is also an undocumented feature regarding global communication to a "published" object, but this feature is not officially supported. The distinctive feature of Odyssey is its design to accommodate multiple transport mechanisms. Currently, Odyssey supports Java RMI, Microsoft Distributed Component Object Model (DCOM), and CORBA Internet Inter-ORB Protocol (IIOP). However, the current release of Odyssey does not add new or distinctive features from its Telescript predecessor, and the mobile agent model is not yet stable.

1.4 SOME APPLICATIONS

All the above mobile agent platforms are targeted at providing execution environment and programming support for developing applications. Primitive language-level operations required by programmers for developing agent-based applications are identified. They are (1) *basic agent management functions*, such as creation, dispatching, cloning, and migration; (2) *agent-to-agent communication and synchronization functions*; (3) *agent monitoring and control functions*, such as status queries, recall, and termination of agents; (4) *fault tolerance functions,* such as check pointing, exception handling, and audit trails; and (5) *security-related functions*, such as encryption, authentication, signing, and data sealing.

As mentioned before, many mobile agent–based applications have been studied. Readers can find surveys on various types of applications [7, 8]. Based on earlier mobile agent platforms, several new platforms have been developed to meet the requirements of newly emerging computing technologies and applications, including mobile computing, ad hoc networking, and ubiquitous or pervasive computing [6, 16–20].

For distributed and network computing, mobile agent technology has been used to design both system functions and applications. This book includes excellent tutorial and advanced materials that cover a wide range of topics. Here, we just describe one of the typical mobile agent applications that can help reduce network communication cost. The mobile agent is particularly attractive as a promising technology for information retrieval in large-scale distributed systems like the Internet. The mobile agent acts as task-specific executable code traveling the relevant information source nodes to retrieve data. Several approaches have been proposed with both experimental and analytical evaluations [21–23].

More recently, the mobile agent has been used in designing dynamic and ad hoc systems. It enables the system to have the ability to deal with the uncertainty in a dynamic environment. For example, works have been reported [24, 25] on using mobile agents for monitoring, traffic detection, and management in highly dynamic distributed systems. Other examples include using mobile agents for ad hoc networks [6, 26].

Mobile agents are also being used for developing applications for wireless sensor networks (WSNs). Various operations and system functions in WSNs can be designed and implemented using mobile agents, which can greatly reduce the communication cost, especially over low-bandwidth links. Efficient data dissemination and data fusion in sensor networks using mobile agents have been proposed [4, 27, 28]. Solutions to location tracking in sensor networks using mobile agents are also proposed [29]. Location tacking aims to monitor the roaming path of a moving object. There are two primary challenges: no central control mechanism and backbone network in such environment and the very limited wireless communication bandwidth. A mobile agent can assist in tracking such a mobile object by choosing to migrate in the sensor closest to the object. For programming support, mobile agent–based WSN middleware has been developed as a better foundation for rapidly developing flexible applications for WSNs [5, 30]. Also WSN-based structural health monitoring applications use mobile agent–based network middleware [31] to enhance flexibility and to reduce raw data transmission. Design of wireless sensor networks for structural health monitoring presents a number of challenges, such as adaptability and the limited communication bandwidth. In [31], an integrated wireless sensor network consisting of a mobile agent–based network middleware and distributed high computational power sensor nodes has been developed. The mobile agent middleware is built on a mobile agent system called Mobile-C that allows a sensor network to move computational programs to the data source. With mobile agent middleware, a sensor network is able to adopt newly developed diagnosis algorithms and make adjustments in response to operational or task changes.

1.5   OVERVIEW OF THE BOOK

As briefly described in the previous sections, there exists many applications to benefit from mobile agent technology such as e-commerce, information retrieval, process coordination, mobile computing, personal assistance, and network management. Still more and more applications are switching to use mobile agents due to their flexibility and adaptability. Also their abilities of asynchronous and autonomous execution make connectionless execution possible, which might be extremely valuable in the mobile computing context. By moving computation to data rather than data to computation, mobile agents can also reduce the flow of raw data in the network and therefore overcome network latency, which is especially critical to real-time applications. Additionally, other distinguishable features, such as fault tolerance, natural heterogeneity, and protocol encapsulation enhance the utilization and application horizon of the mobile agent technology over traditional approaches.

This book focuses on cutting-edge research and applications of mobile agent technology in the areas of networking and distributed computing. The book is divided into four parts: (1) introduction, (2) principles of applying mobile agents to networking and distributed computing, (3) mobile agents techniques as applied to networking and distributed computing, and (4) design and evaluation.

The first part introduces the idea of mobile agents and discusses their potential as an important tool in networking and distributed computing.

The second part will show how to apply mobile agents to networking and distributed computing. In this part, we cover mobile agent communication, coordination, and cooperations as well as mobile agent security mechanisms. Agents must communicate with each other in order to solve problems together. *Communication* has been viewed between agents as planned actions that are not aimed at changing the environment; rather the aim is to change the beliefs and intentions of the agent to whom the message is sent. This implies that social agents should have a framework with which to analyze each other's behavior. Mobile agent *coordination* is mainly required for distributed programs consisting of a team of cooperating agents, where each agent is responsible for performing part of a common, global task. Teams of mobile agents are likely to become the means to implement several distributed and networked applications in the future. For example, one possible application is the search for some information in the network to be performed in parallel by a group of agents that will not visit the same host more than once. *Cooperation* between a collection of mobile agents is required for exchanging information or for engaging in cooperative task-oriented behaviors. In addition to the advantages of the mobile agent, using a cooperating mobile agent allows us to provide clear and useful abstractions in building network services through the separation of different concerns. Furthermore, mobile agents can be used to perform intrusion detection. With mobile agent technology, the collection nodes, internal aggregation nodes, and command and control nodes do not have to continuously reside on the same physical machine. For example, a mobile agent may function as an aggregation node and move to whatever physical location in the network is best for its purposes.

The third part of the book, describes in detail the techniques of mobile agents in networking. Especially we discuss the applications of agents in network routing, resource and service discovery, distributed control, distributed databases and transaction processing, and wireless and mobile computing. Mobile agents can have interesting applications at the network infrastructure layer. Agents can adapt the network infrastructure to changing needs over time and can facilitate network routing. Mobile agents can also dynamically discover resources they need to accomplish their tasks. When an agent arrives at a site, it should be able to discover the services offered at that site or things it could do. Distributed control using mobile agents is a useful approach for load balancing, deadlock detection, mutual exclusion, and so on. Characteristics of mobile agents make them useful in achieving load balance in the whole system. We also present some distributed algorithms using mobile agent systems for mutual exclusion, deadlock detection, consensus, and so on. Using transactions for managing large data collections will guarantee the consistency of data records when multiple users or processes perform concurrent operations on them. Owing to the heterogeneous and autonomous environment that the mobile agents operate in and their typical longevity, agent-based transactions have specific requirements. We discuss those requirements and possible recovery mechanisms. With the advent of mobile wireless communications and the growth of mobile computing devices, such as laptop computers, personal digital assistants (PDAs), and cell/smart phones, there is a growing demand for mobile agent–based mobile computing middleware and the mobile agent platforms for wireless hand-held devices and pervasive computing [32].

In the final part, we will discuss the means of measuring performances of mobile agent systems during the development of agent code, such as capturing the overhead of local agent creation, point-to-point messaging, and overhead for agent roaming. We can a keep track of the execution-related performances of mobile agents, such as the migration performance.

REFERENCES

1. A. Lingnau, O. Drobnik, and P. Domel, An HTTP-based infrastructure for mobile agents, *WWW J., Proceedings 4th International WWW Conference*, Vol. 1, Dec. 1995, pp. 461–471.

2. K. Rothermel and R. Popescu-Zeletin, Eds., Mobile agents, Lecture Notes in Computer Science, 1219, Springer, 1997.

3. J. Cao, Y. Sun, X. Wang, and S. K. Das, Scalable load balancing on distributed web servers using mobile agents, *J. Parallel Distrib. Comput.*, 63(10):996–1005, Oct. 2003.

4. M. Chen, S. Gonzalez, and V. C. M. Leung, Applications and design issues for mobile agents in wireless sensor network, *IEEE Wireless Commun.*, 14(6):20–26, Dec. 2007.

5. C.-L. Fok, G.-C. Roman, and C. Lu, Agilla: a mobile agent middleware for self-adaptive wireless sensor networks, *ACM Trans. Auton. Adapt. Syst.*, 4(3), July 2009.

6. J. Park, H. Yong, and E. Lee, A mobile agent platform for supporting ad hoc network environment, *Int. J. Grid Distrib. Comput.*, 1(1), 2008.

7. D. Milojicic, Mobile agent applications, *IEEE Concurrency*, July–Sept. 1999.

8. A. Outtagarts, Mobile agent-based applications: A survey, *IJCSNS Int. J. Comput. Sci. Network Security*, 9(11), Nov. 2009.

9. V. A. Pham and A. Karmouch, Mobile software agents: An overview, *IEEE Commun. Mag.*, 36(7):26–37, July 1998.

10. H. S. Nwana and N. Azarmi, Eds., Software agents and soft computing: Towards enhancing machine intelligence, Lecture Notes AI Series, 1198, Springer, 1997.

11. J. Vitek and C. Tschudin, Eds., Mobile object systems: Towards the programmable internet, Lecture Notes in Computer Science, 1222, Springer, 1997.

12. J. White, Prospectus for an open simple agent transfer protocol. White paper, General Magic, online, 1996.

13. A. Piszcz, A brief overview of software agent technology. White paper, The MITRE Corporation, McLean, VA, 1998.

14. GMD Fokus, Mobile agent system interoperability facilities specification, OMG TC Document orbos/97-10-05, Nov. 1997. (OMG homepage - www.omg.org)

15. C. Baumer, M. Breugst, S. Choy, and T. Magedanz, Grasshoper: a universal agent platform based on OMG MASIF and FIPA Standards, http://www.ikv.de/products/grasshopper.html.

16. J. Cao, D. C. K. Tse, A. T. S. Chan, PDAgent: A platform for developing and deploying mobile agent-enabled applications for wireless devices, in *Proceedings of 2004 International Conference on Parallel Processing (ICPP'2004)*, Montreal, Quebec, Canada, Aug. 2004, pp. 510–517.

17. F. Bagci, J. Petzold, W. Trumler, and T. Ungerer, Ubiquitous mobile agent system in a P2P-Network, paper presented at the UbiSys-Workshop at the Fifth Annual Conference on Ubiquitous Computing, Seattle, WA, Oct. 12–15, 2003.

18. M. Kumar, B. Shirazi, S. K. Das, B. Sung, D. Levine, and M. Singhal, PICO: A middleware framework for pervasive computing, *IEEE Pervasive Comput.*, 2(3):72–79, July–Sept. 2003.

19. J. R. Kim and J. D. Huh, Context-aware services platform supporting mobile agents for ubiquitous home network, in *Proceedings of the 8th International Conference on Advanced Communication Technology (ICACT'2006)*, Phoenix Park, Gangwon-Do, Korea, February 20–22, 2006.

20. G. S. Kim, J. Kim, H.-j. Cho, W.-t. Lim, and Y. I. Eom, Development of a lightweight middleware technologies supporting mobile agents, Lecture Notes in Computer Science, Vol. 4078, 2009.

21. S. Pears, J. Xu, C. Boldyreff, Mobile agent fault tolerance for information retrieval applications: An exception handling approach, *in Proceedings of the 6th International Symposium on Autonomous Decentralized Systems (ISADS'03)*, 2003.

22. W. Qu, M. Kitsaregawa, and K. Li, Performance analysis on mobile-agent based parallel information retrieval approaches, *in Proceedings of 2007 IEEE International Conference on Parallel and Distributed Systems*, Dec. 5–7, 2007.

23. W. Qu, W. Zhou, and M. Kitsaregawa, An parallel information retrieval method for e-commerce, *Int. J. Comut. Syst. Sci. Eng.*, 5:29–37, 2009.

24. B. Chen, H. H. Cheng, and J. Pelen, Integrating mobile agent technology with multi-agent systems for distributed traffic detection and management, *Transport. Res. Part-C.*, 17, 2009.

25. J. Ahn, Fault tolerant mobile-agent based monitoring mechanism for highly dynamic distributed networks, *Int. J. Comput. Sci. Iss.*, 7(3):1–7, May 2010.

26. G. Stoian, Improvement of handoff in mobile WiMAX network using mobile agent, *in Latest Trends in Computers*, Vol. 1, WSEAS Press, 2010, pp. 300–305.

27. Q. Hairong, S. Iyengar, and K. Chakrabarty, Multiresolution data integration using mobile agents in distributed sensor networks, *IEEE Trans. Syst. Man Cybernet.*, 31(3):383–391, August 2001.

28. Q. Wu, N. S. V. Rao, and J. Barhen, On computing mobile agent routes for data fusion in distributed sensor networks, *IEEE Trans. Knowledge Data Eng.*, 16(6):740–753, June 2004.

29. Y.-C. Tseng, S.-P. Kuo, H.-W. Lee, and C.-F. Huang, Location tracking in a wireless sensor network by mobile agents and its data fusion strategies, *Comput. J.*, 47(4):448–460, July 2004.

30. C.-L. Fok, G.-C. Roman, and C. Lu, Mobile agent middleware for sensor networks: An application case study, *Proceedings of the 4th International Symposium on Information Processing in Sensor Networks (IPSN)*, Los Angeles, CA, 2005, pp. 382–287.

31. B. Chen and W. Liu, Mobile agent computing paradigm for building a flexible structural health monitoring sensor network, *Comput.-Aided Civil Infrastruct. Eng.*, 25(7):504–516, October 2010.

32. Y. Feng, J. Cao, I. Lau, Z. Ming, and J. Kee-Yin Ng, A component-level self-configuring personal agent platform for pervasive computing, *Int. J. Parallel, Emergent Distrib. Syst.* 26(3):223–238, June 2011.

PART II
Principles of Applying Mobile Agents

2 Mobile Agent Communications

JIAN LU and XINYU FENG

State Key Laboratory for Novel Software Technology at Nanjing University, Nanjing, Jiangsu, P.R. China

2.1 INTRODUCTION

Mobile agents are regarded as the future of distributed computing. They are promising to offer a unified and scalable framework for applications in widely distributed heterogeneous open networks, such as electronic commerce, parallel computing and information retrieval. Among essential features of mobile agents, communication is a fundamental ability that enables mobile agents to cooperate with each other by sharing and exchanging information and partial results and collectively making decisions.

Much work on agent communication languages, such as Knowledge Query and Manipulation Language (KQML) [1, 2] or Foundation for Intelligent Physical Agents (FIPA) Agent Communication Language (ACL) [3], has been proposed. However, in this chapter, we do not introduce the common semantic layer for knowledge sharing. Instead we focus our discussion on the underlying transportation layer of interagent communication and are concerned solely with the delivery of opaque application data to a target agent, which is closer to the tradition of research on distributed systems.

Although process communication has been a cliché in distributed systems research, the presence of mobility raises a number of new challenges in designing message delivery protocols for effective and efficient communications between mobile agents. In designing such a protocol, two fundamental issues must be addressed: (1) tracking the location of the target mobile agent and (2) delivering the message reliably to the agent. To solve these two problems, a wide range of message delivery protocols have been proposed for agent tracking and reliable message delivery. Some representative research works are introduced and discussed in this chapter.

Mobile Agents in Networking and Distributed Computing, First Edition.
Edited by Jiannong Cao and Sajal K. Das.
© 2012 John Wiley & Sons, Inc. Published 2012 by John Wiley & Sons, Inc.

The rest of this chapter is organized as follows. Section 2.2 argues the importance of remote communications between mobile agents. Section 2.3 presents an analysis of the requirements of message delivery protocols for mobile agents. Section 2.4 describes several schemes for communication between mobile agents. Section 2.5 introduces the mailbox-based framework and describes some of its meaningful special cases. The final section provides some concluding remarks and some recent works for future reading.

2.2   IMPORTANCE OF REMOTE COMMUNICATION BETWEEN MOBILE AGENTS

The typical use of a mobile agent paradigm is for bypassing a communication link and exploiting local access to resources on a remote server. Thus one could argue that, all in all, communication with a remote agent is not important and a mobile agent platform should focus instead on the communication mechanisms that are exploited locally, that is, to get access to the server or to communicate with the agents that are colocated on the same site. Many mobile agent systems provide mechanisms for local communication, using some sort of meeting abstraction as initially proposed by Telescript [4], event notification for group communication [5, 6], or, more recently, tuple spaces [7, 8].

Nevertheless, as we will argue, remote interagent communication is also a fundamental facility in mobile agent platforms. Its necessity can be shown from the following aspects:

1. The mobile computing paradigm makes good compensation to traditional distributed computing, which is based on message passing or remote procedure call (RPC), but it cannot completely replace the traditional computing mode. Although the mobility of agents has the potential benefits of reducing network traffic and overcoming network latency, these benefits are obtained at the expense of transmitting the state and code of the mobile agent across the network. If the network traffic caused by agent migration is larger than the cost of sending requirements to remote services and receiving the results from the remote server, message passing will be more efficient in terms of network traffic than agent mobility. Experiments [9–11] have shown that in many scenarios the most efficient way is to combine message passing with agent mobility.

2. Mobile agent platforms can be used as general-purpose distributed computing middleware, which combine mobility with message passing naturally. Since mobile agents are generally computer and transport layer independent (dependent on only their execution environments), mobile agent technology can also be used to deploy distributed systems easily over a heterogeneous network, in which an agent can be an encapsulated component of the system and not necessarily mobile (called a stationary agent). From this point of view, message passing between remote agents, the most popular communication mode in traditional distributed

computing, should be an indispensable mechanism in mobile agent platforms.

3. In many applications cooperating mobile agents need to exchange information and partial results, collectively making decisions while migrating around the network. For instance, when using mobile agents for information retrieval over the network, it is efficient for cooperating agents to share the partial results while each of them searches their subarea, so that the search space can be considerably reduced [12]. In another example, a mobile agent could visit a site and perform a check on a given condition. If the condition is not satisfied, the agent could register an event listener with the site. This way, while the mobile agent is visiting other sites and before reporting its results, it could receive notifications of state changes in the sites it has already visited and decide whether a second visit is warranted. Another application can be find in [13, 14].

2.3 REQUIREMENTS ANALYSIS OF COMMUNICATION BETWEEN MOBILE AGENTS

In this chapter, we choose message passing as the communication mechanism we adapt to mobility, because it is a basic and well-understood form of communication in a distributed system. This incurs no loss of generality because more complex mechanisms such as RPC and method invocation are easily built on top of message passing.

Although the interprocess message has been a cliché in distributed systems research, agent mobility raises a number of new challenges in designing message delivery mechanisms for effective and efficient communications between mobile agents.

1. **Location Transparency** Since a mobile agent has its autonomy to move from host to host, it is unreasonable, if not impossible, to require that agents have a priori knowledge about their communication peers' locations before they send messages. Therefore, the first requirement of a practical mobile agent communication protocol is to allow mobile agents to communicate in a location-transparent way, that is, an agent can send messages to other agents without knowing where they reside physically. The message delivery protocol, therefore, is required to keep track of the location of mobile agents.

2. **Reliability** A desirable requirement for any communication mechanism is reliability. Programming primitives that guarantee that the data sent effectively reach the communication target, without requiring further actions by the programmer, simplify greatly the development task and lead to applications that are more robust.

In traditional distributed systems, reliability is typically achieved by providing some degree of tolerance to faults in the underlying communication link or in the communicating nodes. However, fault tolerance techniques are not sufficient to ensure reliability in systems that exhibit mobility. Because mobile agents are typically allowed to move freely from one host to another according to some a priori unknown pattern, it is difficult to ensure that the data effectively reach the mobile agent before it moves again. If this condition is not guaranteed, data loss may occur. Thus, the challenge to reliable communication persists even under the assumption of an ideal transport mechanism, which itself guarantees only the correct delivery of data from host to host despite the presence of faults. It is the sheer presence of mobility, and not possibility of faults, that undermines reliability.

3. **Efficiency** The cost of a protocol is characterized by the number of messages sent, size of the messages, and distance traveled by the messages. An efficient protocol should attempt to minimize all these quantities. More specifically, a protocol should efficiently support two operations: "migration," which facilitates the move of an agent to a new site, and "delivery," which locates a specified agent and delivers a message to it. The objective of minimizing the overhead of these two operations results in conflicting requirements [15]. To illustrate this trade-off, consider the two extreme strategies, namely full-information strategy, in which every host in the network maintains complete up-to-date information about the whereabouts of every agent, and the no-information strategy, which does not require any update of information for mobile agents during migration. Clearly, the former strategy makes the delivery operation cheap, but the migration operation becomes very expensive because it is necessary to update the location information at every host. With the latter strategy, the migration operation has a zero cost but the delivery operation has a very high overhead because it requires searching over the whole network. In general, a protocol should perform well for any or some specific communication and migration pattern, achieving a balance of the trade-off between the costs of migration and delivery.

4. **Asynchrony** Here, the term *asynchrony* includes two aspects of meanings: asynchronous migration and asynchronous execution. The former means that the agent can freely migrate to other hosts whenever necessary. Although coordination of message forwarding and agent migration is necessary to guarantee reliable message delivery, agent mobility should not be overconstrained by frequent and tight synchronization. The latter means the agent is independent of the process that created it and the agent home can be disconnected as soon as the agent is transferred. The mobile agent's asynchronous execution should not be restricted by heavily relying on the agent home for locating the agent and delivery of every message to the agent. In one word, since asynchrony is regarded as an important advantage of the mobile agent paradigm [16, 17], it is desirable that the protocol can keep the asynchrony of both migration and execution so that little or no offset of the merits of mobile agent technology will be introduced.

5. **Adaptability** Different applications may have different requirements and thus different emphasis on the above issues. In some applications, asynchrony is favored and thus the agent home should not be relied as the sole location server. In other applications, reliability is more important so synchronization is

needed. Different interagent communication and agent migration patterns may also have different implications on the update and search cost. Although protocols can be designed for specific applications to achieve optimal performance, it is desirable to have an adaptive protocol in a general-purpose mobile agent system, which can suit as many kinds of applications as possible.

These requirements can be served as a guideline for the design of the mobile agent communication protocols, which will be presented in the following sections.

2.4 SEVERAL SCHEMES FOR COMMUNICATION BETWEEN MOBILE AGENTS

As analyzed above, the message delivery protocol for mobile agents should satisfy the requirements of location transparency, reliability, efficiency, asynchrony, and adaptability. Location transparency and reliability are two basic requirements of an effective protocol. More specifically, to satisfy these two requirements, the message delivery protocol should be able to:

1. Identify communicating agents in a globally unique fashion. The agent ID should not change whenever the agent migrates to other hosts.
2. Map the ID of the receiver agent to its current address. To deliver messages to an agent, the underlying transport layer must know the current address of the receiver. Since the agent ID does not contain the location information of the agent, the mobile agent platform should support agent-tracking mechanisms which map the agent ID to its current location.
3. Deliver the message reliably to its target agent. This process can be done either in parallel with agents tracking or in a second phase after the address has been received. In both cases, the message delivery scheme should overcome the message loss or chasing problem caused by agent migration.

In what follows, we provide a review of related work on these three design issues. Work to meet other requirements, such as efficiency and adaptability, is also surveyed.

2.4.1 Naming Scheme

There are two basic requirements of the naming scheme of mobile agents:

1. Since a mobile agent can migrate from one host to another, the agent should be identified in a globally unique fashion.
2. To let the mobile agents communicate in a location-transparent way, the agent ID should remain unchanged during its life cycle, even if the physical address of the agent has changed.

The usual way to identify a mobile agent is to append the name of the agent's origin host (i.e., agent home) with its title (a free-form string used to refer to this agent) [18, 19]. The name of the agent home can be either its Internet Protocal (IP) address or its uniform resource locator (URL). In both cases no two hosts should have the same name. Thus it is impossible for agents born at different agent platforms to have the same ID. For agents created at the same host, the host is responsible for managing the name space to ensure that each agent created in it has a unique title. Since the name of the agent home and the title of the agent will not be affected by the physical location of the agent, the ID of the agent will remain unchanged during its life cycle (we assume that the name of the agent home will not change during the agent's life cycle).

We adopt this naming scheme in this chapter.

2.4.2 Tracking Mechanisms

The task of the tracking mechanisms is to obtain the current location of mobile agents. With the presence of mobility in distributed systems, many mobile unit tracking schemes have been proposed in the last several years in different contexts, including mobile agents, mobile and wireless communications, and wide-area distributed systems. According to the organization of location servers, the major schemes can be categorized into central server, forwarding pointers, broadcast, and hierarchical location directory.

1. **Central-Server Schemes** The central-server scheme makes use of a location server to keep track of the physical location of a mobile object. There are several variations. For example, Mobile IP [20], which is designed for routing IP packets to mobile hosts, uses the home server scheme, where a mobile host registers its care-of-address with its home agent every time it moves. All the IP packets to the host are sent to the home agent, which forwards the packets to the host. This scheme is also used in IS-41 [21] for personal communication service (PCS) as well as in mobile agent systems [6, 22] and distributed systems.

The central-server scheme is simple to implement and has less communication overhead for locating a mobile object. However, it incurs large overhead of updating the locations and delivering messages. The server can be a bottleneck of performance if the number of mobile objects is growing and communication and migration are frequent. It can also be a single point of failure. The scheme does not support locality of mobile object migration and communication, that is, migration and communication involve the cost of contacting the server, which can be far away. This is the well-known triangle routing problem [23]. Cache-based strategies [24, 25] are proposed to avoid the triangle routing problem. If a cache miss occurs, the server is contacted for a new location. In the Internet Mobile Host Protocol (IMHP) [23], packets are forwarded along the forwarding address left by the mobile host if a cache "miss" occurs. All these protocols do not handle message loss caused by mobility.

2. **Forwarding-Pointer Schemes** In the *forwarding-pointer*-based schemes for tracking mobile objects [26–28], each host on the migration path of an object keeps a forwarding pointer pointing to the next host on the path. Each sender knows the home of the target object. Messages are sent to the agent home and forwarded to the target object along the forwarding pointers.

The forwarding-pointer scheme is also easy to implement and incurs no location registration overhead. However, the scheme cannot guarantee message delivery because a message may follow a mobile object, which frequently migrates, leading to a race condition. Furthermore, it is not practical for a large number of migrations to distinct hosts (a chain of pointers is growing, increasing the cost of message delivery). Some path compression methods can be used to collapse the chain, for example, movement based and search based. In the former case the mobile object would send backward a location update after performing a number of migrations, in the latter case after receiving a number of messages (i.e., after a number of message delivery operations occur). For instance, a search-based path compression technique is proposed in [26]. After a message is routed to the target object along the chain, an Update_Entry message is sent back along the chain and forwarding pointers kept in the nodes of the chain are updated. A similar algorithm has been used in Emerald [29], where the new forwarding address is piggybacked onto the reply message in the object invocation.

3. **Broadcast Schemes** There are three variants of the broadcast scheme, that is, query broadcast, data broadcast, and notification broadcast. The first two are proposed from the perspective of the message sender. In the query broadcast scheme, the message sender sends a query message to all the hosts in the system for the location of the receiver. After receiving the response from the host at which the receiver resides, the sender sends the message to the location obtained from the response. In the data broadcast scheme the sender broadcasts the message directly to all the hosts in the system. The third is proposed from the perspective of the receiver. After migration the mobile agent broadcasts its new location to all the hosts in the system.

Broadcast schemes have less reliance on the agent home for agent tracking or message forwarding; thus they can maintain the disconnected operation ability of mobile agents. They can be implemented in the local Internet domain or the local Ethernet. Broadcasts can be accomplished efficiently in bus-based multiprocessor systems. They are also used in radio networks. However, because of the large communication overhead, it is impractical to broadcast in large-scale networks [30].

4. **Hierarchical Schemes** In the hierarchical schemes, a treelike hierarchy of servers forms a location directory [similar to a domain name system (DNS)]. Each region corresponds to a subtree in the directory. For each agent there is a unique path of forwarding pointers that starts from the root and ends at the leaf that knows the actual address of the agent. Messages to agents are forwarded along this path. These kinds of schemes and their variants have been

used to track mobile users [15, 30, 31], objects [32], and agents [18, 33]. Krishna et al. [31] explored different update and search strategies that can be used in the hierarchical scheme.

The hierarchical scheme scales better than forwarding pointers and central servers. It supports locality of mobile object migration and communication. However, the hierarchy is not always easy to construct, especially in the Internet environment. The hierarchical scheme itself cannot guarantee message delivery. Messages might also chase their recipients under this scheme.

Readers are referred to [34, 35] for excellent surveys of the above techniques.

2.4.3   Efforts on Reliable Message Routing

Tracking mechanisms can map the agent's location-independent ID to its current location. However, they are not sufficient to guarantee message delivery. As discussed above, even though the sender knows the current location of the receiver, the receiver may migrate to other hosts during message transmission. Various approaches have been proposed to overcome message loss caused by migration of recipients.

1. **Forwarding-Pointer Scheme** The idea is the same with the one introduced in Section 2.4.2. Before migration, the mobile object leaves a pointer in its current host pointing to the target host. When a message is sent to an obsolete address of the recipient (this address can be obtained by any of the tracking schemes introduced in Section 2.2), the message is routed along the forwarding pointer. The forwarding-pointer scheme is often used in combination with address caching. A case in point is IMHP [23].

Although messages can be routed along the forwarding pointer, there is not an upper bound of the number of hops a message takes before it reaches the recipient. If the recipient migrates frequently, the message may keep chasing the recipient and could not be received until the death of the recipient. Therefore, the forwarding pointer can only partially overcome the message loss caused by agent mobility and cannot guarantee reliable message delivery (message routing).

2. **Resending-Based Scheme** To implement reliable message delivery for mobile objects, resending-based Transmission Control Protocol (TCP)–like protocols [36–38] are proposed. If a message is missed because of the migration of the recipient, the sender can detect the message loss and resend the message to the new address of the recipient. Using TCP-like slide window mechanism, these protocols can not only overcome message loss caused by both migration of recipients and faults of the network but also maintain the first in–first out (FIFO) order of message delivery. However, as in the forwarding-based scheme, when the recipient migrates frequently, there is no upper bound of the number of message resending. Therefore, it cannot satisfy our requirement of reliability.

3. **Broadcast** If the sender maintains an obsolete address of the recipient and the message sent to that address could not be delivered to the recipient, the message will be broadcasted to all the hosts in the system. This idea is similar to the data broadcast mentioned in Section 2.4.2, but it is used only when the

communication failure occurs because of the recipient's migration. In Emerald [29] broadcast is used to find an object if a node specified by a forwarding pointer is unreachable or has stale data. According to Murphy [39], however, the simple broadcast cannot avoid message loss caused by object mobility. Murphy proposed a snapshot-based broadcast scheme to guarantee reliable delivery of messages to highly mobile agents. The protocol can also be extended for group communication for mobile agents.

   4. **Synchronization between Message Passing and Target Migration** From the perspective of concurrency control, the message loss or chasing problem in the message delivery process is caused by concurrent and asynchronous access to the location information of the target agent. The mobile agent migration and the message delivery processes can be regarded as two kinds of database operations. The migration of the target agent changes its actual address, which can be regarded as a "write" operation of the location information. The message delivery process needs the target agent's actual address, which in fact is a "read" operation of the location information. Strategies are proposed to synchronize the message passing and target migration so that messages can reach the target agent within a bounded number of hops.

   One widely used synchronization strategy is implemented as follows. Before migration the mobile agent informs all the hosts (usually the home of the agent or a central message-forwarding server) that might send messages to its current address and waits for an ACK from each host (containing the number of messages sent from the host). It then waits for these messages due to arrive. After migration it tells these hosts it has finished moving. During migrations (after sending the ACK) the host suspends message forwarding. Variations of the strategy are proposed. For instance, if FIFO message order is maintained in the underlying transport layer, the ACK message does not need to contain the number of messages sent from the host and the agent can leave for the target host as soon as it has collected all the ACK messages. In the Mogent system [19], a synchronous home server–based protocol is proposed to track mobile agents and guarantee message delivery.

   The synchronization scheme can guarantee that messages be routed to its target agent with a bounded number of hops. However, the agent has to wait for all the ACK messages from message-forwarding servers. If there are multiple servers that might forward messages to the agent, the constraint on the mobile agent migration is prohibitive. In this thesis we also use this kind of synchronization scheme to realize reliable message delivery. However, using a mailbox-based scheme, only the migration of the agent's mailbox is constrained. The agent can move freely about the network.

2.4.4 Adaptive Protocols

To suit for different mobility patterns, many adaptive algorithms have been proposed in the field of personal communication, including timer-based, movement-based, distance-based, and state-based location-updating algorithms [40, 41]. In these algorithms, mobile users decide whether to update their location

information according to different factors. To optimize the location management cost on a per-user basis, a selective location update strategy for PCS users is
proposed [42, 43]. When a mobile user enters a new location area (LA), it can
choose whether to update its location information or not. According to its own
mobility model and call arrival pattern, each mobile user has its update strategy
$Su = \{ui\}$, consisting of a set of binary decision variables ui (to update or not to
update) for all LAs.

In [44] a tracking agent is used for location tracking and message forwarding
for cooperating agents. It is dynamically generated when cooperation starts
and is killed after the cooperation is finished. The coordinates of the center of
the cooperating agents are set to the average coordinates of each agent. If the
distance between the center and the tracking agent is large enough, the tracking
agent will migrate to the center so that the communication latency between
cooperating agents can be decreased. The authors, however, did not discuss
how a new agent could find an existing tracking agent in order to join the
cooperation by sharing the tracking agent with others.

2.5   MAILBOX-BASED FRAMEWORK FOR DESIGNING MOBILE AGENT MESSAGE DELIVERY PROTOCOLS

The design of a mobile agent (MA) message delivery protocol mainly addresses
two issues: (1) tracking the location of the target mobile agent and (2) delivering
the message to the agent. The framework [45] discussed in this section is based
on the concept of a mailbox associated with a mobile agent. Its flexibility and
adaptability come from the decoupling between a mobile agent and its mailbox,
allowing for the separation between the above two different issues. We discuss
the design space within the mailbox-based framework and identify the relevant
parameters and various protocols that can be derived as special cases.

2.5.1   System Model and Assumptions

Each mobile agent has a mailbox which buffers the messages sent to it. As
shown in Figure 2.1, if an agent wants to send a message to another agent,

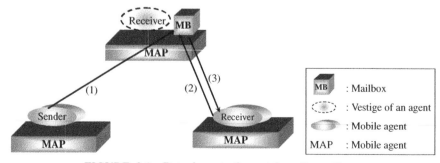

FIGURE 2.1   Detachment of agent from its mailbox.

it simply sends the message to the receiver's mailbox (step 1). Later the receiver receives the message from its mailbox using either pull or push (steps 2 and 3).

The mailbox is logically one part of the agent, but it can be detached from its owner in the sense that the agent can migrate to a new host while leaving its mailbox at a previous host along its migration path. Thus the communication between agents is divided into two steps: (i) the transmission of a message from the sender to the receiver's mailbox and (ii) the delivery of the message from the mailbox to its owner agent. Since the mailbox is also a mobile object (we do not call it another mobile agent dedicated to message delivery because it has no autonomy to decide its migration), step (i) is identical to the inter—mobile agent communication; thus it can be realized by any existing message delivery strategies. Notice, however, that for a frequently migrating agent its mailbox can migrate at a much lower frequency. When to migrate the mailbox is a parameter of protocol design. The second step, that is, the delivery of the message from the mailbox to its owner agent, raises new issues to be discussed in detail later.

The term migration path is used in the following discussion to denote the list of hosts that a mobile agent or its mailbox has visited in sequence. For instance, as shown in Figure 2.2, a mobile agent is migrating through hosts $h_1, h_2, \ldots ,$ h_5 sequentially. It takes its mailbox while moving to h_1, h_3, and h_5. We say h_1, $h_2, \ldots , h_5$ are all on the migration path of the agent, while h_1, h_3, and h_5 are on the migration path of the mailbox. By definition, we know that the set of hosts on the migration path of the mailbox is a subset of those on the migration path of the mailbox's owner agent. The home of the mobile agent, that is, the origin host of the agent, is the first host on the migration paths of both the agent and its mailbox.

We assume that mobile agent communication is largely asynchronous. This is reasonable because, with mobile agents roaming the Internet, it is rare that two agents use synchronous communication to talk to each other. The large and unpredicted message delays on the Internet, which can easily become on the order of several seconds, also prohibit frequent use of synchronous communication in a mobile agent application. We also assume that our framework

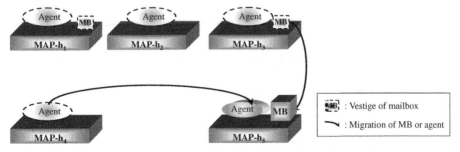

FIGURE 2.2   Migration paths of mobile agent and its mailbox.

is built on top of a reliable network communication layer, which guarantees that messages will not be lost during transmission and will be delivered between hosts. Finally, we do not deal with host failures and assume that no message is lost due to host crash.

2.5.2 Three-Dimensional Design Model

In our framework, choices can be made in three aspects of designing a protocol that best suits the specific requirement of an application: mailbox migration frequency, mailbox-to-agent message delivery, and synchronization of message forwarding with object migration.

1. *Frequency of Mailbox Migration* By frequency of mailbox migration, we mean the number of mailbox migrations during the life cycle of the agent and the time when these migrations happen. The migration frequency of the agent's mailbox can be categorized as follows:

   - *No Migration (NM)* In this case, the mobile agent moves alone and never takes its mailbox. The mailbox is left at the agent home during the agent's life cycle. All the messages are sent to the agent's home and the agent obtains messages from its home using one of the mailbox-to-agent message delivery modes described below. The cost for tracking the mailbox is zero, but the message delivery cost is high because all the messages must be forwarded by the agent's home. The triangle routing [23] increases the communication overhead.

   - *Full Migration (FM)* The mailbox is part of the data of the mobile agent and migrates with the agent all the time. The cost of message delivery between the mailbox and the agent is zero, but it is difficult to track the mailbox. If the agent (and the mailbox) migrates frequently, there is a trade-off between the number of messages that could be lost and how much communication overhead will be introduced to guarantee message delivery.

   - *Jump Migration (JM)* Between the above two extreme cases, the mobile agent determines whether to take its mailbox dynamically before each migration. To make the decision, an agent can consider such factors as the number of messages it will receive at its target host and the distance between the target host and the host where its mailbox currently resides. If an agent seldom receives messages from others at its target host, it does not need to take its mailbox to the new host. On the other hand, if an agent expects to receive messages frequently from others and its target host is far away from the host where its mailbox currently resides, it will be expensive to leave the mailbox unmoved and to fetch messages from the remote mailbox. In this case the agent should migrate to the target host together with its mailbox. Under the jump migration mode, the protocol can work more flexibly based on a decision that best suits particular agent migration and the interagent communication pattern, reducing the cost of both "tracking" and "delivery" operations.

2. *Mailbox-to-Agent Message Delivery* As mentioned before, messages destined to an agent are all sent to the agent's mailbox and the agent receives the messages later by either a push or a pull operation.

   • *Push (PS)* The mailbox keeps the address of its owner agent and forwards every message to it. In this way real-time message delivery can be implemented and the message query cost is avoided. However, the agent must notify the mailbox of its current location after every migration. If the agent migrates frequently but communicates with other agents only at a small number of hosts on the migration path, most of the location registration messages (for the purpose of message delivery) would be superfluous and introduce large migration overhead.

   • *Pull (PL)* The agent keeps the address of its mailbox and retrieves messages from the mailbox whenever needed. The mailbox does not need to know the agent's current location, and therefore the location registration is avoided. On the other hand, the agent has to query its mailbox for messages. The polling messages would increase the message delivery overhead. Moreover, in the pull mode the message may not be processed in real time.

3. *Synchronization of Message Forwarding and Agent/Mailbox Migration* With the help of the proposed framework, users can choose whether or not they need reliable message delivery. If higher reliability is required, we increase the degree of synchronization in order to overcome message loss. The synchronization is performed either for coordinating the message forwarding by the host and the migration of the destination mailbox (denoted by SHM) or for coordinating the message forwarding from an agent's mailbox and the migration of the agent (denoted by SMA), or both (called full synchronization and denoted by FS). We use NS to denote the extreme case where no synchronization is performed.

The synchronization between the message-forwarding object (mobile agent server or the mailbox) and the moving object (mailbox or mobile agent) can be realized in the following way. Before migration, the moving object sends Deregister messages to all objects that might forward messages to it and waits for the ACK message from each object (containing the number of messages forwarded from the object). It then waits for these messages due to arrive. After migration it informs of its arrival to all the message-forwarding objects by sending them Register messages. The state change of the moving object is shown in Figure 2.3. Messages can be forwarded to the mobile object when in *stationary* and *waiting* states and must be blocked when it is in the *moving* state.

The above three aspects can be used to develop a three-dimensional model, as shown in Figure 2.4. Each aspect represents one dimension in the model, showing a spectrum of different degrees of constraints for that

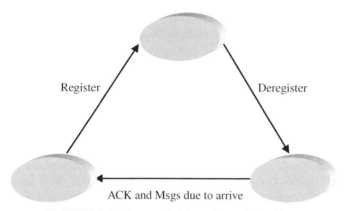

FIGURE 2.3 State switching of a mobile object.

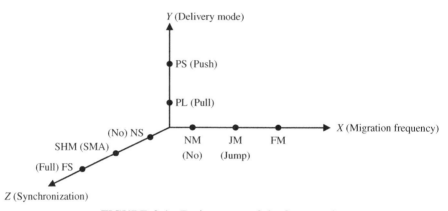

FIGURE 2.4 Design space of the framework.

dimension. The three dimensions are orthogonal. That is, each aspect may be discussed independently of one another, and a property in one dimension can logically have various combinations with the properties in the other dimensions. For different applications with different requirements in the three aspects, the required degree of properties can be different. Message delivery protocols can be described by combining parameters in the orthogonal dimensions.

2.5.3 Parameter Combinations

The three-dimensional model introduces a taxonomy of mobile agent message delivery protocols. In this section, we describe a classification of these protocols according to different parameter combinations. A string of the format XX-YY-ZZ is used to express a protocol, where XX stands for NM, JM, or FM; YY for PL or PS; and ZZ for NS, SHM, SMA, or FS. The overall configuration of a

TABLE 2.1   Parameter Combinations and Corresponding Protocols

Protocols	Location Registration	Reliability
NM-PS-NS	Yes (agent → mailbox)	No
NM-PS-SMA	Yes (agent → mailbox)	Yes
NM-PL-NS	No	Yes
FM-*-NS	No	No
JM-PL-NS	No	No
JM-PS-NS	Yes (agent → mailbox)	No
FM-*-SHM	Yes (mailbox → host)	Yes
JM-PL-SHM	Yes (mailbox → host)	Yes
JM-PS-FS	Yes (agent → mailbox, mailbox → host)	Yes

protocol has a special value for each of the three parameters. Most combinations have plausible applications. However, brevity precludes a discussion of the full range of protocols that can be derived, and we study here only those combinations with the most popular features. Table 2.1 shows the different protocols derived from our mailbox-based framework, with the description of their location registration modes and whether they can satisfy the required reliability. An asterisk in a string denotes a "don't care" state where multiple values are applicable.

1. *Home Server–Based Protocols* All the protocols under the NM mode adopt a home server approach. In this case the agent's home acts as the message-forwarding server.

The NM-PS-NS protocol is identical to the Mobile IP used in mobile computing [20]. The agent registers its current location with its mailbox residing at its home. Messages are sent to the mailbox. The mailbox pushes messages to its owner agent. This protocol does not guarantee message delivery. If the agent migrates during the message forwarding, the message will be lost.

To ensure reliable message delivery, synchronization between agent migration and message forwarding from the mailbox (SMA) is needed. This produces the NM-PS-SMA protocol, a synchronized version of Mobile IP. In the NM-PL-NS protocol the agent pulls messages from its mailbox; therefore the message delivery can be guaranteed without using synchronization.

The home server–based protocols are simple and work well for small-to-medium systems where the number of agents is small. However, the triangle routing will increase the communication overhead, especially when the range of agent distribution is large. In a system with a large number of agents and frequent interagent communication, the home may become a performance bottleneck and a single point of failure. Furthermore, a mobile agent's ability for asynchronous execution is constrained because of the dependence of the agent home as a message-forwarding server.

2. *Forwarding-Pointer-Based Delivery Protocols* The FM-*-NS protocol adopts the forwarding-pointer strategy. Each host on the mailbox migration

path keeps a forwarding pointer to the successive host in the path. The sender caches the location of the target mailbox obtained before. If there is no such address in the cache, the sender uses the home of the receiver agent as the cached address of the mailbox. Messages are sent to the cached address directly. If a cache miss occurs, messages will be forwarded along the forwarding pointers. When the mailbox receives the message and finds that the sender has outdated knowledge of its address, it notifies the sender its current location and the sender updates the cached address. Since the mailbox is bound with its owner agent in the FM mode, there is no remote interaction between the mailbox and the agent.

The JM-PL-NS and JM-PS-NS protocols are similar to the FM-*-NS protocol except that the mailbox migrates in the JM mode, which can be regarded as a kind of path compression technique. When the agent and its mailbox are at different hosts, the pull and push modes are used respectively by the agent to get messages from its mailbox.

There is no location update cost in the forwarding-pointer scheme. The sender sends messages to the cached address of the target agent; therefore the workload of the agent home is decreased. Even if the cache is outdated, messages can still be routed to the target agent along the path. But if one host on the migration path fails, the target agent can no longer be reached. The most serious problem with this scheme is that it cannot guarantee the reliability of message delivery because many messages may keep chasing the target agent if the agent migrates frequently.

In the FM-*-NS protocol, the multihop path could degrade the communication performance significantly. In both JM-PL-NS and JM-PS-NS protocols, the mailbox of an agent migrates less frequently than the agent. Thus the message-forwarding path is shorter and the communication overhead is reduced. Moreover, the chasing problem may be less probable to occur.

3. *Distributed Registration-Based Protocols* Synchronization is used in the distributed registration-based schemes to guarantee reliable message delivery. Before migrating, the *mailbox* informs all the hosts on its migration path and waits for ACK messages from them. After arriving at the target host, the mailbox registers its new address to hosts on the migration path.

The sender sends messages to its cached address, say h_k. If the mailbox has moved away, h_k (apparently h_k is on the migration path of the mailbox) forwards the messages to the current address of the mailbox and notifies the sender of the new address. This scheme is similar to the synchronized home server scheme (NM-PS-SMA), but the role of the agent home is distributed to all the hosts on the migration path. It can also be regarded as a forwarding-pointer scheme with a migration-based path compression technique, that is, the agent updates all the pointers on its migration path after one or several migrations.

Using the FM-*-SHM protocol, the overhead for synchronization and location registration would be unaffordable if the agent migrates frequently.

But if the mailbox migrates in the JM mode, both the time of mailbox registration and the number of hosts on the mailbox's migration path are reduced. In the JM-PL-SHM protocol, the pull mode is used by the agent to obtain messages from its mailbox. Synchronization between message sending from hosts and migration of the mailbox (SHM) is necessary to guarantee message delivery. In the JM-PS-FS protocol, the mailbox pushes each incoming message to its owner agent. Therefore, in addition to SHM, synchronization between message pushing from the mailbox and the migration of the owner agent (SMA) is also needed to achieve reliable message delivery. That is why FS is used in the protocol.

The JM-PL-SHM protocol is described in detail in [46], which shows that the synchronization scheme is effective to guarantee reliable message delivery. Messages are forwarded at most once to reach the receiver's mailbox and no chasing problem exists. The impact of the mailbox migration frequency on the performance of the protocol is also analyzed. We observed that by properly deciding on the migration frequency of the mailbox the protocol could be designed to achieve better trade-off balance between the costs of migration and delivery.

2.6 CONCLUDING REMARKS AND FURTHER RESEARCH

Communication is an essential ability for mobile agents to collaborate with others by information exchanging and knowledge sharing; however, mobility brings new challenges to interagent message passing. In this chaper, we first analyzed the necessity of remote message passing in mobile agent systems and presented requirements of message delivery protocols for mobile agents. We believe mobile agent message delivery protocols should satisfy the requirements of location transparency, reliability, asynchrony, efficiency, and adaptability.

After a comprehensive review of related work, a very general framework has been introduced in detail for designing message delivery protocols in mobile agent systems. The framework uses a flexible and adaptive mailbox-based scheme which associates each mobile agent with a mailbox while allowing the decoupling between them. With different combinations of mailbox migration frequency, the message delivery mode between the mailbox and the agent, and the synchronization mode between message forwarding and migration, the framework not only covers several previously known location management strategies, such as the home server scheme and forwarding-pointer scheme, but also provides the designer with the possibility to define protocols better suited for specific applications. It has the following advantages:

1. It can be used to describe and evaluate various mobile agent communication protocols.
2. It can help users to clearly specify their requirements.

3. It can help users design a flexible, adaptive protocol which can be customized to meet their requirements.

In recent years, some further research approaches have been proposed for communication between mobile agents. Some representative works are sketched as follows.

In Cao et al. [47], the relay communication model is abstracted and identified from existing algorithms and the two possible approaches, namely push and pull, are explored to design adaptive and reliable message delivery protocols. In this model, there are three roles involved, namely sender agents, relay stations, and receiver agents. Each receiver agent in turn has one or more relay stations, which can be its home server, proxy, or a mailbox. Communication between agents is divided into two steps: (1) transmission of a message from the sender to the receiver's relay station and (2) delivery of the message from the relay station to the receiver agent. To send the messages, the message sender will first obtain the address of the target agent's relay station and then send messages to it. Later, the receiver agent can obtain messages from its relay station. There exist two well-known approaches, namely, push and pull, for forwarding messages from the relay station to mobile agents. In the push mode, the relay station maintains the location of the mobile agent and forwards incoming messages to it. In the pull mode, on the other hand, the agent knows the address of its relay station and queries it periodically for messages. This study shows that push and pull modes have complementary properties in terms of agent mobility constraints, communication overhead, support of real-time message processing, and the relay station's resilience to failures and flexibility and concludes that specific applications can select different message delivery approaches to achieve the desired level of performance and flexibility.

Zhong et al. [48], on the other hand, argue that mailbox-based asynchronous persistent communication mechanisms are not sufficient for certain distributed applications like parallel computing. Synchronous transient communication is provided as complementary services that make cooperative agents work more closely and efficiently. Then, a connection migration mechanism in support of synchronous communication between agents is proposed. This reliable connection migration mechanism allows mobile objects in communication to remain connected during their migration. This mechanism supports concurrent migration of both endpoints of a connection and guarantees exactly-once delivery for all transmitted data.

Hsiao et al. [49] point out that communication between mobile agents often relies on an infrastructure of the discovery server to resolve the location of the agents and such an agent discovery infrastructure becomes very sophisticated while considering scalability and reliability. Then they propose a novel substrate called ARMADA that support reliable and scalable agent to agent communication without relying on an auxiliary discovery infrastructure. The ARMADA is based on a scalable and reliable peer-to-peer (P2P) storage overlay that implements distributed hash tables and takes advantage of the self-configuration and self-healing features provided by the underlying P2P overlay network to support communications among agents running on top of the overlay.

Ahn [50] presents three problems of the forwarding-pointer-based approach when it is applied to a large-scale mobile agent system. First, it may lead to a very high message delivery cost whenever each message is sent to a mobile agent. Second, it requires a large size of storage where agent location information is maintained. The third drawback is that even if, among all service nodes on a forwarding path of a mobile agent, only one fails, any message destined to the agent cannot be delivered to it. Then, a new fault-tolerant and efficient mobile agent communication mechanism based on forwarding pointers is proposed. This mechanism enables each mobile agent to keep its forwarding pointer only on a small number of its visiting nodes in an autonomous manner. Consequently, every mobile agent's migration route has been considerably shortened and the time for forwarding each message to the agent is much smaller.

REFERENCES

1. T. Finin et al., KQML as an agent communication language, in N. R. Adam, B. K. Bhargava, and Y.Yesha (Eds.), *Proceedings of the 3rd International Conference on Information and Knowledge Management (CIKM94)*, ACM, New York, NY, USA, Dec. 1994, pp. 456–463.

2. Y. Labrou and T. Finin, A proposal for a new KQML specification, Technical Report CS-97-03, Computer Science and Electrical Engineering Department, University of Maryland Baltimore County, Baltimore, MD 21250, Feb. 1997. Available at http://www.csee.umbc.edu/csee/research/kqml/papers/kqml97.pdf.

3. FIPA ACL, available: http://www.fipa.org.

4. J. E. White, Telescript technology: Mobile agents, in D. Milojicić, F. Douglis, and R. Wheeler (Eds.), *Mobility: Processes, Computers, and Agents*. ACM Press/Addison-Wesley, New York, NY, USA, 1999.

5. J. Baumann et al., Communication concepts for mobile agent systems, in K. Rothermel and R. Popescu-Zeletin (Eds.), *Mobile Agents: 1st International Workshop MA'97*, Lecture Notes in Computer Science, Vol. 1219, Springer, Apr. 1997, pp. 123–135.

6. D. B. Lange and M. Oshima, *Programming and Deploying Java Mobile Agents with Aglets*, Addison-Wesley, Reading, MA, 1998.

7. G. Cabri, L. Leonardi, and F. Zambonelli, Reactive tuple spaces for mobile agent coordination, in L. Rohermel and F. Hohl (Eds.), *Mobile Agents: 2nd International Workshop MA'98*, Lecture Notes in Computer Science, Vol. 1477, Springer, Sept. 1998, pp. 237–248.

8. G. P. Picco, A. L. Murphy, and G.-C. Roman, LIME: Linda meets mobility, in D. Garlan (Ed.), *Proceedings of the 21st International Conference on Software Engineering*, May 1999, ACM, New York, NY, USA, 1999, pp. 368–377.

9. T. Chia and S. Kannapan, Strategically mobile agents, in K. Rothermel and R. Popescu-Zeletin (Eds.), *Mobile Agents: 1st International Workshop MA'97*, Lecture Notes in Computer Science, Vol. 1219, Springer, Apr. 1997, pp. 149–161.

10. R. S. Gray et al., Mobile-Agent versus client/server performance: Scalability in an information-retrieval task, in G. P. Picco (Ed.), *Mobile Agents: 5th International Conference, MA 2001,* Lecture Notes in Computer Science, Vol. 2240, Dec. 2001, pp. 229–243.

11. M. Straβer and M. Schwehm, A performance model for mobile agent systems, in *Proceedings of International Conference on Parallel and Distributed Processing Techniques and Applications,* Vol. II, Las Vegas, CSREA, New York, July 1997, pp. 1132–1140.

12. J. Baumann and N. Radouniklis, Agent groups for mobile agent systems, in H. König, K. Geihs, and T. Preuss (Eds.), *Proceedings of the 1st International Working Conference on Distributed Applications and Interoperable Systems (DAIS'97),* Cottbus, Germany, Champman & Hall, London, Sept. 1997, pp. 74–85.

13. J. Cao, G. H. Chan, W. Jia, and T. Dillon, Checkpointing and rollback of wide-area distributed applications using mobile agents, in *Proceedings of IEEE 2001 International Parallel and Distributed Processing Symposium (IPDPS2001),* IEEE Computer Society Press, San Francisco, Apr. 2001, pp. 1–6.

14. T. K. Shih, Agent communication network—a mobile agent computation model for Internet applications, in *Proceedings of 1999 IEEE International Symposium on Computers and Communications.* Red Sea, Egypt, 1999, pp. 425–431.

15. B. Awerbuch, and D. Peleg, Online tracking of mobile users, *J. ACM,* 42(5): 1021–1058, Sept. 1995.

16. D. Chess, C. Harrison, and A. Kershenbaum, Mobile agents: Are they a good idea? in J. Vitek and C. Tschudin (Eds.), *Mobile Object Systems: Towards the Programmable Internet,* Lecture Notes in Computer Science, Vol. 1222, Springer, Feb. 1997, pp. 25–45.

17. D. B. Lange and M. Oshima, Seven good reasons for mobile agents, *Commun. ACM,* 42(3):88–89, Mar. 1999.

18. W. V. Belle, K. Verelst, and T. D'Hondt, Location transparent routing in mobile agent systems—Merging name lookups with routing, in *Proceedings of the 7th IEEE Workshop on Future Trends of Distributed Computing Systems,* Cape Town, South Africa, 1999, pp. 207–212.

19. X. Tao et al., Communication mechanism in Mogent system, *J. Software,* 11(8): 1060–1065, 2000.

20. C. E Perkins, IP mobility support, in *RFC2002,* Oct. 1996. Available at http://rfc-ref.org/RFC-TEXTS/2002/index.html.

21. M. D. Gallagher and R. A. Snyder, *Mobile Telecommunication Networking with IS-41,* McGraw-Hill, New York, 1997.

22. D. Milojicic et al., MASIF: The OMG mobile agent system interoperability facility, in L. Rohermel and F. Hohl (Eds.), *Mobile Agents: 2nd International Workshop MA'98,* Lecture Notes in Computer Science, Vol. 1477, Springer, Sept. 1998, pp. 50–67.

23. C. Perkins, A. Myles, and D. B. Johnson, IMHP: A mobile host protocol for the Internet, *Computer Networks and ISDN Systems,* 27(3):479–491, Dec. 1994.

24. Y. Lin, Determining the user locations for personal communications services networks, *IEEE Trans. Vehic. Technol.,* 43(3):466–473, Aug. 1994.

25. H. Harjono, R. Jain, and S. Mohan, Analysis and simulation of a cache-based auxiliary user location strategy for PCS, in *Proceedings of IEEE Conference on Networks for Personal Communication*, Long Branch, NJ, USA, Mar. 1994, pp. 1–5.

26. J. Desbiens, M. Lavoie, and F. Renaud, Communication and tracking infrastructure of a mobile agent system, in Hesham El-Rewini (Eds.), *Proceedings of the 31st Annual Hawaii International Conference on System Sciences*, Vol. 7, Kohala Coast, HI, USA, Jan. 1998, pp. 54–63.

27. L. Moreau, Distributed directory service and message routing for mobile agents, Science of Computer Programming, 39(2–3):249–272, Mar 2001.

28. Objectspace, Objectspace voyager core technology, available: http://www.object space.com.

29. E. Jul, H. Levy, N. Hutchinson, and A. Black, Fine-grained mobility in the emerald system, *ACM Trans. Computer Syst.*, 6(1):109–133, Feb. 1988.

30. K. Ratnam, I. Matta, and S. Rangarajan, A fully distributed location management scheme for large PCS networks, *J. Interconnection Networks*, 2(1):85–102, 2001.

31. P. Krishna, N. H. Vaidya, and D. K. Pradhan, Location management in distributed mobile environments, in *Proceedings of the 3rd International Conference on Parallel and Distributed Information Systems (PDIS)*, Austin TX, USA, Sep. 1994, pp. 81–88.

32. M. van Steen, F. Hauck, P. Homburg, and A. Tanenbaum, Locating objects in wide-area systems, *IEEE Commun. Mag.*, Jan. 1998, pp. 104–109.

33. S. Lazar, I. Weerakoon, and D. Sidhu, A scalable location tracking and message delivery scheme for mobile agents, in *Proceedings of the 7th IEEE International Workshops on Enabling Technologies: Infrastructure for Collaborative Enterprises* 1998 (*WET ICE'98*), Stanford, CA, USA, Jun 1998, pp. 243–248.

34. P. T. Wojciechowski, Algorithms for location-independent communication between mobile agents, Technical Report 2001/13, Operating Systems Laboratory, Swiss Federal Institute of Technology (EPFL), Mar. 2001. Available at: http://infoscience.epfl.ch/record/52380/files/IC_TECH_REPORT_200113.pdf.

35. E. Pitoura and G. Samaras, Locating objects in mobile computing, *IEEE Trans. Knowledge Data Eng.*, 13(4):571–592, 2001.

36. M. Ranganathan, M. Bednarek, and D. Montgomery, A reliable message delivery protocol for mobile agents, in D. Kotz and F. Mattern (Eds.), *Proceedings of ASA/MA2000, Linear Notes in Computer Science*, Vol. 1882, Springer, Sept. 2000, pp. 206–220.

37. T. Okoshi et al., MobileSocket: Session layer continuous operation support for java applications, *Trans. Inform. Process. Soc. Jpn.*, 1(1):1–13, 1999.

38. A. Bakre and B. R. Badrinath, I-TCP: Indirect TCP for mobile hosts, Technical Report DCS-TR-314, Rutgers University, New Brunswick, NJ, Oct. 1994.

39. A. Murphy and G. P. Picco, Reliable communication for highly mobile agents, in *Agent Systems and Architectures/Mobile Agents*, 5(1):81–100, Mar 2002.

40. A. Bar-Noy, I. Kessler, and M. Sidi, Mobile users: To update or not to update? *ACM/Baltzer Wireless Networks*, 1(2): 175–185, July 1995.

41. V. Wong and V. Leung, Location management for next generation personal communication networks, *IEEE Network*, 14(5):18–24, Sept./Oct. 2000.

42. S. K. Das and S. K. Sen, A new location update strategy for cellular networks and its implementation using a genetic algorithm, in *Proceedings of the 3rd ACM/IEEE Conference on Mobile Computing and Networking (MobiCom'97)*, Budapest, Hungary, ACM New York, NY, USA, Sept. 1997, pp. 185–194.

43. S. K. Sen, A. Bhattacharya, and S. K. Das, A selective location update strategy for PCS users, *ACM/Baltzer J. Wireless Networks*, special issue on selected Mobicom'97 papers, 5(5):311–326, Oct. 1999.

44. G. Kunito, Y. Okumura, K. Aizawa, and M. Hatori, Tracking agent: A new way of communication in a multi-agent environment, in *Proceedings of IEEE 6th International Conference on Universal Personal Communications*, Vol. 2, San Diego, CA, USA, Oct. 1997, pp. 903–907.

45. J. Cao, X. Feng, J. Lu, and S. Das, Mailbox-based scheme for mobile agent communications, *IEEE Computer* (IEEE Computer Society Publication), 35(9): 54–60, Sept. 2002.

46. X. Feng, J. Cao, J. Lu, and H. Chan, An efficient mailbox-based algorithm for message delivery in mobile agent systems, in G. P. Picco, (Ed.), *Mobile Agents: 5th International Conference, MA 2001*, Lecture Notes in Computer Science, Vol. 2240, Dec. 2001, pp. 135–151.

47. J. Cao, X. Feng, J. Lu, H. Chan, and S. Das, Reliable message delivery for mobile agents: push or pull?, *IEEE Trans. Syst. Man. Cybernet, Part A*, 34(5):577–587, Sept. 2004.

48. X. Zhong and C. Xu, A reliable connection migration mechanism for synchronnous transient communication in mobile codes, in *Proceedings of 2004 International Conference on Parallel Processing*, Montreal, Quebec, Canada, Aug. 15–18, 2004, Vol. 1. pp. 431–438.

49. H. Hsiao, P. Huang, C. King, and A. Banerjee, Taking advantage of the overlay geometrical structures for mobile agent communications, in *Proceedings of the 18th International Parallel & Distributed Processing Symposium*, Santa Fe, NM, Apr. 2004.

50. J. Ahn, Fault-tolerant and scalable communication mechanism for mobile agents, in C. Aykanat et al. (Eds.), *Proceedings of the 19th International Symposium on Computer and Information Science*, Kemer-Antalya, Turkey, Oct. 27–29, 2004, Lecture Notes in Computer Science, Vol. 3280, Springer, Berlin, Heidelberg 2004, pp. 533–542.

3 Distributed Security Algorithms for Mobile Agents

PAOLA FLOCCHINI

School of Electrical Engineering and Computer Science, University of Ottawa, Canada.

NICOLA SANTORO

School of Computer Science, Carleton University, Canada.

3.1 INTRODUCTION

Mobile agents have been extensively studied for several years by researchers in artificial intelligence and in software engineering. They offer a simple and natural way to describe distributed settings where mobility is inherent and an explicit and direct way to describe the entities of those settings, such as mobile code, software agents, viruses, robots, and Web crawlers. Further, they allow to immediately express notions such as selfish behavior, negotiation, and cooperation arising in the new computing environments. As a programming paradigm, they allow a new philosophy of protocol and software design, bound to have an impact as strong as that caused by object-oriented programming. As a computational paradigm, mobile agent systems are an immediate and natural extension of the traditional message-passing settings studied in distributed computing.

For these reasons, the use of mobile agents is becoming increasingly popular when computing in networked environments, ranging from the Internet to the data grid, both as a theoretical computational paradigm and as a system-supported programming platform.

In networked systems that support autonomous mobile agents, a main concern is how to develop efficient agent-based *system protocols*, that is, to design protocols that will allow a team of identical simple agents to cooperatively perform (possibly complex) system tasks. Examples of basic tasks are *wakeup, traversal, rendezvous,* and *election.* The coordination of the agents

Mobile Agents in Networking and Distributed Computing, First Edition.
Edited by Jiannong Cao and Sajal K. Das.

necessary to perform these tasks is not necessarily simple or easy to achieve. In fact, the computational problems related to these operations are definitely nontrivial, and a great deal of theoretical research is devoted to the study of conditions for the solvability of these problems and to the discovery of efficient algorithmic solutions [1–10].

At an abstract level, these environments can be described as a collection of autonomous mobile *agents* (or *robots*) located in a graph *G*. The agents have limited computing capabilities and private storage, can move from node to neighboring node, and perform computations at each node according to a predefined set of behavioral rules called *protocol,* the same for all agents. They are *asynchronous,* in the sense that every action they perform (computing, moving, etc.) takes a finite but otherwise unpredictable amount of time. Each node of the network, also called a *host,* may provide a storage area called *whiteboard* for incoming agents to communicate and compute, and its access is held in fair mutual exclusion. The research concern is on determining what tasks can be performed by such entities, under what conditions, and at what cost. In particular, a central question is to determine what minimal hypotheses allow a given problem to be solved.

At a practical level, in these environments, *security* is the most pressing concern and possibly the most difficult to address. Actually, even the most basic security issues, in spite of their practical urgency and the amount of effort, must still be effectively addressed [11–15].

Among the severe security threats faced in distributed mobile computing environments, two are particularly troublesome: *harmful agent* (that is, the presence of malicious mobile processes) and *harmful host* (that is, the presence at a network site of harmful stationary processes).

The former problem is particularly acute in unregulated noncooperative settings such as the Internet (e.g., e-mail-transmitted viruses). The latter exists not only in those settings but also in environments with regulated access and where agents cooperate toward common goals (e.g., sharing of resources or distribution of a computation on the grid). In fact, a local (hardware or software) failure might render a host harmful. In this chapter we consider security problems of both types and concentrate on two security problems, one for each type: *locating a black hole* and *capturing an intruder*. For each we discuss the computational issues and the algorithmic techniques and solutions.

We first focus (in Section 3.2) on the issue of *host attacks*, that is, the presence in a site of processes that harm incoming agents. A first step in solving such a problem should be to identify, if possible, the harmful host, that is, to determine and report its location; following this phase, a "rescue" activity would conceivably be initiated to deal with the destructive process resident there. The task to identify the harmful host is clearly dangerous for the searching agents and, depending on the nature of the harm, might be impossible to perform. We consider a highly harmful process that disposes of visiting agents upon their arrival, leaving no observable trace of such a destruction. Due to its nature, the site where such a process is located is called a *black hole*.

The task is to unambiguously determine and report the location of the black hole. The research concern is to determine under what conditions and at what cost mobile agents can successfully accomplish this task. The searching agents start from the same safe site and follow the same set of rules; the task is successfully completed if, within a finite time, at least one agent survives and knows the location of the black hole.

We then consider (in Section 3.3) the problem of *agent attacks*, that is, the presence of a harmful mobile agent in the system. In particular, we consider the presence of a *mobile virus* that infects any visited network site. A crucial task is clearly to decontaminate the infected network; this task is to be carried out by a team of antiviral system agents (the *cleaners*), able to decontaminate visited sites, avoiding any recontamination of decontaminated areas. This problem is equivalent to the one of capturing an intruder moving in the network.

Although the main focus of this chapter is on security, the topics and the techniques have a much wider theoretical scope and range. The problems themselves are related to long-investigated and well-established problems in automata theory, computational complexity, and graph theory. In particular, the *black-hole search* problem is related to the classical problems of *graph exploration* and *map construction* [1, 7, 9, 16–22]. With whiteboards, in the case of dispersed agents (i.e., when each starts from a different node), these problems are in turn computationally related (and sometimes equivalent) to the problems of *rendezvous* and *election* [2, 5, 6, 23–25]. The *network decontamination* problem is instead related to the classical problem known as *graph search* [e.g., 26–30], which is in turn closely related to standard graph parameters and concepts, including tree width, cut width, path width, and, last but not least, graph minors [e.g., 31–34].

The chapter is organized as follows. In the next section we will discuss the black-hole search problem, while the network decontamination and intruder capture problems will be the subject of Section 3.3.

3.2 BLACK-HOLE SEARCH

3.2.1 The Problem and Its Setting

The problem posed by the presence of a harmful host has been intensively studied from a programming point of view [35–37]. Obviously, the first step in any solution to such a problem must be to *identify*, if possible, the harmful host, that is, to determine and report its location; following this phase, a "rescue" activity would conceivably be initiated to deal with the destructive process resident there. Depending on the nature of the danger, the task to identify the harmful host might be difficult, if not impossible, to perform.

Consider the presence in the network of a black hole (BH): a host where resides a stationary process that *disposes* of visiting agents upon their arrival, leaving *no observable* trace of such a destruction. Note that this type of highly harmful host is not rare; for example, the undetectable crash failure of a site in an asynchronous network turns such a site into a black hole. The task is to unambiguously determine and report the location of the black hole by a team of mobile agents. More precisely, the black-hole search (shortly BHS) problem is solved if at least one agent survives and all surviving agents know the location of the black hole.

The research concern is to determine under what conditions and at what cost mobile agents can successfully accomplish this task. The main complexity measures for this problem are the *size* of the solution (i.e., the number of agents employed) and the *cost* (i.e., the number of moves performed by the agents executing a size-optimal solution protocol). Sometimes bounded *time* complexity is also considered.

The searching agents usually start from the same safe site (the *homebase*). In general, no assumptions are made on the time for an agent to move on a link, except that it is finite; that is, the system is asynchronous. Moreover, it is usually assumed that each node of the network provides a storage area called a whiteboard for incoming agents to communicate and compute, and its access is held in fair mutual exclusion.

One can easily see that the black-hole search problem can also be formulated as an *exploration* problem; in fact, the black hole can be located only after all the nodes of the network but one have been visited and are found to be safe. Clearly, in this exploration process some agents may disappear in the black hole). In other words, the black-hole search problem is the problem of exploring an unsafe graph. Before proceeding we will first (briefly) discuss the problem of *safe exploration*, that is, of exploring a graph without any black hole.

3.2.2 Background Problem: Safe Exploration

The problem of exploring and mapping an unknown but *safe* environment has been extensively studied due to its various applications in different areas (navigating a robot through a terrain containing obstacles, finding a path through a maze, or searching a network).

Most of the previous work on exploration of unknown graphs has been limited to single-agent exploration. Studies on exploration of *labeled* graphs typically emphasize minimizing the number of moves or the amount of memory used by the agent [1, 7, 17, 21, 22]. Exploration of *anonymous* graphs is possible only if the agents are allowed to mark the nodes in some way, except when the graph has no cycles (i.e., the graph is a tree [9, 18]). For exploring arbitrary anonymous graphs, various methods of marking nodes have been used by different authors. Pebbles that can be dropped on nodes have been proposed first in [16], where it is shown that any strongly connected directed graph can be explored using just one pebble (if the size of the graph is known), and using $O(\log \log n)$ pebbles otherwise. Distinct markers have been used, for example, in [38] to explore unlabeled undirected graphs. Yet another approach, used by Bender and Slonim [39] was to employ two cooperating agents, one of which would stand on a node, while the other explores new edges. Whiteboards have been used by Fraigniaud and Ilcinkas [19] for exploring directed graphs and by Fraigniaud et al. [18] for exploring trees. In [9, 19, 20] the authors focus on minimizing the amount of memory used by the agents for exploration (they however do not require the agents to construct a map of the graph).

There have been few results on exploration by more than one agent. A two-agent exploration algorithm for directed graphs was given in [39], whereas Fraigniaud et al. [18] showed how k agents can explore a tree. In both these cases, the agents start from the same node and they have distinct identities. In [6] a team of dispersed agents explores a graph and constructs a map. The graph is anonymous but the links are labeled with sense of direction; moreover the protocol works if the size n of the network or the number of agents k is coprime and it achieves a move complexity of $O(km)$ (where m is the number of edges). Another algorithm with the same complexity has been described in [23], where the requirement of sense of direction is dropped. In this case the agents need to know either n or k, which must be coprime. The solution has been made "effective" in [24], where effective means that it will always terminate, regardless of the relationship between n and k reporting a solution whenever the solution can be computed, and reporting a failure message when the solution cannot be computed.

The map construction problem is actually equivalent to some others basic problems, such as *agent election, labeling,* and *rendezvous.* Among them rendezvous is probably the most investigated; for a recent account see [2, 25].

3.2.3 Basic Properties and Tools for Black-Hole Search

We return now to the black-hole search problem and discuss first some basic properties and techniques.

3.2.3.1 Cautious Walk

We now describe a basic tool [40] that is heavily employed when searching for a black hole. In order to minimize the number of agents that can be lost in the black hole, the agents have to move *cautiously*. More precisely, we define as *cautious walk* a particular way of moving on the network that prevents two different agents to traverse the same link when this link potentially leads to the black hole.

At any time during the search for the black hole, the ports (corresponding to the incident links) of a node can be classified as *unexplored* (no agent has been sent/received via this port), *explored* (an agent has been received via this port), or *dangerous* (an agent has been sent through this port but no agent has been received from it). Clearly, an explored port does not lead to a black hole; on the other hand, both unexplored and dangerous ports might lead to it.

The main idea of cautious walk is to avoid sending an agent over a dangerous link while still achieving progress. This is accomplished using the following two rules:

1. No agent enters a dangerous link.
2. Whenever an agent a leaves a node u through an unexplored port p (transforming it into dangerous), upon its arrival to node v and before proceeding somewhere else, a returns to u (transforming that port into explored).

Similarly to the classification adopted for the ports, we classify nodes as follows: At the beginning, all nodes except the homebase are unexplored; the first time a node is visited by an agent, it becomes explored. Note that, by definition, the black hole never becomes explored. Explored nodes and edges are considered safe.

3.2.3.2 Basic Limitations
When considering the black-hole search problem, some constraints follow from the asynchrony of the agents (arising from the asynchrony of the system, that is, the impossibility to distinguish the BH from a slow node). For example [40]:

- If G has a cut vertex different from the homebase, then it is impossible for asynchronous agents to determine the location of the BH.
- It is impossible for asynchronous agents to determine the location of the BH if the size of G is not known.
- For asynchronous agents it is impossible to verify if there is a BH.

As a consequence, the network must be 2-connected; furthermore, the existence of the black hole and the size of G must be common knowledge to the agents.

As for the number of searching agents needed, since one agent may immediately wander into the black hole, we trivially have:

- At least two agents are needed to locate the BH.

How realistic is this bound? How many agents suffice? The answers vary depending on the a priori knowledge the agents have about the network and on the consistency of the local labelings.

3.2.4 Impact of Knowledge

3.2.4.1 Black-Hole Search without a Map
Consider first the situation of *topological ignorance*, that is, when the agents have no a priori knowledge of the topological structure of G (e.g., do not have a

map of the network). Then any generic solution needs at least $\Delta + 1$ agents, where Δ is the maximal degree of G, even if the agents know Δ and the number n of nodes of G.

The goal of a black-hole search algorithm $\mathcal{P}$ is to identify the location of the BH; that is, within finite time, at least one agent must terminate with a map of the entire graph where the homebase, the current position of the agent, and the location of the black hole are indicated. Note that termination with an exact map in finite time is actually impossible. In fact, since an agent is destroyed upon arriving to the BH, no surviving agent can discover the port numbers of the black hole. Hence, the map will have to miss such an information. More importantly, the agents are asynchronous and do not know the actual degree d (BH) of the black hole (just that it is at most Δ). Hence, if an agent has a local map that contains $N - 1$ vertices and at most Δ unexplored edges, it cannot distinguish between the case when all unexplored ports lead to the black hole and the case when some of them are connected to each other; this ambiguity cannot be resolved in finite time or without the agents being destroyed. In other words, if we require termination within finite time, an agent might incorrectly label some links as incident to the BH; however, the agent needs to be wrong only on at most $\Delta - d(\text{BH})$ links. Hence, from a solution algorithm $\mathcal{P}$ we require termination by the surviving agents within finite time and creation of a map with just that level of accuracy.

Interestingly, in any *minimal* generic solution (i.e., using the minimum number of agents), the agents must perform $\Omega(n^2)$ moves in the worst case [41]. Both these bounds are *tight*. In fact, there is a protocol that correctly locates the black hole in $O(n^2)$ moves using $\Delta + 1$ agents that know Δ and n [41].

The algorithm essentially performs a collective "cautious" exploration of the graph until all nodes but one are considered to be safe. More precisely, the agents cooperatively visit the graph by "expanding" all nodes until the black hole is localized, where the expansion of a node consists of visiting all its neighbors. During this process, the homebase is used as the cooperation center; the agents must pass by it after finishing the expansion of a node and before starting a new expansion. Since the graph is simple, two agents exploring the links incident to a node are sufficient to eventually make that node "expanded." Thus, in the algorithm, at most two agents cooperatively expand a node; when an agent discovers that the node is expanded, it goes back to the homebase before starting to look for a new node to expand. The whiteboard on the homebase is used to store information about the nodes that already have been explored and the ones that are under exploration. If the black hole is a node with maximum degree, there is nothing to prevent Δ agents disappearing in it.

3.2.4.2 *Black-Hole Search with Sense of Direction*

Consider next the case of topological ignorance in systems where there is *sense of direction* (SD); informally, sense of direction is a labeling of the ports that allows the nodes to determine whether two paths starting from a node lead to the same node using only the labels of the ports along these paths (for a survey on sense of direction see [42]). In this case, two agents suffice to locate the black hole, regardless of the (unknown) topological structure of G. The proof of [41] is constructive, and the algorithm has a $O(n^2)$ cost. This cost is optimal; in fact, it is shown that there are types of sense of direction that, if present, impose an $\Omega(n^2)$ worst-case cost on any generic two-agent algorithm for locating a black hole using SD. As for the topological ignorance case, the agents perform an exploration. The algorithm is similar to the one with topological ignorance (in fact it leads to the same cost); sense of direction is however very useful to decrease the number of casualties. The exploring agents can be only two: A node that is being explored by an agent is considered "dangerous" and, by the properties of sense of direction, the other agent will be able to avoid it in its exploration, thus ensuring that one of the two will eventually succeed.

3.2.4.3   Black-Hole Search with a Map

Consider the case of *complete topological knowledge* of the network; that is, the agents have a complete knowledge of the edge-labeled graph G, the correspondence between port labels and the link labels of G, *and* the location of the source node (from where the agents start the search). This information is stronger than the more common *topological awareness* (i.e., knowledge of the class of the network, but not of its size or of the source location, e.g., being in a mesh, starting from an unknown position).

Also in this case two agents suffice [41]; furthermore, the cost of a minimal protocol can be reduced in this case to $O(n \log n)$, and this cost is worst-case optimal. The technique here is quite different and it is based on a partitioning of the graph in two portions which are given to the two agents to perform the exploration. One will succeed in finishing its portion and will carefully move to help the other agent finishing its own.

Informally, the protocol works as follows. Let G_{ex} be the explored part of the network (i.e., the set of safe nodes); initially it consists only of the homebase h. Agents a and b partition the unexplored area into disjoint subgraphs G_a (the working set for a) and G_b (the working set for b) such that for each connected component of G_a and G_b there is a link connecting it to G_{ex} (this partitioning can always be done). Let T_a and T_b be trees spanning G_a and G_b, respectively, such that $T_a \cap G_b = T_b \cap G_a = \emptyset$. (The graphs G_a and G_b are not necessarily connected—the trees T_a and T_b are obtained from the spanning forests of G_a and G_b by adding edges from G_{ex} as necessary but avoiding the vertices of the opposite working set.)

Each agent then traverses its working set using cautious walk on the corresponding spanning tree. In this process, it transforms unexplored nodes into safe ones.

Let a be the first agent to terminate the exploration of its working set; when this happens, a goes to find b. It does so by first going to the node w where the working sets were last computed using an optimal path and avoiding G_b, then following the trace of b, and finally reaching the last safe node w' reached by b.

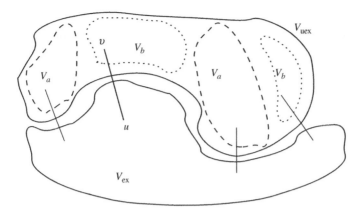

FIGURE 3.1   Splitting the unexplored subgraph G_{uex} into G_a and G_b.

Agent a then computes the new subgraph G_{uex} containing all nonsafe nodes (Figure 3.1). If G_{uex} contains a single node, that node is the black hole. Otherwise a computes the new working sets for itself and b; it leaves a note for b at the current node w' indicating the new working set G_b for b and goes to explore its new assigned area avoiding the (new) working set of b. When (if) b returns to w', it finds the note and starts exploring its new working set. Note that, at any time, an agent is either exploring its working set or looking for the other agent to update the workload or is destroyed by the black hole.

3.2.4.4   Topology-Sensitive Universal Protocols

Interestingly, it is possible to considerably improve the bound on the number of moves without increasing the team size. In fact, there is a recent *universal* protocol, *Explore and Bypass,* that allows a team of *two* agents with a map of the network to locate a black hole with cost $O(n + d \log d)$, where d denotes the diameter of the network [43]. This means that, without losing its universality and without violating the worst-case $\Omega(n \log n)$ lower bound, this algorithm allows two agents to locate a black hole with $\Theta(n)$ cost in a very large class of (possibly unstructured) networks: those where $d = O(n/\log n)$.

The algorithm is quite involved. The main idea is to have the agents explore the network using cooperative depth-first search of a spanning tree T. When further progress using only links of T is blocked, the blocking node is appropriately bypassed and the process is repeated. For efficiency reasons, the bypass is performed in different ways depending on the structure of the unexplored set U and on the size of its connected components. The overall exploration is done in such a way that:

1. The cost of the cooperative depth-first search is linear in the number of explored vertices.
2. Bypassing a node incurs an additional overhead of $O(d)$ which can be charged to the newly explored vertices if there are enough of them.

3. If there are not enough unexplored vertices remaining for bypassing to be viable, the remaining unexplored graph is so small $[O(d)]$ that applying the general $O(n \log n)$ algorithm would incur an $O(d \log d)$ additional cost [which is essentially optimal, due to the lower bound of $\Theta(n \log n)$ for rings].

Importantly, there are many networks with $O(n/\log n)$ diameter in which the previous protocols [41, 44] fail to achieve the $O(n)$ bound. A simple example of such a network is the *wheel*, a ring with a central node connected to all ring nodes, where the central node is very slow: Those protocols will require $O(n \log n)$ moves.

3.2.4.5 Variations with a Map
A very simple algorithm that works on any topology (a priori known by the agents) is shown in [45].

Let C be a set of simple cycles such that each vertex of G is covered by a cycle from C (Figure 3.2). Such a set of cycles with some connectivity constraint is called *open-vertex cover* by cycles. The algorithm is based on the precomputation of such an open-vertex cover by cycles of a graph. The idea is to explore the graph G by exploring the cycles C.

The algorithm uses the optimal number of agents (two). If an agent is blocked on an edge e (because either the transmission delay on e is very high or it leads to the BH), the other agent will be able to bypass it, using the cycle containing e, and continue the exploration. The number of moves depends on the choice of the cover and it is optimal for several classes of networks. These classes include all Abelian Cayley graphs of degree 3 and more (e.g., hypercubes, multidimensional

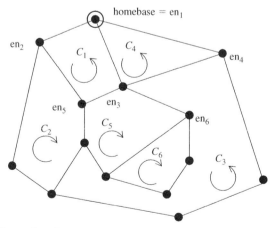

FIGURE 3.2 Example of an open vertex cover $C = \{C_1, C_2, C_3, C_4, C_5, C_6\}$, for graph G. The cycle directions are shown, as well as the entry nodes en_i for each cycle C_i.

tori, etc,), as well as many non-Abelian cube graphs (e.g., cube-connected-cycles, butterfly, wrapped-butterfly networks). For some of these networks, this is the only algorithm achieving such a bound.

3.2.5 Special Topologies

A natural question to ask is whether the bounds for arbitrary networks with full topological knowledge can be improved for networks with special topologies by topology-dependent proptocols.

3.2.5.1 Rings

The problem has been investigated and its solutions characterized for *ring* networks [40]. For these networks, at least two agents are needed, and a $\Omega(n \log n)$ lower bound holds on the number of moves regardless of the number of agents [40].

An agent and move optimal solution exists based on a partitioning of the ring and a nonoverlapping exploration by the agents. The solution is similar to (and simpler than) the one for the known arbitrary topology case. Initially the agents use the whiteboard to differentiate their tasks: each taking charge of (cautiously) exploring roughly half of the ring. One of the two agents will necessarily succeed (say agent A); by that time, the other (agent B) is either still exploring or trapped into the black hole. The successful agent A follows the safe trace of B; at the last safe node found following the trace, A writes on the whiteboard a message for B, indicating that it will now take charge of half of the area still to be explored. In this way, if B is not in the black hole and returns to the node (as imposed by cautious walk), it will find the message and will act accordingly. Notice that the size of the ring must be known for the algorithm to work; furthermore, without knowledge of the size, the problem is unsolvable. The key point of the algorithm's correctness is that the agents are always exploring disjoint areas; hence one of them will always complete its exploration. Since the unexplored area to be explored is halved in each stage, the total number of stages is $O(\log n)$; the amount of moves in each stage is linearly proportional to the size of the explored area; therefore, the total number of moves is $O(n \log n)$. The time complexity of this solution is also $O(n \log n)$.

Interestingly, increasing the number of agents, the number of moves cannot decrease, but the time to finish the exploration does [40]. For example, suppose n agents $x_1, x_2, \ldots, x_n$ are available. By accessing the whiteboard they can assign to themselves different tasks: for example, agent x_i could take care of exploring the node at distance i (clockwise: if there is no orientation, a similar trick would work). To explore node u at distance i, agent x_i moves to visit the nodes that precede u clockwise and the one that precedes u counterclockwise. Only one agent will be successful because all the others will terminate in the black hole either when moving clockwise or when moving counterclockwise. Notice that, in their exploration, the agents do not need to move with cautious walk. Clearly the agents can perform their tasks concurrently and the time complexity is $\Omega(n)$. Indeed, there exists an optimal trade-off between time complexity and number of agents.

Notice that the lower bound for rings implies an $\Omega(n \log n)$ lower bound on the worst-case cost complexity of any *universal* protocol.

The ring has been investigated also to perform another task: rendezvous of k anonymous agents dispersed in the ring in spite of the presence of a black hole. The problem is studied in [46] and a complete characterization of the conditions under which the problem can be solved is established. The characterization depends on whether k or n is unknown (at least one must be known for any nontrivial rendezvous). Interestingly, it is shown that, if k is unknown, the rendezvous algorithm also solves the black-hole location problem, and it does so with a bounded-time complexity of $\Theta(n)$; this is a significant improvement over the $O(n \log n)$ bounded-time complexity of [40].

3.2.5.2 Interconnection Networks

The $\Omega(n \log n)$ lower bound for rings does not necessarily generalize to other topologies. Sometimes the network has special properties that can be exploited to obtain a lower cost network-specific protocol. For example, two agents can locate a black hole with only $O(n)$ moves in a variety of highly structured interconnection networks such as *hypercubes*, square *tori* and *meshes, wrapped butterflies*, and *star graphs* [44].

The protocol achieving such a bound is based on the novel notion of *traversal pairs* of a network which describes how the graph will be explored by each agent and will be used by an agent to avoid "dangerous" parts of the network. The algorithm proceeds in logical rounds. In each round the agents follow a usual cooperative approach of dynamically dividing the work between them: The unexplored area is partitioned into two parts of (almost) equal size. Each agent explores one part without entering the other one; exploration and avoidance are directed by the traversal pair. Since the parts are disjoint, one of them does not contain the black hole and the corresponding agent will complete its exploration. When this happens, the agent (reaches the last safe node visited by the other agent and there) partitions whatever is still left to be explored leaving a note for the other agent (should it be still alive). This process is repeated until the unexplored area consists of a single node: the black hole. In addition to the protocol and its analysis, in [44] there is also the algorithm for constructing a traversal pair of a biconnected graph.

3.2.6 Using Tokens

As we have seen, the problem of asynchronous agents exploring a dangerous graph has been investigated assuming the availability of a whiteboard at each node: Upon gaining access, the agent can write messages on the whiteboard and can read all previously written messages, and this mechanism has been used by the agents to communicate and mark nodes or/and edges. The whiteboard is indeed a powerful mechanism of interagent communication and coordination.

Recently the problem of locating a black hole has been investigated also in a different, weaker model where there are no whiteboards at the nodes. Each agent has instead a bounded number of tokens that can be carried, placed on a node or on a port, or removed from it; all tokens are identical (i.e., indistinguishable) and no other form of marking or communication is available [47–49]. Some natural questions immediately arise: Is the BHS problem still solvable with this weaker mechanism and if so under what conditions and at what cost? Notice that the use of tokens introduces another complexity measure: the number of tokens. Indeed, if the number of tokens is unbounded, it is possible to simulate a whiteboard environment; hence the question immediately arises of how many tokens are really needed.

Surprisingly, the black-hole search problem in an unknown graph can be solved using only this weaker tool for marking nodes and communicating information. In fact, it has been shown [47] that $\Delta + 1$ agents with a single token each can successfully solve the black-hole search problem; recall that this team size is *optimal* when the network is unknown. The number of moves performed by the agents when executing the protocol is actually polynomial. Not surprisingly, the protocol is quite complex. The absence of a whiteboard, in fact, poses serious limitations to the agents, which have available only a few movable bits to communicate with each other.

Special topologies have been studied as well, and in particular, the case of the *ring* has been investigated in detail in [48]. There it has been shown that the two-agent $\Theta(n \log n)$ move strategies for black-hole search in rings with whiteboards can be successfully employed also without whiteboards by carefully using a bounded number of tokens. Observe that these optimal token-based solutions use only $O(1)$ tokens in total, whereas the protocols using whiteboards assumed at least $O(\log n)$ dedicated bits of storage at each node. Further observe that any protocol that uses only a constant number of tokens implies the existence of a protocol (with the same size and cost) that uses only whiteboards of constant size; the converse is not true.

In the case of arbitrary networks, it has been recently shown that, with a map of the graph, a $\Theta(n \log n)$-moves solution exists for 2 agents with a single token each, which can be placed only on nodes (like in classical exploration algorithms with pebbles) [49]. These results indicate that, although tokens appear as a weaker means of communication and coordination, their use does not negatively affect solvability and it does not even lead to a degradation of performance.

3.2.7 Synchronous Networks

The black-hole search problem has been studied also in synchronous settings where the time for an agent to traverse a link is assumed to be unitary.

When the system is synchronous, the goals and strategies are quite different from the ones reviewed in the previous sections. In fact, when designing an algorithm for the asynchronous case, a major problem is that an agent cannot wait at a node for another agent to come back; as a consequence, agents must always move and have to do it carefully. When the system is synchronous, on the other hand, the strategies are mostly based on waiting the right amount of time before performing a move. The algorithm becomes the determination of the shortest traversal schedule for the agents, where a traversal schedule is a sequence of actions (move to a neighboring node or stay at the current node). Furthermore, for the black-hole search to be solvable, it is no longer necessary that the network is two-node connected; thus, black-hole search can be performed by synchronous agents also in trees.

In synchronous networks tight bounds have been established for some classes of trees [50]. In the case of general networks the decision problem corresponding to the one of finding the optimal strategy is shown to be non-deterministic polynomial time (NP) hard [51, 52] and approximation algorithms are given in [50] and subsequently improved in [52, 53]. The case of multiple black holes has been very recently investigated in [54], where a lower bound on the cost and close upper bounds are given.

3.2.8   Rendezvous in Spite of Black Hole

In networks where a black hole is present, the primary task is clearly that of determining the location of the black hole. In addition to the BHS problem, the research has also focused on the computability of other tasks in the presence of a black hole, that is, on the design of *black-hole resilient* solutions to problems. In particular, the focus has been on the rendezvous problem: k anonymous agents are dispersed in the network and must gather at the same node; the location of the rendezvous is not fixed a priori. As mentioned in Section 3.2.2, rendezvous (with whiteboards) is equivalent to the exploration and leader election problems, and it has been extensively studied in networks without black holes. In the presence of a black hole, the problem changes drastically. In fact, it is impossible to guarantee that all agents will gather since some agents might enter the black hole. Thus the main research concern is to determine how many agents can be guaranteed to rendezvous and under what conditions.

A solution strategy would be to first determine the location of the black hole and then perform a rendezvous in the safe part of the network. This strategy requires solving the BHS problem in a setting quite different from that examined by the investigators so far; in fact, the agents would not start from the same safe node, the homebase, but are instead *scattered* in the network. To date, the only solutions to the BHS problem when the agents are scattered are for ring networks using whiteboards [40] or tokens [48].

Interestingly, the overall problem can be solved without necessarily having to locate the black hole. The overall problem has been first investigated when the k anonymous agents are dispersed in a ring, and a complete characterization of the conditions under which the problem can be solved is established in [46]. The characterization depends on whether k or n is unknown (at least one must be known for any nontrivial rendezvous). Interestingly, it is shown that, if k is unknown, the rendezvous algorithm also solves the black-hole location problem, and it does so with a bounded-time complexity of $\Theta(n)$; this is a significant improvement over the $O(n \log n)$ bounded-time complexity of [40].

The problem has recently been investigated in arbitrary networks in the presence not only of one or more black holes but also of *black links,* that is, edges that destroy traversing agents without leaving any trace [55]. A complete characterization of the conditions necessary for solvability has been provided, and a protocol has been designed that allows rendezvous if these conditions hold and otherwise determines impossibility [40].

3.3   INTRUDER CAPTURE AND NETWORK DECONTAMINATION

3.3.1   The Problem

A particularly important security concern is to protect a network from unwanted and possibly dangerous intrusions. At an abstract level, an intruder is an alien process that moves on the network to sites unoccupied by the system's agents, possibly "contaminating" the nodes it passes by. The concern for the severe damage intruders can cause has motivated a large amount of research, especially on detection [56–58]. From an algorithmic point of view, the concern has been almost exclusively on the intruder capture problem, that is, the problem of enabling a team of system agents to stop the dangerous activities of the intruder by coming in direct contact with it. The goal is to design a protocol, to be executed by the team of agents, enabling them to capture the intruder. To make the protocol as flexible as possible, the intruder is assumed to be very powerful: arbitrarily fast and possibly aware of the positions of all the agents; on the contrary, the agents could be arbitrarily slow and unable to sense the intruder presence except when encountering it (i.e., on a node or on a link).

In this setting, the intruder capture problem is equivalent (both from a computational and a complexity point of view) to the problem of network decontamination. In this problem, the nodes of the network are initially *contaminated* and the goal is to *clean* (or decontaminate) the infected network. The task is to be carried out by a team of antiviral system agents (the cleaners). A cleaner is able to decontaminate an infected site once it arrives there; arriving at a clean site, clearly no decontamination operation needs to be performed by the cleaner. A decontaminated site can become *recontaminated* (e.g., if the virus returns to that site in the absence of a cleaner); the specifications of the conditions under which this can happen is called a recontamination rule. The most common recontamination rule is that, if a node without an agent on it has a contaminated neighbor, it will become (re-)contaminated. A solution protocol will then specify the strategy to be used by the agents; that is, it specifies the sequence of moves across the network that will enable the agents, upon all being injected in the system at a chosen network site, to decontaminate the whole network, possibly avoiding any recontamination.

3.3.2 Background Problem: Graph Search

A variation of the decontamination problem described above has been extensively studied in the literature under the name graph search [26–30].

The graph search problem has been first discussed by Breisch [59] and Parson [30, 60]. In the graph-searching problem, we are given a "contaminated" network, that is, whose links are all contaminated. Via a sequence of operations using "searchers," we would like to obtain a state of the network in which all links are simultaneously clear. A *search step* is one of the following operations: (1) place a searcher on a node, (2) remove a searcher from a node, and (3) move a searcher along a link. There are two ways in which a contaminated link can become clear. In both cases, a searcher traverses the link from one extremity u to the other extremity v. The two cases depend on the way the link is preserved from recontamination: Either another searcher remains in u or all other links incident to u are clear. The goal is to use as few searchers as possible to decontaminate the network. A *search strategy* is a sequence of search steps that results in all links being simultaneously clear. The *search number* $s(G)$ of a network G is the smallest number of searchers for which a search strategy exists. A search strategy using $s(G)$ searchers in G is called minimal.

The decision problem corresponding to the computation of the search number of a graph is NP hard [29] and NP completeness is shown in [28, 61]. In particular, Megiddo et al. [29] gave a $O(n)$ time algorithm to determine the search number of n-node trees and a $O(n \log n)$ time algorithm to determine a minimal search strategy in n-node trees. Ellis, Sudborough, and Turner [26] linked $s(G)$ with the vertex separation vs(G) of G (known to be equal to the pathwidth of G [62]). Given an n-node network $G = (V, E)$, vs(G) is defined as the minimum taken over all (one-to-one) linear layouts $L : V \to \{1, \ldots, n\}$, of vs$_L$ (G), the latter being defined as the maximum, for $i = 1, \ldots, n$, of the number of vertices $x \in V$ such that $L(x) \leq i$ and there exists a neighbor y of x such that $L(y) > i$. Ellis et al. showed that vs$(G) \leq s(G) \leq$ vs$(G) + 2$ and $s(G) =$ vs(G'), where G' is the 2-augmentation of G, that is, the network obtained from G by replacing every link $\{x, y\}$ by a path $\{x, a, b, y\}$ of length 3 between x and y. They also showed that the vertex separation of trees can be computed in linear time, and they gave an $O(n \log n)$ time algorithm for computing the corresponding layout. It yields another $O(n)$ time algorithm returning the search number of trees and an $O(n \log n)$ *time* algorithm returning a minimal search strategy in trees.

Graph searching has many other applications (see [63]), including pursuit–evasion problems in a labyrinth [60], decontamination problems in a system of tunnels, and mobile computing problems in which agents are looking for a hostile intruder [64]. Moreover, the graph-searching problem also arises in very large scale integration (VLSI) design through its equivalence with the gate matrix layout problem [62]. It is hence not surprising that it gave rise to numerous papers. Another reason for this success is that the problem and its several variants (nodesearch, mixedsearch, t-search, etc.) are closely related to standard graph parameters and concepts, including tree width, cut width, path width, and, last but not least, graph minors [31]. For instance, Makedon and Sudborough [33] showed that $s(G)$ is equal to the cut width of G for all networks of maximum degree 3. Similarly, Kiroussis and Papadimitriou showed that the node search number of a network is equal to its interval width [32] and to its vertex separator plus 1 [27]. Seymour and Thomas [34] showed that the t-search number is equal to the tree width plus 1. Takahashi, Ueno, and Kajitani [65] showed that the mixed-search number is equal to the proper path width. Bienstock and Seymour [61] simplified the proof of Lapaugh's result [28] stating that there is a minimal search strategy that does not recontaminate any link (see also [66]). Thilikos [67] used graph minors to derive a linear-time algorithm that checks whether a network has a search number at most 2. For other results on graph searching, the reader is referred to the literature [68–72]. Contributions to related search problems can be found elsewhere [64, 73–78].

In graph searching, there has been a particular interest in monotone strategies [28, 61, 79]. A strategy is monotone if after a node (link) is decontaminated it will not be contaminated again. An important result from Lapaugh has shown that monotonicity does not really change the difficulty of the graph search problem; in fact, it has been shown in [28] that for any graph there exists a monotone search strategy that uses the minimum number of agents.

Let us stress that in the classical graph search problem the agents can be arbitrarily moved from a node "jumping" to any other node in the graph.

The main difference in the setting described in this chapter is that the agents, which are pieces of software, *cannot be removed from the network;* they can only move from a node to a *neighboring* one (*contiguous search*). This additional constraint was introduced and first studied in [4] resulting in a *connected-node search* where (i) the removal of agents is not allowed and (ii) at any time of the search strategy the set of clean nodes forms a connected subnetwork. With the connected assumption the nature of the problem changes considerably and the classical results on node and edge search do not generally apply.

In the case of connected graph search usually more agents are required to decontaminate a network G. It has been shown that for any graph G with n nodes the ratio between the connected search number $\mathrm{csn}(G)$ and the regular search number $\mathrm{sn}(G)$ is always bounded. More precisely, it is known that $\mathrm{csn}(G)/\mathrm{sn}(G) \leq \log n + 1$ [80], and, for a tree T, $\mathrm{csn}(T)/\mathrm{sn}(T) \leq 2$ [81]. Monotonicity also plays an important role in connected graph searches. It has been shown that, as in the more general graph search problem, a solution allowing recontamination of nodes cannot reduce the optimal number of agents required to decontaminate trees [4]. On the other hand, unlike the classical graph search, there exist graphs for which any given monotone-connected graph search strategy requires more searchers than the optimal non-monotone-connected search strategy [82]. The decision problem corresponding to the computation of the connected search number of a graph is NP hard, but it is not known whether there exists a yes certificate that is checkable in polynomial time.

As we will survey below, the problem has been studied mostly in specific topologies. Also the arbitrary topology has been considered; in this case, some heuristics have been proposed [83] and a move-exponential optimal solution has been given in [84].

In this chapter we use decontamination to refer to the connected monotone node search as defined in [4].

3.3.3   Models for Decontamination

Initially, all agents are located at the same node, the homebase, and all the other nodes are contaminated; a decontamination strategy consists of a sequence of movements of the agents along the edges of the network. The agents can communicate when they reside on the same node.

Starting from the classical model employed in [4] (called the local model), additional assumptions have sometimes been added to study the impact that more powerful agents' or system's capabilities have on the solutions of our problem:

1. In the *local model* an agent located at a node can "see" only local information, such as the state of the node, the labels of the incident links, and the other agents present at the node.
2. *Visibility* is the capability of the agent to see the state of its neighbors; that is, an agent can see whether a neighboring node is guarded and whether it is clean or contaminated. Notice that, in some mobile agent systems, the visibility power could be easily achieved by "probing" the state of neighboring nodes before making a decision.
3. *Cloning* is the capability, for an agent, to clone copies of itself.
4. *Synchronicity* implies that local computations are instantaneous, and it takes one unit of time (one step) for an agent to move from a node to a neighboring one.

The efficiency of a strategy is usually measured in terms of number of agents, amount of decontamination operations, number of moves performed by the agents, and ideal time.

We say that a cleaning strategy is *monotone* if, once a node is clean, it will never be contaminated again. The reason to avoid recontamination derives from the requirement to minimize the amount of decontamination performed by the system agents: If recontamination is avoided, the number of decontamination operations performed is the smallest possible, one per network site. All the results reported here are for monotone strategies.

3.3.4   Results in Specific Topologies

3.3.4.1   Trees
The tree has been the first topology to be investigated in the local model [4]. Notice that, for a give tree T, the minimum number of agents needed depends

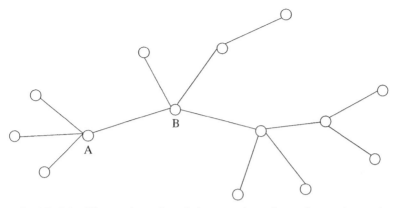

FIGURE 3.3   The number of needed agents depends on the starting node.

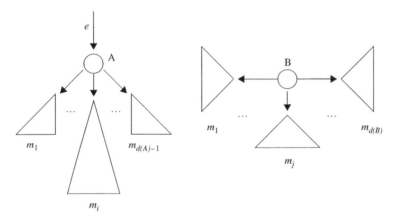

FIGURE 3.4   Determining the minimum number of cleaners.

on the node from which the team of agents starts. Consider for example the tree shown in Figure 3.3. If the team starts from node A, then two agents suffice. However, the reader can verify that at least three agents are needed if they start from node B.

In [4], the authors describe a simple and efficient strategy to determine the minimum number of agents necessary to decontaminate an arbitrary given tree from any initial starting node. The strategy is based on the following two observations.

Consider a node A. If A is not the starting node, the agents will arrive at A for the first time from some link e (see Figure 3.4). Let $T_1(A), \ldots, T_i(A), \ldots,$ $T_{d(A)-1}$ be the subtrees of A from the other incident links, where $d(A)$ denotes the degree of A; let m_i denote the number of agents needed to decontaminate $T_i(A)$ once the agents are at A, and let $m_i \le m_{i+1}$, $1 \le i \le d(A) - 2$. The first observation is that to decontaminate A and all its other subtrees without recontamination the number $m(A, e)$ of agents needed is

$$m(A,e) = \begin{cases} m_1 & \text{if } m_1 > m_2 \\ m_{1+1} & \text{if } m_1 = m_2 \end{cases}$$

Consider now a node B and let $m_j(B)$ be the minimum number of agents needed to decontaminate the subtree $T_j(B)$ once the agents are at B and let $m_j \leq m_{j+1}$, $1 \leq j \leq d(B)$. The second observation is that to decontaminate the entire tree starting from B the number $m(B)$ of agents needed is

$$m(B) = \begin{cases} m_1 & \text{if } m_1 > m_2 \\ m_{1+1} & \text{if } m_1 = m_2 \end{cases}$$

Based on these two properties, the authors show in [4] how determination of the optimal number of agents can be done through a saturation where appropriate information about the structure of the tree is collected from the leaves and propagated along the tree until the optimal is known for each possible starting point. The most interesting aspect of this strategy is that it yields immediately a decontamination protocol for trees that uses exactly that minimum number of agents. In other words, the technique of [4] allows to determine the minimum number of agents and the corresponding decontamination strategy for every starting network, and this is done exchanging only $O(n)$ short messages [or, serially, in $O(n)$ time].

The trees requiring the largest number of agents are *complete binary trees*, where the number of agent is $O(\log n)$; by contrast, in the *line* two agents are always sufficient.

3.3.4.2   *Hypercubes*

It has been shown in [85] that to decontaminate a hypercube of size n, $\Theta(n/\sqrt{\log n})$ agents are necessary and sufficient. The employ of an optimal number of agents in the local model has an interesting consequence; in fact, it implies that $\Theta(n/\sqrt{\log n})$ is the search number for the hypercube in the classical model, that is, where agents may "jump."

In the algorithm for the local model one of the agents acts as a *coordinator* for the entire cleaning process. The cleaning strategy is carried out on the broadcast tree of the hypercube. The cleaning strategy broadcast tree of the hypercube; see Figure 3.5. The main idea is to place enough agents on the homebase and to have them move, level by level, on the edges of the broadcast tree, leaded by the coordinator in such a way that no recontamination may occur. The number of moves and the ideal time complexity of this strategy are indicated in Table 3.1.

The visibility assumption allows the agents to make their own decision regarding the action to take solely on the basis of their local knowledge. In fact, the agents are still moving on the broadcast tree, but they do not have to follow the order imposed by the coordinator. The agents on node x can proceed to

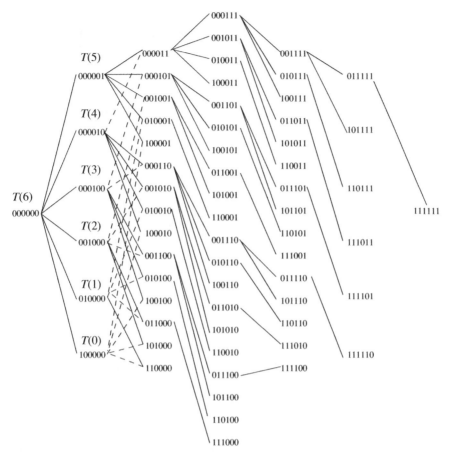

FIGURE 3.5 The broadcast tree T of the hypercube H_6. Normal lines represent edges in T, dotted lines (only partially shown) the remaining edges of H_6.

TABLE 3.1 Decontamination of Hypercube

Model	Agents	Time	Moves
Local	$(\star)O(n/\sqrt{\log n})$	$O(n \log n)$	$O(n \log n)$
Local, cloning, synchronicity	$n/2$	$(\star) \log n$	$(\star) n - 1$
Visibility	$n/2$	$(\star) \log n$	$O(n \log n)$
Visibility and cloning	$n/2$	$(\star) \log n$	$(\star) n - 1$

Note: The star indicates an optimal bound.

clean the children of x in the broadcast tree when they see that the other neighbors of x are either clean or guarded. With this strategy the time complexity is drastically reduced (since agents move concurrently and independently) but the number of agents increases. Other variations of those two models have been studied and summarized in Table 3.1.

A characterization of the impact that these additional assumptions have on the problem is still open. For example, an optimal move complexity in the local model with cloning has not been found, and it is not clear whether it exists; when the agents have visibility, synchronicity has not been of any help, although it has not been proved that it is indeed useless; the use of an optimal number of agents in the weaker local model is obtained at the expense of employing more agents, and it is not clear whether this increment is necessary.

3.3.4.3   Chordal Rings

The local and the visibility models have been the subject of investigation also in the chordal ring topology in [86].

Let $C(\langle d_1 = 1, d_2, \ldots, d_k \rangle)$ be a chordal ring network with n nodes and link structure $\langle d_1 = 1, d_2, \ldots, d_k \rangle$, where $d_i < d_{i+1}$ and $d_k \leq \lfloor n/2 \rfloor$. In [86] it is first shown that the smallest number of agents needed for the decontamination depends not on the size of the chordal ring but solely on the *length* of the longest chord. In fact, any solution of the contiguous decontamination problem in a chordal ring $C(\langle d_1 = 1, d_2, \ldots, d_k \rangle)$ with $4 \leq d_k \leq \sqrt{n}$ requires at least $2d_k$ searchers ($2d_k + 1$ in the visibility model).

In both models, the cleaning is preceded by a deployment stage after which the agents have to occupy $2d_k$ consecutive nodes. After the deployment, the decontamination stage can start. In the local model, nodes x_0 to $x_{d_k - 1}$ are constantly guarded by one agent each, forming a window of d_k agents. This window of agents will shield the clean nodes from recontamination from one direction of the ring while the agents of the other window are moved by the coordinator (one at a time starting from the one occupying node x_{d_k}) along their longest chord to clean the next window in the ring. Also in the case of the chordal ring, the visibility assumption allows the agents to make their own decision solely on the basis of their local knowledge: An agent moves to clean a neighbor only when this is the only contaminated neighbor.

Figure 3.6 shows a possible execution of the algorithm in a portion of a chordal ring $C(\langle 1, 2, 4 \rangle)$. Figure 3.6a shows the guarded nodes (in black) after the deployment phase. At this point, the nodes indicated in the figure can independently and concurrently start the cleaning phase, moving to occupy their only contaminated neighbor. Figure 3.6b shows the new state of the network if they all move (the arrows indicate the nodes where the agents could move to clean their neighbor).

The complexity results in the two models are summarized in Table 3.2.

Consistent with the observations for the hypercube, also in the case of the chordal ring the visibility assumption allows to drastically decrease the time complexity (and in this case also the move complexity). In particular, the strategies for the visibility model are optimal in terms of both number of agents and number of moves; as for the time complexity, visibility allows some concurrency (although it does not bring this measure to optimal as was the case for the hypercube).

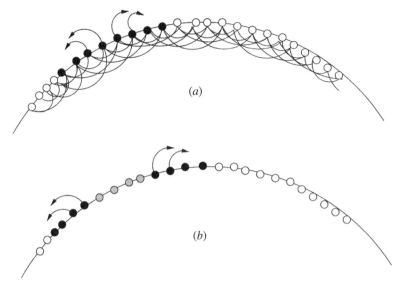

FIGURE 3.6   A chordal ring $C(\langle 1, 2, 4\rangle)$. (*a*) The agents are deployed and four of them (the ones pointed by an arrow) could move to clean the neighbor. (*b*) Four agents have moved to clean their only contaminated neighbor and four more (the ones pointed by an arrow) could now move.

TABLE 3.2   Results for Chordal Ring

Model	Agents	Time	Moves
Local	$2d_k + 1$ ($\star$)	$3n - 4d_k - 1$	$4n - 6d_k - 1$
Visibility	$2d_k$ ($\star$)	$\left\lceil \dfrac{n - 2d_k}{2(d_k - d_k - 1)} \right\rceil$	$n - 2d_k$ ($\star$)

Note: The star indicates an optimal bound.

3.3.4.4   Tori

A lower bound for the torus has beed derived in [86]. Any solution of the decontamination problem in a torus $T(h, k)$ with $h, k \geq 4$ requires at least 2 min $\{h, k\}$ agents; in the local model it requires at least 2 min$\{h, k\} + 1$ agents. The strategy that matches the lower bound is very simple. The idea is to deploy the agents to cover two consecutive columns and then keep one column of agents to guard from decontamination and have the other column move along the torus. The complexity results are summarized in Table 3.3. As for the other topologies, visibility decreases time and slightly increases the number of agents. In the case of the torus it is interesting to notice that in the visibility model all three complexity measures are optimal.

TABLE 3.3   Results for Two-Dimensional Torus with Dimensions $h, k, h \le k$

Model	Agents	Time	Moves
Local	$2h + 1$ $(\star)$	$hk - 2h$	$2hk - 4h - 1$
Visibility	$2h$ $(\star)$ $(\star)$	$\left\lceil \dfrac{1}{2} \dfrac{k-2}{2} \right\rceil (\star)$	$hk - 2h$ $(\star)$ $(\star)$

Note: The star indicates an optimal bound.

TABLE 3.4   Results for d-Dimensional Torus $T(h_1, h_2, \ldots, h_d)$

Model	Agents	Time	Moves
Local	$2(N/h_d) + 1$	$N - 2(N/h_d)$	$2N - 4(N/h_d) - 1$
Visibility	$2(N/h_d)$	$(\lceil h_d - 2 \rceil)/2$	$N - 2(N/h_d)$

Finally, these simple decontamination strategies can be generalized to d-dimensional tori (although the lower bounds have not been generalized). Let $T(h_1, \ldots, h_d)$ be a d-dimensional torus and let $h_1 < h_2 \le \cdots \le h_d$. Let N be the number of nodes in the torus and let $H = N/h_d$. The resulting complexities are reported in Table 3.4.

3.3.5   Different Contamination Rules

In [87] the network decontamination problem has been considered under a new model of *neighborhood-based immunity* to recontamination: a clean node, after the cleaning agent has gone, becomes recontaminated only if a weak majority of its neighbors are infected. This recontamination rule is called *local immunization*. Luccio et al. [87] studied the effects of this level of immunity on the nature of the problem in tori and trees. More precisely, they establish lower bounds on the number of agents necessary for decontamination and on the number of moves performed by an optimal-size team of cleaners and propose cleaning strategies. The bounds are tight for trees and for synchronous tori; they are within a constant factor of each other in the case of asynchronous tori. It is shown that with local immunization only $O(1)$ agents are needed to decontaminate meshes and tori, regardless of their size; this must be contrasted with, for example, the $2 \min\{n, m\}$ agents required to decontaminate an $n \times m$ torus without local immunization [86]. Interestingly, among tree networks, binary trees were the worst to decontaminate without local immunization, requiring $\Omega(\log n)$ agents in the worst case [4]. Instead, with local immunization, they can be decontaminated by a *single* agent starting from a leaf or by two agents starting from any internal node.

A different kind of immunity has been considered in [88]: one disinfected, a node is immune to contamination for a certain amount of time. This *temporal immunity* has been investigated and characterized for tree networks [88].

3.4 CONCLUSIONS

Mobile agents represent a novel powerful paradigm for algorithmic solutions to distributed problems; unlike the message-passing paradigm, mobile agent solutions are naturally suited for dynamic environments. Thus they provide a unique opportunity for developing simple solutions to complex control and security problems arising in ever-changing systems such as dynamic networks. While mobile agents per se have been extensively investigated in the artificial intelligence, software engineering, and specification and verification communities, the algorithmic aspects (problem solving, complexity analysis, experimental evaluation) are very limited. It is only recently that researchers have started to systematically explore this new and exciting distributed computational universe. In this chapter we have described some interesting problems and solution techniques developed in this investigation in the context of security. Our focus has been on two security problems: locating a black hole and capturing an intruder. For each we have described the computational issues and the algorithmic techniques and solutions. These topics and techniques have a much wider theoretical scope and range. In particular, the problems themselves are related to long-investigated and well-established problems in automata theory, computational complexity, and graph theory.

Many problems are still open. Among them:

- The design of solutions when the harmful host represents a *transient danger*, in other words, when the harmful behavior is not consistent and continuous but changes over time.
- The study of *mobile harm,* that is, of pieces of software that are wandering around the network possibly damaging the mobile agents encountered in their path.
- The study of *multiple attacks*, in other words, the general harmful host location problem when dealing with an arbitrary, possibly unknown, number of harmful hosts present in the system. Some results have been recently obtained [54, 55]

ACKNOWLEDGMENTS

This work has been supported in part by the Natural Sciences and Engineering Research Council of Canada under the Discovery Grant program, and by Dr. Flocchini's University Research Chair.

REFERENCES

1. S. Albers and M. Henzinger, Exploring unknown environments, *SIAM J. Comput.*, 29:1164–1188, 2000.

2. S. Alpern and S. Gal, *The Theory of Search Games and Rendezvous*, Springer, New York, 2002.

3. B. Awerbuch, M. Betke, and M. Singh, Piecemeal graph learning by a mobile robot, *Inform. Comput.*, 152:155–172, 1999.

4. L. Barrière, P. Flocchini, P. Fraigniaud, and N. Santoro, Capture of an intruder by mobile agents, in *Proceedings of the 14th ACM Symposium on Parallel Algorithms and Architectures (SPAA)*, Winnipeg, 2002, pp. 200–209.

5. L. Barrière, P. Flocchini, P. Fraigniaud, and N. Santoro, Can we elect if we cannot compare? in *Proceedings of the 15th ACM Symposium on Parallel Algorithms and Architectures (SPAA)*, San Diego, 2003, pp. 200–209.

6. L. Barrière, P. Flocchini, P. Fraigniaud, and N. Santoro, Rendezvous and election of mobile agents: Impact of sense of direction, *Theory Comput. Syst.*, 40(2): 143–162, 2007.

7. X. Deng and C. H. Papadimitriou, Exploring an unknown graph, *J. Graph Theory*, 32(3):265–297, 1999.

8. A. Dessmark, P. Fraigniaud, and A. Pelc, Deterministic rendezvous in graphs, *Algorithmica*, 46:69–96, 2006.

9. K. Diks, P. Fraigniaud, E. Kranakis, and A. Pelc, Tree exploration with little memory, *J. Algorithms*, 51:38–63, 2004.

10. E. Kranakis, D. Krizanc, N. Santoro, and C. Sawchuk, Mobile agent rendezvous in a ring, in *Proceedings of the 23rd International Conference on Distibuted Computing Systems (ICDCS)*, Providence, 2003, pp. 592–599.

11. N. Borselius, Mobile agent security, *Electron. Commun. Eng. J.*, 14(5):211–218, 2002.

12. D. M. Chess, Security issues in mobile code systems, in *Mobile Agent Security*, G. Vigna (Ed.), Lecture Notes in Computer Science, Vol. 1419, Springer, London, 1998, pp. 1–14.

13. M. S. Greenberg, J. C. Byington, and D. G. Harper, Mobile agents and security, *IEEE Commun. Mag.*, 36(7):76–85, 1998.

14. R. Oppliger, Security issues related to mobile code and agent-based systems, *Computer Commun.*, 22(12):1165–1170, 1999.

15. K. Schelderup and J. Ones, Mobile agent security—Issues and directions, in *Proceedings of the 6th International Conference on Intelligence and Services in Networks*, Lecture Notes in Computer Science, Vol. 1597, Barcelona, 1999, pp. 155–167.

16. M. Bender, A. Fernandez, D. Ron, A. Sahai, and S. Vadhan, The power of a pebble: Exploring and mapping directed graphs, *Inform. Comput.*, 176(1):1–21, 2002.

17. A. Dessmark and A. Pelc, Optimal graph exploration without good maps, *Theor. Computer Sci.*, 326:343–362, 2004.

18. P. Fraigniaud, L. Gasieniec, D. Kowalski, and A. Pelc, Collective tree exploration, *Networks*, 48(3):166–177, 2006.

19. P. Fraigniaud, and D. Ilcinkas, Digraph exploration with little memory, in *Proceedings of the 21st Symposium on Theoretical Aspects of Computer Science (STACS)*, Montpellier, 2004, pp. 246–257.

20. P. Fraigniaud, D. Ilcinkas, G. Peer, A. Pelc, and D. Peleg, Graph exploration by a finite automaton, *Theor. Computer Sci.*, 345(2–3):331–344, 2005.

21. P. Panaite and A. Pelc, Exploring unknown undirected graphs, *J. Algorithms*, 33:281–295, 1999.

22. P. Panaite and A. Pelc, Impact of topographic information on graph exploration efficiency, *Networks*, 36:96–103, 2000.

23. S. Das, P. Flocchini, S. Kutten, A. Nayak, and N. Santoro, Map construction of unknown graphs by multiple agents, *Theor. Computer Sci.*, 385(1–3):34–48, 2007.

24. S. Das, P. Flocchini, A. Nayak, and N. Santoro, Effective elections for anonymous mobile agents, in *Proceedings of the 17th International Symposium on Algorithms and Computation (ISAAC)*, Kolkata, 2006, pp. 732–743.

25. E. Kranakis, D. Krizanc, and S. Rajsbaum, Mobile agent rendezvous, in *Proceedings of the 13th International Colloquium on Structural Information and Communication Complexity (SIROCCO)*, Chester, 2006, pp. 1–9.

26. J. Ellis, H. Sudborough, and J. Turner, The vertex separation and search number of a graph, *Inform. Comput.*, 113(1):50–79, 1994.

27. L. Kirousis and C. Papadimitriou, Searching and pebbling, *Theor. Computer Sci.*, 47(2):205–218, 1986.

28. A. Lapaugh, Recontamination does not help to search a graph, *J. ACM*, 40(2): 224–245, 1993.

29. N. Megiddo, S. Hakimi, M. Garey, D. Johnson, and C. Papadimitriou, The complexity of searching a graph, *J. ACM*, 35(1):18–44, 1988.

30. T. Parson, The search number of a connected graph, in *Proceedings of the 9th Southeastern Conference on Combinatorics, Graph Theory and Computing*, Utilitas Mathematica, Boca Raton, 1978, pp. 549–554.

31. D. Bienstock and M. Langston, Algorithmic implications of the graph minor theorem, in M.O. Ball, T.L. Magnanti, C.L. Monma, and G.L. Nemhauser (Eds.), *Handbook of Operations Research and Management Science: Network Models*, Vol. 7, Elsevier, Amsterdam, 1995, pp. 481–502.

32. L. Kirousis and C. Papadimitriou, Interval graphs and searching, *Discrete Math.*, 55:181–184, 1985.

33. F. Makedon and H. Sudborough, Minimizing width in linear layout, in *Proceedings of the 10th International Colloquium on Automata, Languages, and Programming (ICALP '83)*, Barcelona, 1983, pp. 478–490.

34. P. Seymour and R. Thomas, Graph searching, and a min-max theorem for treewidth, *J. Comb. Theory, Ser. B*, 58(1):22–33, 1993.

35. F. Hohl, Time limited blackbox security: Protecting mobile agents from malicious hosts, in G. Vigna (Ed.), *Mobile Agent Security*, Lecture Notes in Computer Science, Vol. 1419, Springer, London, 1998, pp. 92–113.

36. T. Sander, and C. F. Tschudin, Protecting mobile agents against malicious hosts, in G. Vigna (Ed.), *Mobile Agent Security*, Lecture Notes in Computer Science, Vol. 1419, Springer, London, 1998, pp. 44–60.

37. J. Vitek, and G. Castagna, Mobile computations and hostile hosts, in D. Tsichritzis (Ed.), *Mobile Objects*, University of Geneva, Geneva, 1999, pp. 241–261.

38. G. Dudek, M. Jenkin, E. Milios, and D. Wilkes, Robotic exploration as graph construction, *Trans. Robot. Autom.*, 7(6):859–865, 1991.

39. M. Bender, and D. K. Slonim, The power of team exploration: Two robots can learn unlabeled directed graphs, in *Proceedings of the 35th Symposium on Foundations of Computer Science (FOCS)*, Santa Fe, 1994, pp. 75–85.

40. S. Dobrev, P. Flocchini, G. Prencipe, and N. Santoro, Mobile search for a black hole in an anonymous ring, *Algorithmica*, 48(1):67–90, 2007.

41. S. Dobrev, P. Flocchini, G. Prencipe, and N. Santoro, Searching for a black hole in arbitrary networks: Optimal mobile agents protocols, *Distrib. Comput.* 19(1):1–19, 2006.

42. P. Flocchini, B. Mans, and N. Santoro, Sense of direction in distributed computing, *Theor. Computer Sci.*, 291:29–53, 2003.

43. S. Dobrev, P. Flocchini, and N. Santoro, Improved bounds for optimal black hole search in a network with a map, in *Proceedings of the 10th International Colloquium on Structural Information and Communication Complexity (SIROCCO)*, Smolenice Castle, 2004, pp. 111–122.

44. S. Dobrev, P. Flocchini, R. Kralovic, G. Prencipe, P. Ruzicka, and N. Santoro, Optimal search for a black hole in common interconnection networks, *Networks*, 47(2):61–71, 2006.

45. S. Dobrev, P. Flocchini, and N. Santoro, Cycling through a dangerous network: A simple efficient strategy for black hole search, in *Proceedings of the 26th International Conference on Distributed computing Systems (ICDCS)*, Lisboa, 2006, p. 57.

46. S. Dobrev, P. Flocchini, G. Prencipe, and N. Santoro, Multiple agents rendezvous in a ring in spite of a black hole, in *Proceedings of the 6th International Symposium on Principles of Distributed Systems (OPODIS)*, La Martinique, 2003, pp. 34–46.

47. S. Dobrev, P. Flocchini, R. Kralovic, and N. Santoro, Exploring a dangerous unknown graph using tokens, in *Proceedings of the 5th IFIP International Conference on Theoretical Computer Science (TCS)*, Santiago, 2006, pp. 131–150.

48. S. Dobrev, N. Santoro, and W. Shi, Using scattered mobile agents to locate a black hole in an unoriented ring with tokens, *Intl. J. Found. Comput. Sci.*, 19(6):1355–1372, 2008.

49. P. Flocchini, D. Ilcinkas, and N. Santoro, Ping pong in dangerous graphs: Optimal black hole search with pebbles, *Algorithmica*, 62(3–4):1006–1033, 2012.

50. J. Czyzowicz, D. Kowalski, E. Markou, and A. Pelc, Searching for a black hole in synchronous tree networks, *Combinator. Prob. Comput.*, 16:595–619, 2007.

51. J. Czyzowicz, D. Kowalski, E. Markou, and A. Pelc, Complexity of searching for a black hole, *Fund. Inform.*, 71(2–3):229–242, 2006; 35–45, 2004.

52. R. Klasing, E. Markou, T. Radzik, and F. Sarracco, Approximation bounds for black hole search problems. *Networks*, 52(4):216–226, 2008.

53. R. Klasing, E. Markou, T. Radzik, and F. Sarracco, Hardness and approximation results for Black Hole Search in arbitrary networks, *Theor. Computer Sci.*, 384(2–3): 201–221, 2007.

54. C. Cooper, R. Klasing, and T. Radzik, Searching for black-hole faults in a network using multiple agents, in *Proceedings of the 10th International Conference on Principle of Distributed Systems (OPODIS)*, Bordeaux, 2006, pp. 320–332.

55. J. Chalopin, S. Das, and N. Santoro, Rendezvous of mobile agents in unknown graphs with faulty links, in *Proceedings of the 21st International Symposium on Distributed Computing (DISC)*, Lemesos, 2007, pp. 108–122.

56. W. Jansen, Intrusion detection with mobile agents, *Computer Communications*, 25(15): 1392–1401, 2002.

57. E. H. Spafford, and D. Zamboni, Intrusion detection using autonomous agents, *Computer Networks*, 34(4):547–570, 2000.

58. D. Ye, Q. Bai, M. Zhang, and Z. Ye, Distributed intrusion detections by using mobile agents, in *Proceedings of the 7th IEEE/ACIS International Conference on Computer and Information Science (ICIS)*, Portland, 2008, pp. 259–265.

59. R. Breisch, An intuitive approach to speleotopology, *Southwestern Cavers* VI(5): 72–78, 1967.

60. T. Parson, Pursuit-evasion in a graph, in *Proceedings of Conference on Theory and Applications of Graphs*, Lecture Notes in Mathematics, Springer, Michigan, 1976, pp. 426–441.

61. D. Bienstock and P. Seymour, Monotonicity in graph searching, *J. Algorithms*, 12:239–245, 1991.

62. N. Kinnersley, The vertex separation number of a graph equals its path-width, *Inform. Process. Lett.*, 42(6):345–350, 1992.

63. F. V. Fomin and D. M. Thilikos, An annotated bibliography on guaranteed graph searching, *Theor. Comput. Sci.*, 399(3):236–245, 2008.

64. I. Suzuki and M. Yamashita, Searching for a mobile intruder in a polygonal region, *SIAM J. Comput.*, 21(5):863–888, 1992.

65. A. Takahashi, S. Ueno, and Y. Kajitani, Mixed searching and proper-path-width, *Theor. Comput. Sci.*, 137(2):253–268, 1995.

66. D. Bienstock, Graph searching, path-width, tree-width and related problems, (A Survey), in *Proceedings of DIMACS Workshop on Reliability of Computer and Communication Networks*, Rutgers University, 1991, pp. 33–49.

67. D. Thilikos, Algorithms and obstructions for linear-width and related search parameters, *Discrete Appl. Math.*, 105:239–271, 2000.

68. R. Chang, Single step graph search problem, *Inform. Process. Lett.* 40(2):107–111, 1991.

69. N. Dendris, L. Kirousis, and D. Thilikos, Fugitive-search games on graphs and related parameters, *Theor. Comput. Sci.*, 172(1–2):233–254, 1997.

70. F. Fomin and P. Golovach, Graph searching and interval completion, *SIAM J. Discrete Math.*, 13(4):454–464, 2000.

71. J. Smith, Minimal trees of given search number, *Discrete Math.*, 66:191–202, 1987.

72. Y. Stamatiou and D. Thilikos, Monotonicity and inert fugitive search games, in *Proceedings of the 6th Twente Workshop on Graphs and Combinatorial Optimization*, Electronic Notes on Discrete Mathematics, Vol. 3, Enschede, 1999, pp. 184.

73. H. Buhrman, M. Franklin, J. Garay, J.-H. Hoepman, J. Tromp, and P. Vitányi, Mutual search, *J. ACM*, 46(4), 517–536, 1999.

74. T. Lengauer, Black-white pebbles and graph separation, *Acta Inform.*, 16(4): 465–475, 1981.

75. S. Neufeld, A pursuit-evasion problem on a grid, *Inform. Process. Lett.*, 58(1):5–9, 1996.

76. I. Suzuki, M. Yamashita, H. Umemoto, and T. Kameda, Bushiness and a tight worst-case upper bound on the search number of a simple polygon, *Inform. Process. Lett.*, 66(1):49–52, 1998.

77. B. von Stengel and R. Werchner, Complexity of searching an immobile hider in a graph, *Discrete Appl. Math.*, 78:235–249, 1997.

78. M. Yamamoto, K. Takahashi, M. Hagiya, and S.-Y. Nishizaki, Formalization of graph search algorithms and its applications, in *Proceedings of the 11th International Conference on Theorem Proving in Higher Order Logics*, Canberra, 1998, pp. 479–496.

79. P. Fraigniaud and N. Nisse, Monotony properties of connected visible graph searching, *Inf. Comput.*, 206(12):1383–1393, 2008.

80. D. Ilcinkas, N. Nisse, and D. Soguet, The cost of monotonicity in distributed graph searching, *Distributed Comput.*, 22(2):117–127, 2009.

81. L. Barrière, P. Fraigniaud, N. Santoro, and D. M. Thilikos, Searching is not jumping, in *Proceedings of the 29th International Workshop on Graph Theoretic Concepts in Computer Science (WG)*, Lecture Notes in Computer Science, Vol. 2880, Elspeet, 2003, pp. 34–45.

82. B. Yang, D. Dyer, and B. Alspach, Sweeping graphs with large clique number, *Discrete Math.*, 309(18):5770–5780, 2009.

83. P. Flocchini, A. Nayak, and A. Shulz, Cleaning an arbitrary regular network with mobile agents, in *Proceedings of the 2nd International Conference on Distributed Computing and Internet Technology (ICDCIT)*, Bhubaneswar, 2005, pp. 132–142.

84. L. Blin, P. Fraigniaud, N. Nisse, and S. Vial, Distributed chasing of network intruders, *Theor. Comput. Sci.*, 399(1–2):12-37, 2008.

85. P. Flocchini, M. J. Huang, and F. L. Luccio, Decontamination of hypercubes by mobile agents, *Networks,* 52(3):167–178, 2008.

86. P. Flocchini, M. J. Huang, and F. L. Luccio, Decontamination of chordal rings and tori using mobile agents, *Int. J. Found. Computer Sci.*, 18(3):547–564, 2007.

87. F. Luccio, L. Pagli, and N. Santoro, Network decontamination in presence of local immunity, *Int. J. Found. Comput. Sci.*, 18(3):457–474, 2007.

88. P. Flocchini, B. Mans, and N. Santoro, Tree decontamination with temporary immunity, in *Proceedings of the 19th International Symposium on Algorithms and Computation (ISAAC)*, Gold Coast, 2008, pp. 330–341.

4 Mobile Agent Coordination

GIACOMO CABRI and RAFFAELE QUITADAMO

University of Modena and Reggio Emilia, Modena, Italy

In this chapter, we will discuss mobile agent coordination, starting with some general considerations about coordination and mobility. By means of a case study, we compare some coordination models and point out that fully uncoupled models better suit dynamic scenarios such as those involving mobility. Therefore, we then focus mainly on uncoupled coordination models (Linda-like models), surveying some proposed coordination systems in the literature and evaluating them with regard to some chosen classification criteria. Finally, the chapter proposes an innovative approach to mobile agent coordination which exploits the powerful concept of roles in multiagent systems.

4.1 INTRODUCTION

In the early days of software agent research, much attention was focused on the development of efficient and easy-to-use agent platforms or systems. Although this has positively contributed to the maturation of this technology, it seemed clear that a step forward was demanded. Design methodologies and coordination patterns were needed in order to enable a larger number of industrial applications using agent technologies [1].

Designing complex applications with software agents implies decomposing the main activity goal into activities (or subgoals) which together will achieve the original goal [2]. Decomposition produces dependencies among the participant agents who should be able to engage in, possibly, complex communications with other agents in order to exchange information or ask their help in pursuing a goal. The latter leads naturally to the notion of a number of agents "cooperating" with each other toward the accomplishment of some common objective.

Mobile Agents in Networking and Distributed Computing, First Edition.
Edited by Jiannong Cao and Sajal K. Das.
© 2012 John Wiley & Sons, Inc. Published 2012 by John Wiley & Sons, Inc.

The need to communicate and cooperate leads to the need for coordinating the activities pursued by agents in order to both

1. simplify the process of building multiagent systems and
2. provide the ability to reuse descriptions of coordination mechanisms and patterns.

Coordination has been defined as the process of "managing dependencies between activities" [3]. In the field of computer science, coordination is often defined as the process of separating computation from communication concerns. Although collaboration and coordination are very important tasks in multiagent applications, embedding all the required logic in agents themselves seems to be an awkward solution, since it does not grant adaptability to environmental changes and does not promote reusability. Furthermore, the choice of an embedded coordination schema does not meet today's agent-based application requirements, which include the capability to operate in open and dynamic environments.

To overtake the above problems, it is fundamental to use a collaboration and coordination approach able to deal with application and developer needs and which can be applied and reused in several scenarios. Coordination basically implies the definition of a *coordination model* and a related *coordination architecture* or a related coordination language. An agent coordination model is a conceptual framework which should cover the issues of creation and destruction of agents, communications among agents, and spatial distribution of agents as well as synchronization and distribution of their actions over time. In particular, the "*coordinables*" are the coordinated entities (or agents) whose mutual interaction is ruled by the model, the *coordination media* are the abstractions enabling the interaction among the agents, and the *coordination laws* are the rules governing the interaction among agents through the coordination media as well as the behavior of the coordination media itself [4]. A coordination architecture is a software infrastructure supporting and implementing a coordination model. A coordination language is "the linguistic embodiment of a coordination model" [5], which should orthogonally combine two languages: one for coordination (interagents) and one for computation (intraagent).

4.2   MOBILE AGENT COORDINATION: GENERAL OVERVIEW

In this paragraph, we will introduce the coordination approaches that have emerged in the mobile agent literature. The general spatial–temporal taxonomy proposed here will be discussed with references to a case study example and is meant as a more comprehensive introduction to the next paragraphs, where we will focus more on recent fully uncoupled coordination models.

4.2.1 Case Study

A promising application of mobile agents is the provision of an intelligent urban mobility system. Modern cities are being more and more disseminated with different types of computational devices, for example, wireless-enabled devices providing tourist information or traffic updates (see Figure 4.1).

The devices installed into the cars can be exploited for helping drivers achieve goals such as:

- Retrieving information about the current street, for instance, about hotels, restaurants, or interesting monuments
- Checking the traffic situation for the presence of accidents or congestions in order to select a different path
- Planning and coordinating their movements with other cars if they are part of organized groups

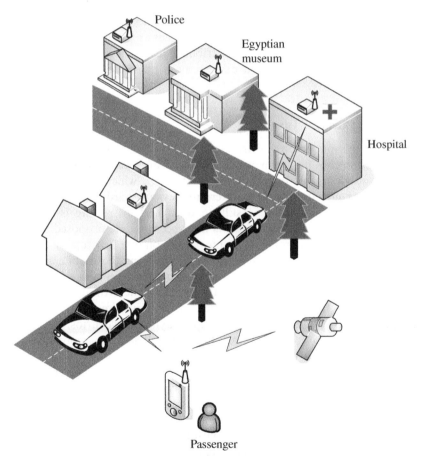

FIGURE 4.1   Intelligent urban mobility system.

To this end, we can assume that (i) cars are provided with a software agent running on some wireless device, like a little PC for vehicles or a personal digital assistant (PDA), in charge of giving the driver information on interesting places to see, what they offer, and where they are; (ii) the city is provided with an adequate embedded network infrastructure to support the provision of such information by mobile agents and to enable their coordination; and (iii) both the vehicles and the city infrastructure elements are provided with a localization mechanism to find out where they are actually located in the city [e.g., the global positioning system (GPS) in Figure 4.1]. With regard to the infrastructure, installed in sensible places or buildings, there will be a wired network of computer hosts each capable of communicating with each other and with the mobile devices located in its proximity via the use of a short-range wireless link.

This system, even in the brevity of the description, is a case study that captures in a powerful way features and constraints of mobile agent applications: It represents a very dynamic scenario with a variable number of cars issuing different requests entering and exiting the system. The above scenario and the associated coordination problems are of a very general nature, being isomorphic to scenarios such as software agents exploring the Web, where mobile software agents coordinate distributed researches on various websites. Therefore, besides the addressed case study, all our considerations are of a more general validity.

4.2.2   A First General Taxonomy

While coordination among distributed, even mobile, objects in closed and stable environments has been widely explored and well understood, coordination of mobile agents operating in the Internet, which is characterized by mutable communication paths, intermittency of hardware and software resources, and heterogeneity, is still an open and interesting issue. Several coordination models for mobile agents have been to date proposed and a first general taxonomy [6] is proposed in this paragraph (see Figure 4.2): It is based on the degree of *spatial* and *temporal coupling* among the mobile agents in a MAS (multiagent system). This taxonomy provides a very high level view of coordination models in literature, useful to follow the evolution of coordination approaches from the first coupled client–server models toward fully uncoupled Linda-like models, more suited to dynamic scenarios.

As mentioned above, the explained taxonomy divides the approaches in the literature according to their degree of spatial and temporal coupling. Spatial coupling requires that agents share a common *name space*. Therefore, agents can communicate by explicitly naming the receiving agents. In order to support this ability, a naming or location service is often used to prevent the need for explicit references from the sender agent to the receiver agent. Temporal coupling implies some form of synchronization between the interacting agents. *Temporally uncoupled models* instead relax this synchronization requirement. Therefore, agents are no longer dependent on meeting and exchanging

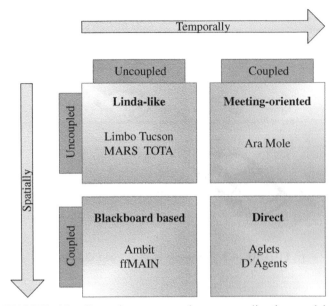

FIGURE 4.2 General taxonomy of agent coordination models.

information with others at specific times, and they do not have to worry as much about other agents' schedules.

Coordination models that exhibit both high spatial and temporal coupling are referenced as *direct coordination* models and are the most implemented ones (many Java-based agent systems, e.g., Aglets [7] and D'Agents [8], adopt this model). From the standpoint of interagent coordination, *direct coordination systems* allow agents to communicate in a peer-to-peer manner and the ease of communication comes from the fact of using agreed-upon communication protocols (messages are formatted in a previously agreed-upon format, e.g., using an Agent Communication Language (ACL) like Knowledge Query and Manipulation Language (KQML) [9]). An example from our case study is reported in Figure 4.3, where driver agents A and B share their references and can directly exchange messages using some agreed-upon protocol.

As concerns the interactions with the hosting environment, direct coordination models use client–server coordination, since the server makes its managed resources available through well-specified interfaces to client agents. In the case of mobile agent systems, this approach is quite inadequate. Since mobile agents move frequently, locating and direct messaging are expensive operations and rely heavily on the stability of the network. Statically located agents, especially those located in the same environment, can benefit from spatially coupled models, since the protocols, locations, and coordination times can be agreed upon a priori and network involvement is minimized.

The *meeting-oriented* category aims at defining spatially uncoupled models, where agents coordinate in the context of "meetings" without needing to

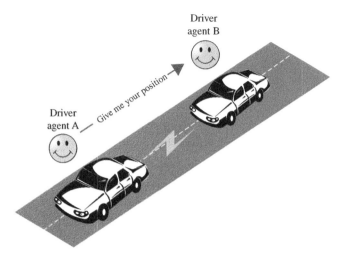

FIGURE 4.3 Example of direct coordination.

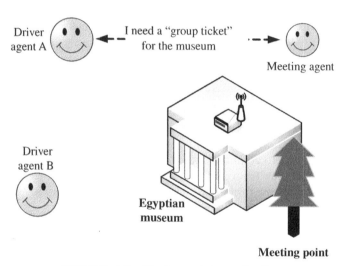

FIGURE 4.4 Meeting-point coordination.

explicitly name the partners involved. Proper meeting points are established and agents join them in order to communicate and synchronize with the other agents involved in the meeting. In Figure 4.4, we have depicted a possible meeting-oriented coordination among a group of driver agents trying to buy a "group ticket" for the Egyptian museum. Drivers may not even know the number and identity of all the group members, but they are requested to simply synchronize with a meeting-point agent that will handle the coordination among the requesters.

A very common design choice of meeting-oriented models is allowing only local interactions among the agents in a certain meeting. Nonlocal communications would introduce unpredictable delays and unreliability problems typical of direct coordination models. Examples of meeting-oriented inspired systems include the Ara mobile agent system [10] and Mole [11]. In the latter system, agents must share a reference to a specific synchronization object, which represents the meeting; accessing one of these synchronization objects, they are implicitly allowed to join the meeting and communicate with the other participants. It must be observed that meeting-oriented coordination systems do not fully achieve spatial uncoupling since agents must share the meeting names at least.

The major drawback of the two previous temporal-coupled models derives from their enforcing synchronization among interacting agents: In many mobile agent applications neither the schedule nor the position of agents is predictable, and this makes synchronization run the risk of missing interactions. All temporal uncoupled models have some shared data space in common, used as a repository for messages. This way, the sender is totally unaware of when the receiver will retrieve the message. The *blackboard-based* coordination model calls such a shared structure a "blackboard." This model can be used, for instance, to coordinate the routes of a group of drivers, avoiding them to follow wrong paths. In the example of Figure 4.5, a driver agent in the past wrote that the current road is wrong for group "GROUP1098" and a second driver agent, belonging to the same group, retrieves the record and changes its path.

Even though the use of the blackboard makes this model temporally uncoupled, it requires that agents must agree on a common message identifier

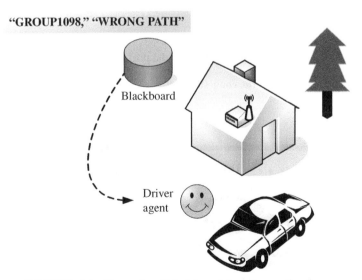

FIGURE 4.5 Example of blackboard-based coordination.

to communicate (i.e., a common namespace and thus spatial coupling). Further, since all interagent communications must be performed via a blackboard, hosting environments can easily monitor messages flowing and thus enforce a more secure execution model than previous ones. Ambit [12] and ffMAIN [13] follow the blackboard-based coordination model.

Tuple-based coordination was introduced in the late 1980s in the form of the Linda coordination language for concurrent and parallel programming [14] and consisted of a limited set of primitives, the coordination primitives, to access a tuple space. Later, in the 1990s, the model received widespread recognition as a general-purpose coordination paradigm for distributed programming. *Linda-like* coordination models use local tuple spaces, just like Blackboard-based models use a blackboard. The key difference is in the behavior of the space and its operations, which are based upon an associative mechanism: The system stores tuples in a certain format and retrieves them using associative pattern matching. In the example of Figure 4.6, the police server provides agents with a shared tuple space used to coordinate with other agents and the driver agent can request records that are related to accidents along this road to avoid wasting time queued in 10 km and change its route accordingly.

The Linda tuple space model allows full uncoupling of the cooperating agents, both spatial and temporal. Linda-like coordination systems well suit mobile agent applications situated in a wide and dynamic environment like the Internet. Equipping each agent with a complete and updated knowledge of its hosting environments and of other agents is hardly feasible. Pattern-matching mechanisms help the programmer deal with uncertainty, heterogeneity, and dynamicity and can contribute to reducing complexity. Since Linda-like

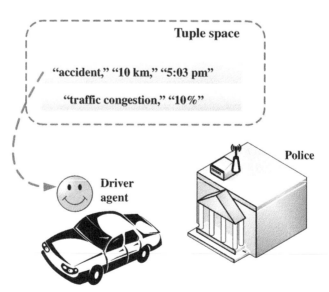

FIGURE 4.6 Using tuple spaces to coordinate car drivers.

coordination models better suit dynamic and unpredictable scenarios, they will be deeply analyzed in the following paragraph.

4.3   LINDA-LIKE COORDINATION MODELS: SOME EVALUATION CRITERIA

The atomic units of interaction in tuple-based coordination models are *tuples*. A tuple is a structured set of typed data items. Coordination activities between application agents (there included synchronization) are performed indirectly via exchange of tuples through a shared tuple space. Linda primitives provide means for agents to manipulate the shared tuple space, thereby introducing *coordination operations*. Two operations are provided to associatively retrieve data from the tuple space: *rd* and *in*, to read or extract, respectively, a tuple from the tuple space. A tuple can be written in the tuple space by an agent performing the out primitive. For instance, the out ("total", 22, l) writes a tuple with three fields: the string "total," the integer 22, and the contents of the program variable l.

The associative mechanism to get tuples from the space is based upon a *matching rule*: rd and in take a template as their argument, and the returned tuple is one matching the template. The full matching requires that the template and the tuple must be of the same length, the field types must be the same, and the values of constant fields have to be identical. For instance, the operation in ("total", ?b, l) looks for a tuple containing the string "total" as its first field followed by a value of the same type as the program variable b and the value of the variable l; the notation ?b indicates that the matching value is to be bound to the variable b after retrieval. Supposing that the above tuple ("total", 22, l) has been inserted in the tuple space, performing the previous in operation triggers the matching rules that associate the value 22 to the program variable b.

Input operations are blocking, that is, they return only when a matching tuple is found, thus implementing indirect synchronization based upon tuples' occurrences. When multiple tuples match a template, one is selected non-deterministically. There are also two nonblocking versions of the previous input operations—inp and rdp—which return true if a matching tuple has been found and false otherwise.

Though originally defined with a closed-system perspective, Linda-like coordination is attractive for programming open mobile agent applications because it offers some key advantages as concerns:

- *Uncoupling* As already stressed, the use of a tuple space as the coordination medium uncouples the coordinating components both in space and time: An agent can perform an out operation independently of the presence or even the existence of the retrieving agent and can terminate its execution before such a tuple is actually retrieved. Moreover, since agents do not have to be in the same place to interact, the tuple space helps abstract from locality issues.

- *Associative Addressing* The template used to retrieve a tuple specifies what kind of tuple is requested, rather than which tuple. This well suits mobile agent scenarios: In a wide and dynamic environment, a complete and updated knowledge of all execution environments and of other application agents may be difficult or even impossible to acquire. As agents would somehow require pattern-matching mechanisms to deal with uncertainty, dynamicity, and heterogeneity, it is worthwhile integrating these mechanisms directly in the coordination model to simplify agent programming and to reduce application complexity.

- *Context Awareness* A tuple space can act as a natural repository of contextual information to let agents get access to information about "what's happening" in the surrounding operational environment.

- *Security and Robustness* A tuple space can be made in charge of controlling all interactions performed via tuples, independently of the identity of involved agents.

- *Separation of Concerns* Coordination languages focus on the issues of coordination only: They are not influenced by characteristics of the host programming language or of the involved hardware architecture. This leads to a clear coordination model, simplifies programming, and intrinsically suits open and dynamic scenarios.

4.3.1 Some Criteria to Compare Tuple-Based Systems

Taking as a starting point the Linda system, a lot of projects have been proposed to extend the original model under different aspects: Many have extended the set of coordination primitives to address specific problems or to enrich the expressiveness of the coordination language; programmability has been often added to customize the behavior of the primitives and make the coordination system more intelligent.

The identification of an extensive set of criteria (features), wide enough to include all possible variations from the Linda model, would turn out to be confusing rather than explanatory. Here we focus on three main classification criteria, taking into account only the most significant systems in the literature:

1. *Location* of Tuple Space with Respect to Agents   The tuple space is usually considered as an external entity made available by a third party and accessed for coordination purposes by mobile agents. Nonetheless, some approaches have been proposed that "internalize" the tuple space concept, allowing single agents to have their internal instance of the tuple space.

2. *Communication scope* of Tuple Space   Given a set of tuple space instances, it may happen that they are only locally connected or are not connected

at all. This means that either a space instance can communicate with only other neighboring spaces or it can communicate only with connected agents. Other kinds of systems exploit a more global concept of tuple space, enabling also long-range, multihop, remote communications between tuple space instances.

3. *Degree of programmability* Allowed Nonprogrammable tuple space models are not able to support any computational activity. All the computations are left to the agents. This kind of system provides only a predefined set of hardwired capabilities without the possibility for the middleware itself or for an agent to change or customize any features. Adding programmability to a tuple space means that the system is able to dynamically download, store, and execute external code. Agents can program the tuple space, not only by reshaping its predefined set of features, but also by injecting new programs and services. These new implanted services can be associated with some triggering conditions (*reactivity*) to let the middleware execute those procedures whenever the proper conditions are met.

The programmability requirement is becoming of paramount importance for most modern Linda-like systems. In a reactive tuple space model, the tuple space transcends its role as a mere "tuple repository" with a built-in and stateless associative mechanism. Instead, reactive spaces can also have their own state and can react with specific actions to accesses performed by mobile agents. Reactions can access the tuple spaces, change their content, and influence the semantics of the agents' accesses.

Tuple space reactivity offers several advantages for mobile agent applications. First, it can be used to enforce local policies to rule the interactions between the agents and the hosting environment and to defend the integrity of the environments from malicious agents; for example, a site administrator can monitor access events and implement specific security policies for its execution environment. Second, using reactions, an agent can more easily adapt the semantics of its interactions to the specific characteristics of the hosting environment: This will result in a much simpler agent programming model compared to the fixed pattern-matching mechanism of Linda-like models and adds "distributed intelligence" to the whole system.

4.4 OVERVIEW OF SOME MODERN TUPLE-BASED APPROACHES

Considering the above-mentioned criteria, in this paragraph we will survey some important Linda-like extensions from the literature. They have been graphically reported in Figure 4.7, where each axis corresponds to one of the chosen features.

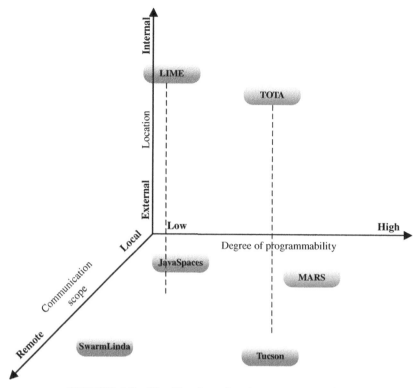

FIGURE 4.7 Classification of tuple space approaches.

4.4.1 JavaSpaces

JavaSpace [15] is a technology developed by Sun aimed at providing distributed repositories of information similar to Linda tuple spaces. JavaSpaces are tuple spaces, where tuples are Java objects: Tuples are classes implementing the Entry interface. The access to the tuple space is defined in terms of Java methods, belonging to the JavaSpace interface. Tuples are stored in their serialized form (i.e., using the standard Java serialization mechanism) and the pattern-matching mechanism relies on the serialized form of the tuples. Other useful features provided by JavaSpaces are:

- The possibility to define a "lease time" for a tuple, that is, the tuple lifetime that, when it is over, removes the tuple from the space
- The support for transactions, to ensure the correctness of the performed operations
- A notifying mechanism, which signals a number of registered listeners when a new tuple is written

Besides these relevant functionalities, JavaSpaces still can be classified as *nonprogrammable*, because there is no way for the Java programmer to inject code to customize the behavior of the tuple space. It has been also evaluated as a system with a *low communication scope*, in the sense that there is no connection among neighbor spaces and interactions can occur only in the local tuple space. The middleware is positioned *externally* with respect to the agents storing and retrieving tuples: Agents simply get a shared reference to the tuple space object but are not supposed to have a personal tuple space.

4.4.2   SwarmLinda

SwarmLinda [16] is an example of a tuple space middleware with a *remote communication scope*. These tuple spaces can interact with each other, meaning that different instances can exchange data and information. Often, this is reached through a peer-to-peer network, where different hosts run an instance of the middleware, and such instances connect to other instances running on other hosts. SwarmLinda is implemented in Java and exploits XML documents to describe tuples. SwarmLinda gets inspiration from concepts of swarm intelligence and multiagent systems, modeling the tuple space as a set of nodes, and provides services (inserting, retrieving, etc.) performed by "ants" that travel across the nodes and search for (or carry on) one or more tuples. An interesting feature of SwarmLinda is its tuple aggregation, based on patterns criteria, thus similar tuples will be closer and kept (possibly) in the same node space. This implies that, while the whole system can be seen as a composition of distributed tuple spaces, it is really a single tuple space with clients connected to different instances but that perceives the system as unique. Such a tuple space is situated externally to the interacting agents and is not programmable with special user-defined code.

4.4.3   LIME

LIME (Linda in a Mobile Environment) is the evolution of the Linda coordination model, which addresses mobility of agents and their tuple spaces. LIME tries to adapt the advantages of the Linda model (in terms of uncoupled interactions) to the dynamicity of a mobile world.

In LIME the tuple space can be considered *internal*, since each mobile entity is associated to an *interface tuple space* (ITS), which can be considered as a personal space where tuples can be stored or retrieved from. When mobile agents meet in the same physical place, their ITSs are automatically merged and a new shared tuple space is available for their coordination. The merged tuple spaces become a single tuple space from the viewpoint of agents, making transparent any details about the actual location of tuples. When a new agent arrives, its ITS is merged with the current one, and the content is automatically recomputed. Agents can also define "private ITSs," that are not shared with other agents, to be exploited to store private information.

Interactions among agents can occur in LIME only in their shared ITS and this make LIME a *local* coordination model from the standpoint of the communication scope. In addition, LIME cannot be evaluated as a *programmable* system, though it provides a limited form of event reaction capability: Due to a new reactive statement, system events can be captured and proper code fragments can be executed in response to each event. Therefore, LIME provides its tuple space with a form of *reactivity*: A mobile entity can program the behavior of its own ITS, so as to provide more flexible control on accesses to it.

4.4.4 TuCSoN

TuCSoN [17] is a coordination model for mobile Internet information agents. TuCSoN exploits the tuple center concept to define an interaction space spread over a collection of Internet nodes and, according to our classification criteria, is clearly characterized by a remote communication scope. In particular, each tuple center is associated to a node and is denoted by a locally unique identifier. A TuCSoN tuple center can be identified by means of either its full absolute name or its relative local name: The former is unique all over the Internet, while the latter is unique only within a single-node name space. The syntax tc@node uniquely identifies the specific tuple center tc hosted by node, while the relative name tc refers to the local tuple center of the node where the mobile agent is currently running. Due to this distributed network of tuple nodes and centers, TuCSoN allows agents to perform tuple center operations in both a *network-transparent* and a *network-aware* fashion: The former allows an agent to remotely look for its next hosting environment, while the latter is typically used by an agent wishing to interact with a node locally.

It must be pointed out that a tuple center is not just a tuple space, since it can support "specification tuples," which define the reaction logic to communicative actions on the tuple space. This make TuCSoN a coordination model with a *high degree of programmability* because it gives the programmer the capability of modifying the standard behavior of the tuple space by programming it without changing the set of primitives. In addition, TuCSoN provides a tuple space external to the agents as concerns the location criteria.

4.4.5 MARS

MARS (Mobile Agent Reactive Spaces) [18] extends the Sun JavaSpaces, implementing *programmable reactive* tuple spaces for Java-based mobile agent applications. The MARS model assumes the existence of one (unnamed) tuple space locally to each execution environment (thus *located externally* to mobile agents), which is independent of other neighboring spaces (a *local communication scope*, following one of the above evaluation criteria). This tuple space represents the only means that agents can use to interact both with the local execution environment and with other agents. As in JavaSpaces, MARS tuples are Java objects whose instance variables represent tuple fields. To access the tuple space, the MARS interface provides some fundamental Linda-like operations, such as read, write, take, readAll, takeAll.

The key feature of MARS relies, however, in the notion of programmable reactive tuple space, which makes it possible to embody computational abilities within the tuple space, assuming specific behaviors in response to access events. Therefore, a MARS tuple space is no longer a mere tuple repository, but an active component with its own state, which can be programmed so as to react to tuple access operations by performing specific reactions. The MARS reaction model complies with the standard tuple space model: Reactions are coded as *metalevel tuples* (i.e., *metatuples*) stored in a local *metalevel tuple space*. Each metatuple has the form (Rct, T, Op, I), meaning that when an agent I invokes the operation Op on a tuple matching T, MARS must trigger the reaction Rct. The reaction itself is a method of a Java object. Whenever an access occurs, the system searches the metalevel tuple space for a matching metatuple. If such a tuple is found, the corresponding reaction object is retrieved and its reaction method is executed. To avoid endless recursion, reactions are not allowed to trigger other reactions in a chain. Metatuples can be stored and retrieved at run time, leading to dynamic installations and uninstallations of reactions, both by the local administrator and by agents.

4.4.6 TOTA

Unlike traditional shared data space models, tuples in the TOTA (Tuples On The Air) middleware [19] are not associated to a specific node (or to a specific data space) of the network (it can be classified as an internal tuple space as concerns the location criteria). Instead, tuples are injected in the network and can autonomously propagate and diffuse in the network accordingly to a specified pattern. Thus, TOTA tuples form a sort of spatially distributed data structure able to express not only messages to be transmitted between application components but also, more generally, some contextual information on the distributed environment. Upon the distributed space identified by the dynamic network of TOTA nodes, each component is capable of locally storing tuples and letting them diffuse through the network. Tuples are injected in the system from a particular node and spread hop by hop accordingly to their propagation rule. In fact, a TOTA tuple is defined in terms of a "content" and a "propagation rule" $T = (C,P)$. The content C is an ordered set of typed fields representing the information carried on by the tuple. The propagation rule P determines how the tuple should be distributed and propagated in the network. In addition, the propagation rules can determine how the content of the tuple should change while it is propagated. This propagation provides a sufficient degree of programmability to the tuple space and makes its communication scope remote since tuples move dynamically through network nodes crossing neighboring spaces.

4.5 ROLES FOR MOBILE AGENT COORDINATION

In this section we show an alternative research direction concerning mobile agent coordination. This approach is based on roles, discussed next.

4.5.1 Roles

Roles have been already exploited in object-oriented approaches, where a role is defined as a set of behaviors common to different entities [20], with the capability to apply them to an entity in order to change its capabilities and behavior. Other approaches promote roles as views of a particular object or entity [21], stressing the similarity between roles in computer programs and those in real life.

Starting from previous work in object-oriented programming, roles have been applied to agents, which after all can be thought as autonomous and active objects, promoting the reuse of solutions and making the definition of coordination scenarios easier. Roles allow not only the agent application developers/designers to model the execution environment, but also allow agents to actively "feel" the environment itself. In other words, roles allow the developer and its agents to perceive the execution environment in the same way.

The importance of the use of roles is supported by the fact that they are adopted in different areas of the computing systems, in particular to obtain uncoupling at different levels. Some examples of such areas are *security*, in which we can recall role-based access control (RBAC) [22], which allows uncoupling between users and permissions, and *computer-supported cooperative work* (CSCW) [23], where roles grant dynamism and separation of duties. Also in the area of software development we can find approaches based on roles, especially in *object-oriented programming* [24, 25], in *design patterns* [20], and in computer-to-human interfaces [26], which remarks the advantages of role-based approaches.

4.5.2 Applying Roles to Agents

Applied to the agent scenario, roles are mainly exploited to define common interactions between agents (e.g., the interactions between the contract-net participants or between auctioneers and bidders in an auction) and promote an organizational view of the system, which well suits agent-oriented approaches [26]. Roles embed all information and capabilities needed in a particular execution environment to communicate, coordinate, and collaborate with other entities and/or agents. Due to this, an agent (and its programmer) does not need to know details about the current execution environment but only needs to know which role to assume and use to interact with (or to exploit) the environment itself. This leads to a separation of issues related to the agent logic and its communication with other entities. The former is embedded in the agent itself, since it defines the agent base behavior, while the latter is embedded in a role and expresses an added behavior. This separation is more emphasized at the development phase, since it is possible to develop agents and roles they are going to use in separated times and ways, leading to a more modular development process. Another advantage of the use of roles is solution reusability. Since roles embed a behavior applied to a specific application context (e.g., collaboration in a MAS system), keeping it separated from agents, a set of roles can be successfully applied to other similar areas, leading not only to a faster development but also to an experience and solution reuse.

4.5.3   Roles and Mobility

Roles are pragmatically useful in mobile agent scenarios. Mobile agents change their execution environment during their life, and they can hardly make assumptions on the hosting environments. Separating the agent logic from the interaction logic, roles allow agents to interact with the hosting environment without knowing details about it.

On the one hand, when a mobile agent arrives at an execution environment, it can rely on the presence of well-known roles that make given capabilities available; the agent can exploit such capabilities to carry out its tasks, disregarding how these capabilities are implemented. On the other hand, hosting environments can implement roles not only in terms of local functionalities but also in terms of local policies. When a mobile agent changes its execution environment and it must carry out the same task in the new environment, it can find and play the same role disregarding implementation details. In general, roles can be exploited to coordinate not only with the hosting environments but also with other entities.

4.5.4   Running Examples

This section presents some application examples where roles are exploited to coordinate agents with other entities.

1. *Bridge* This example shows in particular the coordination between agents and their environments. Let us consider a driver that has to cross a bridge for which a fee is required. When the driver arrives near the bridge, he can assume the *bridgeCrosser* role, which gives him the capability of paying the service and enables the crossing of the bridge. Different bridges can implement the capabilities and the service in different ways: For instance, one bridge can exploit a prepaid card, while another can rely on credit cards; as another example, one bridge can provide a single-use pin that is read by sensors along the bridge, while another bridge can simply raise a tollgate.

2. *Museum* In the latter example we consider tourists that visit a local museum; we suppose that each involved human is supported by an agent that is mobile since it resides on a mobile device. We can define three main roles: *visitor, guide*, and *driver*. The visitor role is played by all the tourists that want to visit the museum; the guide role is played by one person who explains the artworks of the museum; the driver is played by the person who drives the tourist bus. Assigning roles is a first way of coordinating agents. We can figure out a situation where there is only one guide, whose agent is the only one authorized to play the guide role; in another situation, a scholar group can arrange the visit

by letting single students explain to other students the part of the museum they have studied: In this case, the guide role is played by different agents in turn. Moreover, roles can coordinate the movement of groups of tourists by suggesting the room where the guide is. An intragroup coordination is also possible, because different guides present in the same museum can be led to not overlap in the same room; to avoid confusion, this information is sent only to those agents playing the guide role. With regard to policies, some environments can impose that only the guide can talk with the driver.

These simple examples should sketch how roles can be exploited to coordinate agents. In our research activities, we have proposed RoleX [28], a role-based infrastructure implemented in the context of the BRAIN framework [29]. RoleX not only enables the assumption of roles by agents but also provides several mechanisms that can be exploited to coordinate agents playing given roles. For instance, RoleX has been exploited to implement *computational institutions* [30], where it has proved to be a powerful and flexible means to implement abstractions for multiagent systems.

4.6 FUTURE DIRECTIONS IN MOBILE AGENT COORDINATION

This chapter has surveyed mobile agent coordination research in the mainstream literature of software agents. The general trend that clearly emerges is toward *an increasing degree of uncoupling among the coordinable entities* (i.e., mobile agents). In the last few years, new research fields, such as *ubiquitous* and *pervasive computing* [31], are gaining momentum, leading to application scenarios that are tremendously open to components (hardware and software) developed by independent parties and whose evolution pace is often overwhelming. Furthermore, in so-called *software intensive systems* [32], application designers no longer produce software components to be integrated in well-defined contexts but, instead, endow them with increasing levels of autonomy. Components integration is typically not controlled or planned externally but is induced by changes perceived in the environment, leading to self-adaptable systems. Many pervasive applications, for instance, running on sensor networks [33], have been implemented as agent-based programs and several initiatives are trying to use agent research to build adaptable and self-organizing service architectures [34].

The above research definitely proves that *interactions must be treated as first-class "citizens"*; designers can no longer bury them in the component code or make them implicit. Rather, they must externalize and represent them in the architecture. Mobile agent coordination, as highlighted in this chapter, has done much of the latter externalization work and we are convinced it will be a valuable source of inspiration for future distributed component models. However, the next step in coordination research is toward the provision of new powerful means to *coordinate mobile agent interactions without any* (*or with minimal*) *design-level agreement among the participant agents.* Tuple-based coordination models introduce associative mechanisms to retrieve tuples (using pattern matching), but this seems to be no longer enough in extremely dynamic and open environments. Pattern matching implies at least two forms of coupling that have not been considered so far: *syntactic* and *semantic coupling.* The former is related to the syntactic structure of the tuple (e.g., field ordering and typing) and can heavily limit coordination possibilities among independently developed mobile agents: The question is how can two or more agents coordinate if their internal representations of a tuple are syntactically incompatible? The solutions we have surveyed in this chapter assume an implicit design-level agreement on syntax and can thus hardly tackle the stressed dynamicity and openness of modern scenarios. Semantic coupling is still more a subtle issue to deal with, because it poses the problem of the correct interpretation of the tuple itself. When two agents are designed by independent programmers, how can we ensure that the meaning of field A (e.g., "road congestion level") is correctly understood by both the coordinable entities?

Some directions can be given to address the two mentioned problems of syntactic and semantic uncoupling. We think that knowledge will be the key factor of future coordination models. Coordination artifacts will no longer be syntactically constrained tuples, but new forms of "semantically enriched" models will rise [35]. Current research on environments for multiagent systems [36, 37] is pointing in this direction, recognizing the importance of powerful knowledge-enabled environments to drive coordination among mobile agents. The concept of *knowledge networks* [38] is an ongoing example of introducing knowledge as the mediator of agents' interactions in open systems. Coordination would thus be driven by more sophisticated ontology-based knowledge data, other than a simple tuple value. Linda-like tuples will become sort of "knowledge atoms" and coordination will be supported by environments capable of reconciling semantic and syntactic mismatches using well-known semantic inference and reasoning algorithms. This will enable more robust and adaptive configuration patterns as well as mitigate the syntactic and semantic coupling problems described in this concluding section.

REFERENCES

1. D. Deugo, M. Weiss, and E. Kendall, Reusable patterns for agent coordination, in A. Omicini, F. Zambonelli, M. Klusch, and R. Tolksdorf (Eds.), *Coordination of Internet Agents: Models, Technologies, and Applications*, Springer, Berlin, Mar. 2001, Chapter 14, pp. 347–368.

2. J. Jennings and M. Wooldridge, Agent-oriented software engineering, in J. Bradshaw (Ed.), *Handbook of Agent Technology*, AAAI/MIT Press, Palo Alto, CA, USA, 2000.

3. T. W. Malone and K. Crowston, The interdisciplinary study of coordination, *ACM Comput. Surv.*, 26:87–119, 1994.

4. P. Ciancarini, Coordination models and languages as software integrators, *ACM Comput. Surv.*, 28(2):300–302, 1996.

5. N. Carriero, and D. Gelernter, Coordination languages and their significance. *Commun. ACM*, 35(2):97–107, 1992.

6. G. Cabri, L. Leonardi, and F. Zambonelli, Mobile-agent coordination models for internet applications, *IEEE Computer*, 33(2):82–89, 2000.

7. D. Lange, M. Oshima, G. Karjoth, and K. Kosaka, Aglets: Programming mobile agents in Java, in G. Goos, J. Hartmanis, and J. van Leeuwen (Eds.), *Proceedings of International Conference Worldwide Computing and Its Applications (WWCA'97)*, Lecture Notes in Computer Science, Vol. 1274, Springer, Berlin, Tsukuba, Japan, March 10–11, 1997, pp. 253–266.

8. R. S. Gray, G. Cybenko, D. Kotz, R. A. Peterson, and D. Rus, D'Agents: Applications and performance of a mobile-agent system, *Software—Pract. Exper.*, 32(6):543–573, 2002.

9. T. Finin R. Fritzson, D. McKay, and R. McEntire, KQML as an agent communication language, in *Proceedings of the 3rd International Conference on Information and Knowledge Management (CIKM'94)*, ACM New York, NY, USA, Dec. 1994.

10. H. Peine, Ara—Agents for remote action, in *Mobile Agents, Explanations and Examples*, Manning/Prentice-Hall, 1997, pp. 96–161.

11. J. Baumann et al., Mole—Concepts of a mobile agent system, *World Wide Web J.*, 1(3):123–137, 1998.

12. L. Cardelli and D. Gordon, *Mobile Ambients. Foundations of Software Science and Computational Structures*, Lecture Notes in Computer Science, Vol. 1378, Springer, Berlin, 1998, pp. 140–155.

13. P. Domei, A. Lingnau, and O. Drobnik, Mobile agent interaction in heterogeneous environments, in *Proceedings of the 1st International Workshop on Mobile Agents*, Lecture Notes in Computer Science, Vol. 1219, Springer, Berlin, 1997, pp. 136–148.

14. D. Gelernter and N. Carriero, Coordination languages and their significance, *Commun. ACM*, 35(2):96–107, 1992.

15. E. Freeman, S. Hupfer, and K. Arnold, *JavaSpaces Principles, Patterns and Practice*, Addison Wesley, Reading, MA, 1999.

16. A. Charles, R. Menezes, and R. Tolksdorf, On the implementation of SwarmLinda, in *Proceedings of the 42nd Annual Southeast Regional Conference*, Spartanburg, Huntsville, AL, ACM, New York, NY, USA, Nov. 2004.

17. A. Omicini and F. Zambonelli, Coordination for Internet application development, *J. Auton. Agents Multi-Agent Syst.*, 2(3):251–269, 1999.

18. G. Cabri, L. Leonardi, and F. Zambonelli, MARS: A programmable coordination architecture for mobile agents. *IEEE Internet Comput.*, 4(4):26–35, 2000.

19. M. Mamei and F. Zambonelli, Programming pervasive and mobile computing applications with the TOTA middleware, in *Proceedings of the 2nd IEEE International Conference on Pervasive Computing and Communication (Percom2004)*, IEEE Computer Society Press, Orlando, FL, 2004.

20. M. Fowler, Dealing with roles, 1997, available: http://martinfowler.com/apsupp/roles.pdf.

21. D. Baumer, D. Ritchie, W. Siberski, and M. Wulf, The role object pattern, in *Proceedings of the 4th Pattern Languages of Programming Conference (PLoP)*, Monticello, IL, USA, Sept. 1997.

22. R. S. Sandhu, E. J. Coyne, H. L. FeinStein, and C. E. YoumanHayes-Roth, Role-based access control models, *IEEE Computer*, 20(2):38–47, 1996.

23. A. Tripathi, T. Ahmed, R. Kumar, and S. Jaman, Design of a policy-driven middleware for secure distributed collaboration, in *Proceedings of the 22nd International Conference on Distributed Computing System (ICDCS)*, Vienna (A), July 2002.

24. B. Demsky and M. Rinard, Role-based exploration of object-oriented programs, in *Proceedings of International Conference on Software Engineering (ICSE 2002)*, IEEE Computer Society Press, Orlando, FL, May 19–25, 2002.

25. B. B. Kristensen and K. Østerbye, Roles: Conceptual abstraction theory & practical language issues, *Special Issue of Theory and Practice of Object Systems on Subjectivity in Object-Oriented Systems*, 2(3):143–160, 1996.

26. B. Shneiderman and C. Plaisant, The future of graphic user interfaces: Personal role managers, in G. Cockton, S. W. Draper, and G. R. S. Weir (Eds.), *People and Computers IX: Proceedings of Conference on HCI '94*, Press Syndicate of the University of Cambridge, Glasgow, UK 1994.

27. F. Zambonelli, N. R. Jennings, and M. Wooldridge, Organizational rules as an abstraction for the analysis and design of multi-agent systems, *J. Knowledge Software Eng.*, 11(3):303–328, 2001.

28. G. Cabri, L. Ferrari, and L. Leonardi, The RoleX environment for multi-agent cooperation, in *Proceedings of the 8th International Workshop on Cooperative Information Agents (CIA)*, Erfurt, Germany, Lecture Notes in Artificial Intelligence, Vol. 3191, Springer, Berlin, Sept. 2004.

29. G. Cabri, L. Leonardi, and F. Zambonelli, BRAIN: A framework for flexible role-based interactions in multiagent systems, in *Proceedings of 2003 Conference on Cooperative Information Systems (CoopIS)*, Lecture Notes in Computer Science, Vol. 2888, Springer, Berlin, Catania, Italy, Nov. 2003.

30. G. Cabri, L. Ferrari, and R. Rubino, *Building Computational Institutions for Agents with RoleX. Artificial Intelligence & Law*, Vol. 1, Springer, Berlin, 2008, pp. 129–145.

31. D. Estrin, D. Culler, K. Pister, and G. Sukjatme, Connecting the physical world with pervasive networks, *IEEE Pervasive Comput.*, 1(1):59–69, 2002.

32. J. L. Fiadeiro, Designing for software's social complexity, *IEEE Computer*, 40(1): 34–39, 2007.

33. C. L. Fok, G. C. Roman, and C. L. Agilla, A mobile agent middleware for self-adaptive wireless sensor networks, ACM Transactions on Autonomous and Adaptive Systems, Vol. 4, ACM, New York, NY, USA, 2009, pp. 1–26.

34. R. Quitadamo and F. Zambonelli, Autonomic communication services: A new challenge for software agents. *J. Auton. Agents Multi-Agent Syst. (JAAMAS)*, 17(3): 457–475, 2007.

35. L. Tummolini, C. Castelfranchi, A. Ricci, M. Viroli, and A. Omicini, *Exhibitionists and Voyeurs do it better: A Shared Environment Approach for Flexible Coordination with Tacit Messages, Environments for Multi-Agent Systems*, Lecture Notes in Artificial Intelligence, Vol. 3374, Springer, Berlin, 2005, pp. 215–231.

36. E. Platon, M. Mamei, N. Sabouret, S. Honiden, and H. V. D. Parunak, Mechanisms for environments in multi-agent systems: Survey and opportunities, *J. Auton. Agents Multi-Agent Syst.*, 14(1):61–85, 2007.

37. P. Valckenaers, J. Sauter, C. Sierra, and J. A. Rodriguez-Aguilar, Applications and environments for multi-agent systems, *J. Auton. Agents Multi-Agent Syst.*, 14(1): 61–85, 2007.

38. M. Baumgarten et al., Towards self-organizing knowledge networks for smart world infrastructures. *Int. Trans. Syst. Sci. Appl.*, 1(3), pp. 123–133, 2006.

5 Cooperating Mobile Agents

SUKUMAR GHOSH and ANURAG DASGUPTA

University of Iowa, Iowa City, IA 52242, USA

5.1 INTRODUCTION

A mobile agent is a piece of code that migrates from one machine to another. The code (often called the *script*), which is an executable program, executes at the host machine where it lands. In addition to the code, agents carry data values or procedure arguments or results that need to be transported across machines. Compared to messages that are passive, agents are active and can be viewed as messengers.

Mobile agents are convenient tools in distributed systems, at both the applications layer and the middleware level. The promise of mobile agents in bandwidth conservation or disconnected modes of operation is now well accepted. Deploying multiple mobile agents cooperating with one another can add a new dimension to distributed applications. While parallelism is the obvious advantage, the issues of load balancing, agent rendezvous, and fault tolerance play major roles. Among numerous possible applications, we highlight the following four problems, each with a different flavor of cooperation:

- **Mapping of an Unknown Network** Network mapping is also known as the *topology discovery* problem. Making such a discovery using a single mobile agent is equivalent to developing an efficient algorithm for graph traversal. With multiple agents, the challenge is to develop an efficient cooperation mechanism so that the discovery is complete in the fewest number of hops and redundant traversals are avoided.
- **Concurrent Reading and Writing** A distributed data structure has different components mapped to host machines at different geographic locations. As multiple agents concurrently access such a distributed data structure, the reading agent and the writing agent need to properly synchronize their operations so that the semantics of data sharing are preserved.

Mobile Agents in Networking and Distributed Computing, First Edition.
Edited by Jiannong Cao and Sajal K. Das.
© 2012 John Wiley & Sons, Inc. Published 2012 by John Wiley & Sons, Inc.

- **Black-Hole Search** A *black hole* is a node that can potentially capture a visiting agent and thus disrupt an application. Although it implies a malicious intent on the part of the host, a black hole can be as simple as a crashed node. If black holes can be located, then traversal paths can be rerouted without incurring further loss of mobile agents.

- **Stabilization** Transient failures occasionally corrupt the global state of a distributed system, and stabilization is an important technique for restoring normal operation. To stabilize a network a mobile agent patrols the network, and plays the role of traveling repairperson. Multiple agents can expedite the process of stabilization, but in doing so, some synchronization issues need to be resolved. We will address how multiple agents can be deployed for maximum speedup of a stabilizing application.

This chapter has seven sections. Section 5.2 describes the model and the notations. Sections 5.3–5.6 address the four problems highlighted above. Finally, Section 5.7 contains some concluding remarks.

5.2   MODEL

We represent a distributed system by a *connected undirected graph* $G = (V, E)$, where V is the set of nodes representing processes and E is the set of edges representing channels for interprocess communication. Basic interprocess communication uses messages that are received in the same order in which they are sent. Processes do not have access to a global clock.

Whenever appropriate, we will represent the program for each process of a set of rules. Each rule is a guarded action of the form $g \to A$, where g is a Boolean function of the state of that process and those of its neighbors received via messages and A is an action that is executed when g is true. An action by a process involves several steps: receiving a message, updating its own state, and sending a message. The execution of each rule is atomic, and it defines a step of the computation. When more than one rule is applicable, any one of them can be chosen for execution. The scheduler is unfair. A *computation* is a sequence of atomic steps. It can be finite or infinite.

Definition A global state of the system is a tuple $(s(0), s(1), \ldots, s(n))$, where $s(i)$ is the state of process i together with the states of all channels.

Each agent is launched by an initiator node that is also called the agent's *home*. An agent consists of the following six components:

1. The identifier *id* (also called a *label*), usually the same as the initiator's id. The id is unnecessary if there is a single agent but is essential to distinguish between multiple agents in the same system.

2. The agent program: This program is executed when the mobile agent lands on a host machine.

3. The briefcase B containing a set of data variables. It defines the state of the agent computation as well as some key results that have to be carried across nodes.

4. The previous process PRE visited by the agent.

5. The next process to visit NEXT that is computed after every hop.

6. A supervisory program for bookkeeping purposes.

Each hop by an agent is completed in zero time. The agent computation is superimposed on the underlying distributed computation executed by the network of processes. The state of the agent is defined by its control and data variables, and the state of the distributed system consists of the local states of all the processes. When an agent executes a step, it changes its own state and also potentially changes the state of the host on which it executes that step. Unless specified otherwise, the visit of an agent at any node will be treated as an *atomic event*.

Finally, an agent model that involves multiple agents can be either *static* or *dynamic*. In the static model, the number of agents and their homes are known at the beginning of the application, and they remain unchanged throughout the entire life of the application. In the dynamic model, an agent can *spawn* child agents or *kill* them whenever necessary. Unless specified otherwise, we will consider the static model only.

5.3   MAPPING A NETWORK

We assume that an agent is trying to construct the complete map of an undirected connected graph. It has to explore all the nodes and edges of the graph starting from some node. When the agent traverses an edge, it explores the corresponding edge and both of its end nodes. During exploration, the agent keeps track of the visited nodes and the edges so that it recognizes them later. In particular, after reaching an already explored node v incident on an explored edge e, the agent recognizes the location of v and of the other end of e on the partial map it constructs. The agent also knows the number of unexplored edges incident on an explored node but does not know the other ends of these edges.

The goal is to explore all the nodes and the edges of the undirected connected graph with the minimum number of edge traversals. One motivation for visiting all nodes and retrieving data from unknown nodes in a vast network is network maintenance. The agent continuously patrols the network. Also, fault detection in a network requires this type of *perpetual exploration*. Before we discuss the multiagent case, we introduce a few exploration protocols using a single mobile agent.

5.3.1   Exploring Undirected Graphs

The *penalty* of an exploration algorithm running on a graph $G = (V, E)$ is the worst-case number of traversals in excess of the lower bound $|E|$. The total cost of an exploration algorithm running on a graph $G = (V, E)$ is the worst-case number of edge traversal it uses taken over all starting points and all adversary decisions. The adversary has the power of choosing an arbitrary unexplored edge.

Panaite and Pelc [1] provide an exploration algorithm whose *penalty* is $O|V|$ for every graph. In fact, they showed that the penalty never exceeds $3|V|$.

Natural heuristics such as GREEDY and Depth First Search (DFS) fail to achieve the penalty $O|V|$ for all graphs. The following theorem shows the inefficiencies of these two strategies:

Theorem 1 *The penalties of GREEDY and DFS are not linear in the order of the graph.*

At any stage of the algorithm execution, the edges that are already traversed are called *explored*, and the remaining are called *free*. A node is *saturated* if all its incident edges are explored. Otherwise it is free.

The arguments why DFS and GREEDY fail are not hard to see. In both cases, the agent uses unexplored edges as long as possible, and when stuck at a node v, it uses a simple strategy to reach a free node v'. In case of GREEDY, v' is the free node closest to v, while in case of DFS, v' is the most recently visited free node. It turns out that both these choices are too naive. The vision of GREEDY is too local, while DFS does not make sufficient use of the knowledge of the explored subgraph, basing its decisions only on the order of visits.

The algorithm presented by Panaite and Pelc also uses unexplored edges as long as possible. But as opposed to DFS and GREEDY, their algorithm explores the graph in the order given by a dynamically constructed tree. The key difference is the choice of the free node to which the agent relocates after getting stuck. The agent gets back to the node of the dynamically constructed tree at which it interrupts the construction of the tree. In this way the number of traversals through already explored edges is reduced. By following the structure of the dynamic tree, the agent is not distracted from systematic exploration of free nodes situated close to it, which is the case for GREEDY. At the same time, temporal preferences dominate over geographic references that lead to the inefficiency of DFS.

Algorithm EXPLORE in [1] addresses this shortcoming. One of the main features of EXPLORE is agent relocations along a dynamically constructed tree T. Assume that the agent starts at node r of graph G. At any stage of exploration, let H denote the known subgraph of G; T represents a tree in H rooted at r and connecting only the saturated nodes. An overview of this algorithm is given in Figure 5.1. We first define the different procedures that are the building blocks of the algorithm.

Procedure SATURATE(v) performs a traversal that starts and ends at v and saturates v. Procedure EXTEND(T) constructs the new tree T that corresponds to the current H. Procedure NEXT(v) is defined as follows: If $V(T) = V(H)$, then NEXT(v) returns the node v. Otherwise, it returns a node $u \in \text{Ext}(w)$, where w is the first node w' in dfs(T) with Chi(w') $= \emptyset$ and Ext(w') $\neq \emptyset$. Here,

<u>Program Explore</u>
r := the starting node;
$T := (\emptyset, \emptyset)$;
$v := r$;
do $V(T) \neq V(H) \rightarrow$
    SATURATE(v);
    EXTEND(T);
    $v :=$ NEXT(v);
    Relocate to v along a shortest path in H
od

FIGURE 5.1   Algorithm for graph exploration with a single agent.

Chi(w') denotes the set of children of w', that is, neighbors of w' in T different from the parent of w' if it exists, and Ext(w') denotes the set of extensions of w', that is, neighbors of w' in H but not in T.

Panaite and Pelc proved the following theorem:

Theorem 2 *The* penalty *of the proposed algorithm EXPLORE is linear in the order of the graph. It uses $m + O(n)$ edge traversals, for every graph with n nodes and m edges.*

5.3.2   Optimal Graph Exploration

For a given graph G and a given starting node v, a measure of the quality of an exploration algorithm A is the ratio $C(A, G, v)$/opt (G, v) of its cost to that of the optimal algorithm having complete knowledge of the graph. Here Opt(G, v) is the length of the shortest *covering walk* which is the exploration with fewest edge traversals starting from node v. The cost $C(A, G, v)$ is the worst-case number of edge traversals taken over all of the choices of the adversary. The ratio represents the relative penalty paid by the algorithm for the lack of knowledge of the environment. For a given class U of graphs, the number

$$O_U(A) = \sup_{G \in U} \max_v \in G\left(\frac{C(A, G, v)}{\text{opt}(G, v)}\right)$$

is called the *overhead* of algorithm A for the class U of graphs. For a fixed scenario, an algorithm is called *optimal* for a given class of graphs if its overhead for this class is minimal among all exploration algorithms working under this scenario.

Dessmark and Pelc [2] presented optimal exploration algorithms for several classes of graphs. They considered the following three scenarios:

1. The agent has no a priori knowledge of the graph. They called it *exploration without a map.*
2. The agent has an unlabeled isomorphic copy of the graph. This is called an *unanchored map* of the graph.

3. The agent has an unlabeled isomorphic copy of the explored graph with a marked starting node. This is called an *anchored map* of the graph. This scenario does not give the agent any sense of direction, since the map is unlabeled. For example, when the agent starts the exploration of a line, such a map gives information about the length of the line and distances from the starting node to both ends but does not tell which way is the closest end.

In all scenarios, the assumption is that nodes have distinct labels, and all edges at a node v are numbered $1, \ldots, \deg(v)$ in the explored graph. Otherwise it is impossible to explore even the star graph with three leaves, as after visiting the second leaf, the agent cannot distinguish the port leading to the first visited leaf from that leading to the unvisited one. Hence the agent can recognize the already visited nodes and traversed edges. However, it cannot tell the difference between yet-unexplored edges incident on its current position. The actual choice of such unexplored edges is made by the adversary when worst-case performance is being considered. Table 5.1 summarizes the main results of [2]

The table indicates that for the class of all undirected connected graphs DFS is an optimal algorithm for all scenarios except for trees. Without any knowledge, DFS is optimal for trees. Under the scenario with an unanchored map, the optimal overhead is at least $\sqrt{3}$ but strictly below 2. Thus DFS is not optimal in that case. Dessmark and Pelc [2] give an optimal algorithm for trees with an anchored map and show that its overhead is $\frac{3}{2}$. Of the many algorithms described in [2] we choose the simplest case of exploration on lines with an anchored map. This means the agent knows the length n of the line as well as the distances a and b between the starting node and the endpoints. The algorithm for $a \leq b$ is given in Figure 5.2.

Fraigniaud et al. [3] showed that for any K-state agent and any $d \geq 3$ there exists a planar graph of maximum degree d with at most $K + 1$ nodes that the agent cannot explore. They also showed that in order to explore all graphs of diameter D and maximum degree d an agent needs $\Omega(D \log d)$ memory bits, even if the exploration is restricted to planar graphs. This latter bound is tight. So the worst-case space complexity of graph exploration is $\Theta(D \log d)$ bits.

TABLE 5.1 Summary of Results about Optmial Exploration

	Anchored Map	Unanchored Map	No Map
Lines	Overhead $= \frac{7}{5}$, optimal	Overhead $= \sqrt{3}$, optimal	DFS
Trees	Overhead $= \frac{3}{2}$, optimal	Overhead < 2	Overhead $= 2$, lower bound $= \sqrt{3}$
General graphs	DFS, overhead $= 2$, optimal		

Source: From ref. [2].

Algorithm Anchored Line

Let $x = 3a + n$ and $y = 2n - a$

if $x \leq y$ **then**

　　　　go at distance a in one direction, or until an endpoint is reached;

　　　if an endpoint is reached

　　　　　　　then return, go to the other endpoint, and stop

　　else　　return, go to the endpoint, return, go to the other endpoint, and stop

else　go to an endpoint, return, go to the other endpoint, and stop

FIGURE 5.2 Optimal algorithm for exploration on a line graph with an anchored map.

5.3.3 Collective Tree Exploration with Multiple Agents

So far we have considered graph exploration with the help of a single agent. Fraigniaud et al. [4] addressed the problem of collective graph exploration using multiple agents: They considered exploring an n-node tree by k agents, $k > 1$, starting from the root of the tree. The agents return to the starting point at the end of the exploration. Every agent traverses any edge in unit time, and the time of collective exploration is the maximum time used by any agent. Even when the tree is known in advance, scheduling optimal collective exploration turns out to be NP hard. The main communication scenario adopted in the paper is the following: At the currently visited node agents write the information they previously acquired, and they read information (provided by other agents) available at this node. The paper provides an exploration algorithm for any tree with overhead $O(k/\log k)$. The authors prove that if agents cannot communicate at all, then every distributed exploration algorithm works with overhead $\Omega(k)$ for some trees.

The model for a k-agent scenario is a little different than that described earlier for a single agent. The agents have distinct identifiers, but apart from that, they are identical. Each agent knows its own identifier and follows the same exploration algorithm which has the identifier as a parameter. The network is anonymous as before, that is, nodes are not labeled, and ports at each node have only local labels that are distinct integers between 1 and the degree of the node. At every exploration step, every agent either traverses an edge incident on its current position or remains in the current position. An agent traversing an edge knows local port numbers at both ends of the edge.

The communication scenario is termed as exploration with *write–read* communication. In every step of the algorithm every agent performs the following three actions: (a) It moves to an adjacent node. (b) It writes some information in it. (c) It then reads all information available at this node, including its degree. Alternatively, an agent can remain in the current node, in which case it skips the writing action.

Actions are assumed to be synchronous: If A is the set of agents that enter v in a given step, then first all agents from A enter v, then all agents from A write, and then all agents currently located at v (those from A and those that have not moved from v in the current step) read. Two extreme communication scenarios are discussed. In case of exploration without communication, the agents are oblivious of one another. At each step, every agent knows only the route it traversed until that point and the degrees of all nodes it visited. In case of exploration with complete communication, all agents can instantly communicate at each step. In both scenarios, an agent does not know the other endpoints of unexplored incident edges. If an agent decides to traverse such a new edge, the choice of the actual edge belongs to the adversary when the worst-case performance is being considered.

The exploration algorithm described in the paper has an overhead of $O(k/\log k)$. To be precise, the algorithm explores any n-node tree of diameter D in time $O(D + n/\log k)$. The algorithm is described for the stronger scenario first, that is, for exploration with complete communication. It can be simulated in the write−read model without changing time complexity. The paper shows that any algorithm must have overhead of at least $(2-1/k)$ under the complete communication scenario. In order to get overhead sublinear in the number of agents, some communication is necessary. Exploration without communication does not allow any effective splitting of the task among agents.

We outline here the algorithm for exploration with complete communication as described in [4]. Let T_u be the subtree of the explored tree T rooted at node u. Then assume T_u is *explored* if every edge of T_u has been traversed by some agent. Otherwise, it is called *unexplored*. If it is explored, T_u is finished, and either there are no agents in it or all agents in it are in u. Otherwise, it is called *unfinished*. If there is at least one agent in it, T_u is inhabited. Figure 5.3 shows the algorithm.

Algorithm Collective Exploration

Fix a step i of the algorithm and a node v in which some agents are currently located. There are three possible (exclusive) cases.

{**Case 1**} Subtree T_v is finished.

   (Action) if $v \neq r$, then all agents move from v to the parent of v, else all agents from v stop.

{**Case 2**} There exists a child u of v such that T_u is unfinished.

Let $u_1, u_2, \ldots, u_j$ be children of v for which the corresponding trees are unfinished, ordered in increasing order of the local port numbers of v. Let x_1 be the number of agents currently located in T_{ul}. Partition all agents from v into sets $A_1, \ldots, A_j$ of sizes $y_1, \ldots, y_j$, respectively, so that integers $x_1 + y_1$ differ by at most 1. The partition is done in such a way that the indices l for which integers $x_1 + y_1$ are larger by one than for some others, form an initial segment $(1, \ldots, z)$ in $(1, \ldots, j)$. Moreover, sets A_l are formed one-by-one, by inserting agents from v in order of increasing identifiers.

   (Action) All agents from the set A_1 go to u_1, for $l = 1, \ldots, j$.

{**Case 3**} For all children u of v, trees T_u are finished, but at least one T_u is inhabited.

   (Action) All agents from v remain in v.

FIGURE 5.3   Algorithm for collective exploration with complete communication.

5.3.4 Deterministic Rendezvous in Arbitrary Graphs

The rendezvous problem is defined as follows: Two mobile agents located in nodes of an undirected connected network have to meet at some node of the graph. If nodes of the graph are labeled, then agents can decide to meet at a predetermined node, and the rendezvous problem reduces to graph exploration. However, in many practical applications where rendezvous is needed in an unknown environment, such unique labeling of nodes may not be available or limited sensory capabilities of the agents may prevent them from perceiving such labels. As before, we assume that the ports at a node are locally labeled as $(1, 2, \ldots, d)$, where d is the degree of the node.

Agents move in synchronous rounds. In each round, an agent may either remain in the same node or move to an adjacent node. Agents can start up simultaneously or arbitrarily, that is, an adversary can decide the starting times of the agents.

One assumption of deterministic rendezvous is, if agents get to the same node in the same round, they become aware of it, and rendezvous is achieved. However, if they cross each other along an edge, moving in the same round along the same edge in opposite directions, they do not notice each other. So rendezvous is not possible in the middle of an edge. The time used by a rendezvous algorithm for a given initial location of agents in a graph is the worst-case number of rounds since the startup of the later agent until rendezvous is achieved. The worst case is taken over all adversary decisions and over all possible startup times (decided by the adversary) in case of the arbitrary startup scenario.

Each agent knows its own label but does not know the label of the other agents. If agents are identical and execute the same algorithm, then deterministic rendezvous is impossible even in the simplest case when the graph consists of two nodes joined by an edge. If both agents knew each other's labels, then the problem can be reduced to that of graph exploration. The same thing applies if the graph has a distinguished node.

The rendezvous problem in graphs has mostly been studied using randomized methods. Dessmark, Fraigniaud, and Pelc [5] addressed deterministic algorithms for the rendezvous problem, assuming that agents have distinct identifiers and are located at nodes of an unknown anonymous connected graph. Their paper showed that rendezvous can be completed in optimal time $O(n + \log l)$ on any n-node tree, where l is the smaller of the two labels. The result holds even with arbitrary startup. But trees are a special case from the point of view of the rendezvous problem, as any tree has either a central node or a central edge[1], which facilitates the meeting. This technique used for trees cannot be applied to graphs containing cycles.

With simultaneous startup, the optimal time of rendezvous on any ring is $\Theta(D \log l)$ and [5] describes an algorithm achieving that time, where D is the

---

[1]Every tree has one or two centers. In the latter case, the edges joining the two centers serve as a central edge.

initial distance between agents. With arbitrary startup, $\Omega(n + D \log l)$ is a lower bound on the time required for rendezvous on an n-node ring. The paper presents two rendezvous algorithms for the ring with arbitrary startup: an algorithm working in time $O(n \log l)$, for known n and an algorithm polynomial in n, l and the difference between the startup times when n is unknown. The paper also gives an exponential cost algorithm for general graphs, which is later improved. The next section discusses the issue.

5.3.5   Polynomial Deterministic Rendezvous in Arbitrary Graphs

Deterministic rendezvous has previously been shown to be feasible in arbitrary graphs [5] but the proposed algorithm had cost exponential in the number n of nodes and in the smaller identifier l and polynomial in the difference τ between startup times. The main result of the paper by Kowalski and Pelc [6] is a deterministic rendezvous algorithm with cost polynomial in n, τ, and $\log l$. Kowalski and Pelc's algorithm contains a nonconstructive ingredient: Agents use *combinatorial objects* whose existence is proved using a probabilistic method. Nevertheless their rendezvous algorithm is deterministic. Both agents can find separately the same combinatorial object with desired properties, which is then used to solve the rendezvous algorithm. This can be done using a brute-force exhaustive search that may be quite complex, but their model only counts the moves of the agents and not the computation time of the agents. The paper concludes with the open question:

Does there exist a deterministic rendezvous algorithm whose cost is polynomial in n and l (or even in n and $\log l$) but independent of τ?

5.3.6   Asynchronous Deterministic Rendezvous in Graphs

Marco et al. [7] studied the asynchronous version of the rendezvous problem. Note that in the asynchronous setting meeting at a node (which is normally required in rendezvous) is in general impossible. This is because even in a two-node graph the adversary can desynchronize the agents and make them visit nodes at different times. This is why the agents are allowed to meet inside an edge as well.

For the case where the agents are initially located at a distance d on an infinite line, the paper describes a rendezvous algorithm with cost $O(D \cdot |L_{\min}|^2)$ where d is known and $O(D + |L_{\max}|)^3)$ if d is unknown, where $|L_{\min}|$ and $|L_{\max}|$ are the lengths of the shorter and longer labels of the agents, respectively. The authors also describe an optimal algorithm of cost $O(n|L_{\min}|)$ if the size n of the ring is known and of cost $O(n|L_{\max}|)$ if n is unknown. For arbitrary graphs, they show that rendezvous is feasible if an upper bound on the size of the graph is known. They present an optimal algorithm of cost $O(D|L_{\min}|)$ when the topology of the graph and the initial positions of the agents are known to each other. The paper asks two open questions:

1. Is rendezvous with cost $O(D \cdot |L_{\min}|)$ possible for a ring of unknown size?
2. Suppose that a bound M on the number of nodes of the graph is known to both agents. Is there a rendezvous algorithm polynomial in the bound M and in the lengths of the agents' labels?

5.4 CONCURRENT READING AND WRITING

This problem addresses the implementation of a read–write object on a wide-area network. The various components of the object are mapped to different processes over the network. We assume that the reading and writing of the global states of a distributed system are carried out by *reading agents* and *writing agents*, respectively. The notion of a consistent snapshot is available from Chandy and Lamport's seminal paper [8]. Also, Arora and Gouda [9] illustrated how to reset a distributed system to a predefined global state. None of these used mobile agents. Our goal is not only to examine how a single reading or writing agent can perform these operations but also to illustrate how reading and writing operations can be performed when multiple agents are active in the network at the same time (Figure 5.4). Clearly, we plan to treat the network of processes as a *concurrent object* that can be accessed by the read and write operations. There are numerous possible

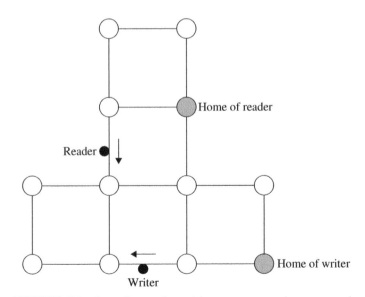

FIGURE 5.4 A reading and a writing agent traversing a network.

applications. The reading and the writing agents can cooperate with one another to implement a consistency model on a distributed data structure. Multiple reading agents may work simultaneously to speed up data retrieval. Multiple reading and writing agents may work together to expedite fault recovery. In parallel programming languages like *Linda* [10], processes communicate with one another via a tuple space. On a network of processes, the tuple space is a part of the network state that can be concurrently read or written by two or more agents. In the area of electronic commerce, multiagent protocols are rapidly growing.

The classical approach to implementing concurrent objects consists of the use of *critical sections* and a mechanism to *mutual exclusion*. This approach involves waiting, since at most one process is allowed in the critical section at any time. A sideeffect is deadlock—if one process is stuck in the critical section, then it indefinitely blocks all processes waiting to enter the critical section. For asynchronous, fault-tolerant distributed systems, it is much more desirable to design a *nonblocking* or *wait-free* [11] solution, where each access by an agent is completed in a finite number of steps, regardless of the speeds of other agents or processes. In this chapter, we will explore nonblocking solutions to the problem of concurrent reading and writing problem using the agent model. The basic correctness condition relevant to our solutions is that, even though the operations of reading and writing by two different agents may overlap, the results of the reading must ensure that the operations are atomic and consistent with what was written by the writing agent(s). Readers might note that similar algorithms have already been proposed for shared registers by Lamport [12], where he showed how to implement atomic registers using weaker versions of registers where the reading and writing operations overlap in time. Below we clarify the semantics of overlapped read and write.

Assume that initially, $\forall i, s(i) = 0$. Let W^k and W^{k+1} denote two consecutive write operations by the writer: W^k updates the local state of every process to 6 and W^{k+1} updates the local state of every process to 5. Let the reading agent take three successive snapshots R^i, R^{i+1}, and R^{i+2} in overlapped time as shown in Figure 5.5.

1. Each read operation must return a consistent value of the global state that will correspond to (i) the previous reset state before the read started,

FIGURE 5.5 Atomic behavior when multiple reads overlap a write.

(ii) the state to which the system is being currently set, or (iii) a state reachable from one of these.

2. No read returns a global state that is "older than", that is, causally ordered before, the global state returned by the previous read.

As a consequence of the second condition, it is okay if both R^i and R^{i+1} return a state $(0,0,0, \ldots)$ or a state reachable from it, but it is not acceptable for R^{i+2} to return $(6,6,6, \ldots)$ and R^{i+1} to return $(5,5,5, \ldots)$. Note that similar conditions are true for atomic registers [13] too.

5.4.1 One-Reader and One-Writer Cases

Before presenting the protocol of concurrent reading and writing, we first describe the individual protocols in the single-reader and single-writer cases. In each case, the agent performs a DFS traversal of the network infinitely often using the protocol described in [14]. The briefcase of the agent is initialized by its home, and at each visited node, the agent executes its designated program before hopping to the next node designated by NEXT := DFS. Ideally, the processes should be oblivious to when the agent traverses and performs the read or write operations. However, in the single-reader case (and in general with the static model of agents), a little help from the visited processes seems unavoidable.

5.4.2 One-Reader Protocol

The snapshot is taken by the reading agent and saved in its briefcase. The agent starts by taking a snapshot of its home process. Thereafter, as the agent visits a node, it records a snapshot state, which ordinarily is the local state of that node. When the agent returns home after each complete traversal, it computes a consistent global state of the system from the states recorded at the individual nodes. For simplicity, we assume that the channels have zero capacity, so the channel states are irrelevant.

At any moment, a message circulating in the system can be of one of the following four types: (1) from an unvisited node to an unvisited node, (2) from a visited node to a visited node, (3) from a visited node to an unvisited node, and (4) from an unvisited node to a visited node. Of these, when a message M propagates from a visited node i to an unvisited node j, there is the potential for a causal ordering between the recordings of $s(i)$ and $s(j)$ by the agent. This is because the following causal chain [record $s(i)$ < send m < receive m < record $s(j)$] may exist. To avoid this, we have to ask process j receiving m to save its current state $s(j)$ into a history variable $h(j)$ before receiving m. It is this saved value that will be returned to the agent as the local state of process j. To detect a message from a visited to an unvisited node, we use the following mechanism:

Each reading agent tags its traversal with a *sequence number* $SEQ \in \{0, 1, 2\}$ that is stored in its briefcase. Before each traversal begins, the home increments

Program for the agent while visiting process i

agent variables SEQ, S; ($S \cdot i$ records the state $s(i)$ of node i)
process variables seq, s, agent_seq, h (initially H is empty);

if SEQ $= $ agent_seq$(i) \oplus_3 1 \wedge$ seq$(i) \neq -1 \rightarrow$
    seq$(i) := $ SEQ; agent_seq$(i) := $ SEQ; $S \cdot i := s(i)$;
$\square$ SEQ $= $ agent_seq$(i) \wedge$ seq$(i) \neq -1 \rightarrow$ skip
$\square$ seq$(i) = -1 \rightarrow$ seq$(i) := $ SEQ; $S \cdot i := h(i)$; delete $h(i)$
fi;
NEXT $:= $ DFS

Program for Process i

do true $\rightarrow$
if message from j: agent_seq$(j) = $ agent_seq$(i) \oplus_3 1 \rightarrow$
    $h(i) := s(i)$; seq$(i) = -1$; agent_seq$(i) := $ agent_seq(j);
fi;
accept the message;
od

FIGURE 5.6   The One-reader protocol.

this value modulo 3. Each node has two variables seq and agent_seq both of which are updated to SEQ when the agent visits that node. The value of agent_seq is appended to every message sent out by a node. Accordingly, when a visited node i sends a message m to an unvisited node j, the condition agent_seq$(i) = $ agent_seq$(j) \oplus_3 1$ will hold. When node j receives the message, it sets seq(j) to -1[2] and it also sets agent_seq(j) to the value of agent_seq(i), and saves $s(j)$ into the history variable $h(j)$. Subsequently, when the reading agent visits j, seq(j) will be reset to the value of SEQ in the briefcase of the agent, and the history variable is deallocated. This leads to the program in Figure 5.6.

We present without proof the following lemma. The proof can be found in [15].

Lemma 3 *Each reading of the global state returns a consistent global state, and if two readings are consecutively taken by the reading agent, then the second read cannot return a reading older than the value returned by the first read.*

5.4.3   One-Writer Protocol

For the writing protocol, we preserve the consistency of the reset operation by disallowing all messages between nodes whose states have been updated and the nodes whose states are yet to be updated [9]. To distinguish between consecutive write operations, we introduce a nonnegative integer variable CLOCK with the writing agent. For the purpose of reset only, a binary value of

---

[2]This flags the agent to record the state from the history.

Program for the writing agent while visiting process i

The agent wants to reset the global state to W
agent variables CLOCK, W;
process variables clock, s;

if (clock(i) < CLOCK) → $s(i) := W \cdot i$;
$$clock(i) := CLOCK$$

fi
NEXT := DFS

FIGURE 5.7 The One-writer protocol

CLOCK will suffice. However, CLOCK will need to have more than two values when we address concurrent reading and writing in the next section. The write will update the value of clock(i) for every process i that it visits. Like the reading agent, the writing agent also traverses the network along a spanning tree. The spanning trees along which the reading and the writing agents traverse the network can be totally independent. The program for the writer is described in Figure 5.7.

When the writing agent returns home, the write operation is over. The home increments CLOCK before the next traversal begins. The write operation does not require the cooperation of the individual nodes, except for the rejection of the messages that originated from nodes with a different value of the local clock. This requires that messages be stamped with the clock value of the sender.

5.4.4 Concurrent Reading and Writing

In the general case when a reading agent and a writing agent carry out their designated tasks in overlapped time, the writer may update the global state to different values during different traversals, and the reader, unaware of when and what the writer is writing, has to capture a consistent snapshot of the global state of the distributed system. There is no relationship between the speeds at which the writer and the reader move around the network. We will use the value of *clock* at the different nodes as the yardstick of progress. The value of clock is updated by the writing agent in all processes, including the home of the reader process. The reader and the writer agents traverse the network following distinct spanning trees, denoted in the algorithm by DFS_R and DFS_W. On occasion, these trees may be identical, but there is no guarantee for it. Since the reader may be slow, the writing agent, in addition to updating the local state and the clock, will record the current state of the visited process into a set *history* for that process that could possibly be used by a slower reader. Each element in the history h is a pair (clock, localstate), and we will designate the entry in the history of process i corresponding to clock j by $h^j(i)$. The saving of the current state becomes unnecessary when seq for the visited node is –1, since the state that will be read by the reader has already been saved by the process while updating seq to –1.

The following two observations are the cornerstones of the algorithm:

Lemma 4 $CLOCK(writer) = k \Rightarrow \forall i\ clock(i) \in \{k-1, k\}$.

If the reading agent looks for copies of local states corresponding to $clock = k$ while the writer is still writing in round K, it is possible that the

Program for the writer while visiting process i

{The writer wants to reset the global state to W}
agent variables CLOCK, W;
process variables clock, s, h;

if $(clock(i) < CLOCK) \rightarrow$
    if $seq(i) \neq -1 \rightarrow h(i) := h(i) \cup (clock(i), s(i))$ **fi**;
    $s(i) := W \cdot i$; $clock(i := CLOCK$
fi;
$NEXT := DFS_W$

Program for the reader while visiting process i

{The reader is trying to assemble a snapshot S}
agent variables SEQ, CLOCK, S;
process variables seq, agent_seq, s, h;

$\forall j < CLOCK - 1$ delete $h^j(i)$;
{Case 1} **if** $clock(i) = CLOCK \rightarrow$

    if $SEQ = agent\_seq(i) \oplus_3 1 \wedge seq(i) \neq -1 \rightarrow$
        $S \cdot i \cdot CLOCK := s(i);\ S \cdot i \cdot (CLOCK - 1) := h^{CLOCK-1}(i)$
    $\square$ $SEQ = agent\_seq(i) \wedge seq(i); \neq -1 \rightarrow$ skip
    $\square$ $seq(i) = -1 \rightarrow S \cdot i \cdot CLOCK := h^{CLOCK}(i);\ S \cdot i \cdot (CLOCK - 1) := h^{CLOCK-1}(i)$
    fi;

$\square$ {Case 2} $clock(i) = CLOCK - 1 \rightarrow$
    if $SEQ = agent\_seq(i) \oplus_3 1 \wedge seq(i) \neq -1 \rightarrow$
        $S \cdot i \cdot (CLOCK - 1) := s(i);\ S \cdot i \cdot CLOCK :=$ null
    $\square$ $SEQ = agent\_seq(i) \wedge seq(i) \neq -1 \rightarrow$ skip
    $\square$ $seq(i) = -1 \rightarrow S \cdot i \cdot (CLOCK - 1) := h^{CLOCK-1}(i);\ S \cdot i \cdot CLOCK :=$ null
    fi;

$\square$ {Case 3} $clock(i) > CLOCK \rightarrow$
    if $SEQ = agent\_seq(i) \oplus_3 1 \rightarrow$
        $S \cdot i \cdot CLOCK := h^{CLOCK}(i);\ S \cdot i \cdot (CLOCK - 1) := h^{CLOCK-1}(i)$
    $\square$ $SEQ = agent\_seq(i) \rightarrow$ skip
    fi;
fi;
$seq(i) := SEQ;\ agent\_seq(i) := SEQ;$
$NEXT := DFS_R$

FIGURE 5.8   The concurrent reading and writing protocol. (From ref. 15.)

reader visits a node whose clock has not yet been updated from $k - 1$ to k. In this case, the reader will construct the snapshot from local states recorded at clock $k - 1$. All entries in the history corresponding to clock lower than $k - 1$ are of no use and can be deleted by the reader. This leads to the following lemma:

Lemma 5 *When the reader makes a traversal after the writer has started writing with a CLOCK k, a consistent snapshot will be assembled from recordings of local states made at clock k only or $k - 1$ only.*

Figure 5.8 shows the final program.

The one-reader case can be easily extended to multiple readers since readers do not interact—each process maintains a separate history. The extension to the multiple-writer case is an open problem. The time complexity for a snapshot or a reset operation is determined by the time for one traversal. An issue of interest is the space complexity per process. Unfortunately, in the present version of the protocol, the space requirement can grow indefinitely when the writing operation is faster than the reading operation. The size of the briefcase for both the reader and the writer scales linearly with the size of the network. Bounding the space complexity when the writer is faster than the reader is an open problem. Also relevant is the issue of implementing various consistency models on the concurrent object.

5.5 FAULT TOLERANCE

When agents traverse an unknown network, they might get trapped in a host, known as a black hole. Once trapped, the agent is lost for all practical purposes, leaving no observable trace of destruction. A black hole need not always be a malicious host—for example, an undetectable crash failure of a host in an asynchronous network can make it a black hole.

An interesting aspect of fault tolerance is finding out a black hole by sacrificing a minimum number of agents. The task is to unambiguously determine and report the location of the black hole, assuming there is only one black hole. The problem is called the *black-hole search* (BHS) problem. More precisely, the BHS is solved if at least one agent survives, and all surviving agents know the location of the black hole within a finite time. Black-hole search is a nontrivial problem. In recent times, the problem has gained renewed significance as protecting an agent from "host attacks" has become a problem almost as pressing as protecting a host from an agent attack.

The problem of efficient BHS has been extensively studied in many types of networks. The underlying assumption in most cases is that the network is totally asynchronous, that is, while every edge traversal by a mobile agent takes a finite time, there is no upper bound on this time.

5.5.1 BHS in Anonymous Ring

Model Dobrev et al. [16] considered the BHS problem in the simplest symmetric topology: an anonymous ring R, that is, a loop network of identical nodes. Each node has two ports, labeled left and right. If this labeling is globally consistent, the ring is oriented; otherwise it is unoriented. Let $0, 1, \ldots, n-1$ be the nodes of the ring in the clockwise direction and node 0 be the home base from where the agents start. Let A denote the set of anonymous mobile agents and $|A| = k$ denote the number of mobile agents. The asynchronous agents are assumed to have limited computing capabilities and bounded storage. They obey the same set of behavioral rules, that is, the protocol. The bounded amount of storage in each node is called its *whiteboard*. Agents communicate by reading from and writing into the whiteboards, and access to a whiteboard is mutually exclusive.

When the anonymous agents start from the same node, they are termed *colocated* agents. Otherwise, when they start from different nodes, they are called *dispersed* agents. The number of agents is the *size* of the fleet, and the total number of moves performed by the agents determines the cost for an algorithm. The following lemmas hold for the BHS problem:

Lemma 6 *At least two agents are needed to locate the black hole.*

If there is only one agent, the BHS problem is unsolvable because the only agent will eventually disappear in the black hole.

Lemma 7 *It is impossible to find the black hole if the size of the ring is not known.*

Lemma 8 *It is impossible to verify whether or not there is a black hole.*

The presence of more than two agents does not reduce the number of moves. It can however be helpful in reducing the time spent by colocated agents to locate the black hole. The number of dispersed agents required to solve the problem depends on whether the ring is oriented or not. If the ring is oriented, then two anonymous dispersed agents are both necessary and sufficient. If the ring is unoriented, three anonymous dispersed agents are both necessary and sufficient [16].

- **Cautious Walk** *Cautious walk* is a basic tool in many BHS algorithms. At any time during BHS, the ports (corresponding to the incident links) of a node can be classified into three types:
- **Unexplored** No agent has moved across this port.
- **Safe** An agent arrived via this port.
- **Active** An agent departed from this port, but no agent has arrived into it.

Both unexplored and active links are potentially dangerous because they might lead an agent to the black hole. Only safe ports are guaranteed to be hazardfree. Cautious walk helps identify safe ports. It is defined by the following two rules:

> **Rule 1** When an agent moves from node u to v via an unexplored port (turning it into active), it immediately returns to u (making the port safe) and only then goes back to v to resume its execution.
>
> **Rule 2** No agent leaves via an active port.

Theorem 9 *In a ring with n nodes, regardless of the number of colocated agents, at least $(n-1)\log(n-1) + O(n)$ moves are needed for solving the BHS problem.*

Sketch of Solving BHS with Two Co-located Agents We present the main idea behind the algorithm. It proceeds in phases. Let E_i and U_i denote the explored and unexplored nodes in phase i, respectively; E_i and U_i partition the ring into two connected subgraphs with the black hole located somewhere in U_i. Divide the unexplored part of the ring between the two agents, assigning to each agent a region of almost equal size. Each agent starts the exploration of the assigned part. Because of the existence of the black hole, only one of them will complete the exploration. When this happens, it will go through the explored part until it reaches the last safe link visited by the other agent. It will then again partition the unexplored area in two parts of almost equal size, leave a message for the other agent (in case it is not in the black hole), and go to explore the newly assigned area. If $|U_{i+1}| = 1$, the surviving agent knows that the black hole is in the single unexplored node, and the algorithm terminates.

The two-agent algorithm is cost optimal. There are algorithms for solving the BHS problem in hypercubes and arbitrary graphs, which also follow the similar idea. We present a few results from [16]:

Theorem 10 *In the worst case, $2n - 4$ time units are needed to find the black hole, regardless of the number of colocated distinct agents available.*

Theorem 11 *The cost of locating the black hole in oriented rings with dispersed agents is at least $\Omega(n \log n)$.*

Theorem 12 *If the agents have prior knowledge of the number of agents k, then the cost of locating the black hole in oriented rings is $\Omega(n \log(n - k))$.*

Theorem 13 *In oriented rings, k agents can locate the black hole in $O((n/\log n)/\log(k - 2))$ time when k is known.*

5.5.2 BHS in Arbitrary Networks

Dobrev et al. studied topology-independent generic solutions [17] for BHS. The problem is clearly not solvable if the graph G representing the network topology is not 2-connected. (BHS in trees requires a change of model that is discussed later.) The cost and size for BHS algorithms are shown to be dependent on the a priori knowledge the agents have about the network and on the consistency of the local labeling. The assumption is that all agents know n, the size of the network.

Model Let $G = (V, E)$ be a simple 2-connected graph, $n = |V|$ be the size of G, $E(x)$ be the links incident on $x \in V$, $d(x) = |E(x)|$ denote the degree of x, and Δ denote the maximum degree in G. If $(x, y) \in E$, then x and y are neighbors of each other. The nodes are anonymous. At each node there is a distinct label called the port number associated to each of its incident links. Let $\lambda_z \in (x, z)$ denote the label associated at x to the link $(x, z) \in E(x)$ and λ_z denote the overall injective mapping at x. The set $\lambda = \lambda_x | x \in V$ of those mappings is called a labeling, and (G, λ) is the resulting edge-labeled graph.

Let $P[x]$ denote the set of all paths with x as a starting point, and let $P[x, y]$ denote the set of paths starting from x and ending in y. Let Λ be the extension of the labeling function λ from edges to paths. A *coding* c of a system (G, λ) is a function such that, $\forall x, y, z \in V$, $\forall \pi_1 \in P[x, y]$, $\forall \pi_2 \in P[x, z]$, $c(\Lambda_x(\pi_1)) = c(\Lambda_x(\pi_2))$ iff $y = z$. For any two paths $\pi_1 \pi_2$ from x to y, $c(\Lambda_x(\pi_1)) = c(\Lambda_x(\pi_2))$. A decoding function d for c is such that $\forall x, y, z \in V$, such that $(x, y) \in E(x)$ and $\pi \in P[y, z]$, $d(\lambda_x(x, y), c(\Lambda_y(\pi))) = c(\lambda_x(x, y)) \circ (\Lambda_y(\pi))$, where $\circ$ is the concatenation operator. The couple (c, d) is called a *sense of direction* for (G, λ). If (c, d) is known to the agents, the agents operate with sense of direction. Otherwise, the agents operate with topological ignorance. The agents have complete topological knowledge of (G, λ) when the following information is available to all agents:

1. Knowledge of the labeled graph (G, λ)
2. Correspondence between port labels and the link labels of (G, λ)
3. Location of the home base in (G, λ)

In Dobrev et al. [17] proved the following results:

Theorem 14 *With topological ignorance, there is an n-node graph G with the highest degree $\Delta \leq n - 4$ such that any algorithm for locating the black hole in arbitrary networks needs at least $\Delta + 1$ agents in G. In addition, if $n - 4 < \Delta < n$, then any such algorithm needs at least Δ agents.*

Theorem 15 *With topological ignorance, there exists a graph G such that any $(\Delta + 1)$ agent algorithm working on all 2-connected n-node networks of maximal degree at most $\Delta \geq 3$ needs $\Omega(n^2)$ moves to locate the black hole in G.*

The authors provide an algorithm that correctly locates the black hole in $O(n^2)$ moves using $\Delta + 1$ agents, where Δ is the highest degree of a node in the graph. They also showed the following:

Theorem 16 *With topological ignorance, if $n - 3 \le \Delta \le n - 1$, Δ agents can locate the black hole with cost $O(n^2)$.*

Theorem 17 *In an arbitrary network with sense of direction, the black hole can be located by two agents with cost $O(n^2)$.*

Theorem 18 *The black hole can be located by two agents with full topological knowledge in arbitrary networks of vertex connectivity 2 with cost $O(n \log n)$, and this is optimal.*

The lower bound of $\Omega(n \log n)$ in general networks does not hold for hypercubes and related networks. Dobrev et al. [18] provided a general strategy that allows two agents to locate the black hole with $O(n)$ moves in hypercubes, cube-connected cycles, star graphs, wrapped butterflies, and chordal rings as well as in multidimensional meshes and tori of restricted diameter. Specifically they proved the following:

Theorem 19 *Two agents can locate the black hole in $O(n)$ moves in all of the following topologies: tori and meshes of diameter $O(n/\log n)$, hypercubes, cube-connected cycles(CCCs), wrapped butterflies, and star graphs.*

In another paper [19], Dobrev et al. showed that it is possible to considerably improve the bound on cost without increasing the size of the agents' team. They presented a universal protocol that allows a team of two agents with a map of the network to locate a black hole with cost $O(n + d \log d)$, where d denotes the diameter of the network. This means that, without losing its universality and without violating the worst-case $\Omega(n \log n)$ lower bound, their algorithm allows two agents to locate a black hole with $\Theta(n)$ cost in a very large class of, possibly unstructured networks, where d $O(n/\log n)$.

5.5.3   BHS in Tree Networks

Model The model for a tree network is different. Obviously the assumption of 2-connectedness is no more valid for trees. Also the network is assumed to be partially synchronous instead of asynchronous. An upper bound on the time of traversing any edge by an agent can be established. Without loss of generality, we normalize this upper bound on edge traversal time to 1.

The partially synchronous scenario allows the use of a time-out mechanism to locate the black hole in any graph with only two agents. Agents proceed along edges of the tree. If they are at a safe node v, one agent goes to the adjacent node and returns, while the other agent waits at v. If after two units of time the first agent does not return, the other one survives and knows the location of the black hole. Otherwise, the adjacent node is safe, and both agents can move to it. This is a variant of the cautious walk. For any network, this version of BHS can be performed using only the edges of its spanning tree.

Clearly, in many graphs, there are more efficient BHS schemes than those operating in a spanning tree of the graph.

Czyzowicz et al. [20] considered a tree T rooted at node s, which is the starting node of both agents. It is assumed that s is not a black hole. Agents have distinct labels. They can communicate only when they meet and not by leaving messages at nodes. There is at most one black hole in the network. Upon completion of the BHS there is at least one surviving agent, and this agent either knows the location of the black hole or knows that there is no black hole in the tree. The surviving agent(s) must return to the root s.

An edge of a tree is *unknown* if no agent has moved yet along this edge (initial state of every edge). An edge is *explored* if either the remaining agents know that there is no black hole incident to this edge, or they know which end of the edge is a black hole. In between meetings, an edge may be neither unknown nor explored when an unknown edge has just been traversed by an agent.

The *explored territory* at step t of a BHS scheme is the set of explored edges. At the beginning of a BHS scheme, the explored territory is empty. A meeting occurs in node v at step t when the agents meet at node v and exchange information which *strictly increases* the explored territory. Node v is called a *meeting point*. In any step of a BHS scheme, an agent can traverse an edge or wait in a node. Also the two agents can meet. If at step t a meeting occurs, then the explored territory at step t is defined as the explored territory after the meeting. The sequence of steps of a BHS scheme between two consecutive meetings is called a *phase*.

Lemma 20 *In a BHS scheme, an unexplored edge cannot be traversed by both agents.*

Lemma 21 *During a phase of a BHS scheme an agent can traverse at most one unexplored edge.*

Lemma 22 *At the end of each phase the explored territory is increased by one or two edges.*

Lemma 23 *Let v be a meeting point at step t in a BHS scheme. Then at least one of the following holds: $v = s$ or v is an endpoint of an edge that was already explored at step $t - 1$.*

Czyzowicz et al. [20] provide an approximation algorithm (Figure 5.9) with ratio $5/3$ for BHS in case of arbitrary trees. The time complexity of the algorithm is linear. It uses the following procedures:

probe(v) One agent traverses edge (p, u) (which is toward node v) and returns to node p to meet the other agent who waits. If they do not meet at step $t + 2$, then the black hole has been found.

split(k, l) One of the agents traverses the path from node m to node k and returns toward node p_l. The other traverses the path from node m to node l

Approximation Algorithm BHS on a Tree

Procedure explore(v) for a general node. Initially $v = s$
Procedure explore(v)
for every pair of unknown edges (v, x), (v, y) incident to v **do**
  split(x, y);
end for
if there is only one remaining unknown edge (v, z) incident to v **then**
  probe(z);
end if
if every edge is explored **then**
  repeat walk(s) **until** both agents are at s
else
  next := relocate(v);
  explore(next)
end if

FIGURE 5.9 Black-hole search on a tree network.

and returns toward node p_k. Let dist(l, k) denote the number of edges in the path from node k to node l. If they do not meet at step $t + \text{dist}(l, k)$, then the black hole has been found.

relocate(v) This function takes as input the current node v where both agents reside and returns the new location of the two agents. If there is an unknown edge incident to a child of v then the agents go to that child. Otherwise, the two agents go to the parent of v.

The authors ask the open question whether there exists a polynomial time algorithm to construct a fastest BHS scheme for an arbitrary tree. More generally, until now it is not known if the problem is polynomial for arbitrary graphs.

5.5.4 Multiple-Agent Rendezvous in a Ring in Spite of a Black Hole

The rendezvous problem requires all the agents to gather at the same node. Both nodes and agents, besides being anonymous, are fully asynchronous. The assumption is that there are k asynchronous, anonymous agents dispersed in a symmetric-ring network of n anonymous sites, one of which is a black hole. Clearly it is impossible for all agents to gather at a rendezvous point, since an adversary can direct some agents toward the black hole. So, Dobrev et al. [21] sought to determine how many agents can gather in the presence of a black hole.

The *rendezvous* problem RV(p) consists of having at least $p \le k$ agents gathering in the same site. There is no a priori restriction on which node will become the rendezvous point. Upon recognizing the gathering point, an agent terminally sets its variable to *arrived*. The algorithm terminates when at least p agents set their *arrived* flag to true. A relaxed version of the rendezvous problem is the *near-gathering* problem $G(p, d)$ that aims at having at least p

TABLE 5.2 A Summary of Results for Rendezvous Problem

	n Unknown,	K Known	n Known,	K Unknown
Oriented	$\forall k > 1$	RV $(k - 1)$	$\forall k > 2$	RV $(k - 2)$
	k odd	RV $(k - 2)$	k odd or n even	RV $(k - 2)$
Unoriented	k even	RV $\left(\dfrac{k - 2}{2}\right)$	k even and n odd	RV $\left(\dfrac{k - 2}{2}\right)$
	$\forall k$	G $(k - 2, 1)$	$\forall k$	G $(k - 2, 1)$

agents within distance d from one another. The summary of the results from the paper is given in Table 5.2.

If k is unknown, then nontrivial[3] rendezvous requires knowledge of the location of the back hole.

Here are some basic results about the rendezvous problem on a ring:

Theorem 24 *In an anonymous ring with a black hole:*

1. *$RV(k)$ is unsolvable.*
2. *If the ring is unoriented, then $RV(k-1)$ is unsolvable.*

Theorem 25 *If k is unknown, then nontrivial rendezvous requires locating the black hole.*

Theorem 26 *If n is not known, then the black-hole location is unsolvable.*

Theorem 27 *Either k or n must be known for nontrivial rendezvous.*

5.5.5 BHS in Asynchronous Rings Using Tokens

Recently Dobrev et al. introduced a *token model* [22] for solving the BHS problem. A token is an atomic entity—agents communicate with one another and with the environment using these tokens. Each agent has a bounded number of tokens that can be carried and placed on a node or removed from it. One or more tokens can be placed at the middle of a node or on a port. All tokens are identical and indistinguishable from one another.

There are no tokens placed in the network in the beginning, and each agent starts with some fixed number of tokens. The basic computational step of an agent, which it executes either when it arrives at a node or upon wake-up, is

1. to examine the node and
2. to modify the tokens and either fall asleep or leave the node through either the left or right port.

---

[3]RV(p) is said to be nontrivial if p is a nonconstant function of k.

The main results show that a team of two agents is sufficient to locate the black hole in a finite time even in this weaker coordination model, and this can be accomplished using only $O(n \log n)$ moves, which is optimal, the same as with the whiteboard model. To achieve this result, the agents need to use only $O(1)$ tokens each. Interestingly, although tokens are a weaker means of communication and coordination, their use does not negatively affect the solvability of the problem or lead to a degradation of performance. On the contrary, it turns out to be better in the sense that, whereas the protocols using whiteboards assume at least $O(\log n)$ dedicated bits of storage at each node, the token algorithm uses only three tokens in total.

An open question is whether it is possible to further reduce the number of tokens and, if so, then what will be the cost in such a scenario.

5.6 STABILIZATION USING COOPERATING MOBILE AGENTS

A distributed system occasionally gets perturbed due to transient failures that corrupt the memory of one or more processes or due to environmental changes that include the joining of a new process or the crash of an existing process. For example, in routing via a spanning tree, if the tree is damaged due to a failure, then some packets will never make it to the destination. A stabilizing system [23] is expected to spontaneously recover from any perturbed configuration to a legal configuration by satisfying the following two properties [9]:

Convergence Starting from an arbitrary initial configuration, the system converges to a legal configuration in a bounded number of steps.

Closure If the system is a legal configuration, then it continues to do so unless a failure or a perturbation occurs.

Traditional stabilizing systems achieve stability using algorithms that use the message-passing or shared-memory model of interprocess communication. In this section, we demonstrate how mobile agents can be used to stabilize a distributed system. The mobile agent creates an extra layer of computation that is superimposed on the underlying distributed computation but does not interfere with it unless a failure occurs. The role of the mobile agent is comparable to that of a repairperson that roams the network, detects illegal configurations, and fixes it by appropriately updating the configuration. The individual processes are oblivious to the presence of the agent. We disregard any minor slowdown in the execution speed of a process due to the sharing of the resources by visiting agents.

At any node, the arrival of an agent triggers the agent program whose execution is atomic. The agent program ends with the departure of the agent from that node or with a waiting phase (in case the agent has to wait at that node for another agent to arrive), after which the execution of the application program at that node resumes. The computation at a node alternates between the agent program and the application program. At any node, the visit of a single agent can be represented by the following sequence of events. We denote an atomic event using $\langle \ \rangle$:

```
agent arrives, ⟨ agent program executed ⟩, agent leaves
```

When a pair of agents I and J meet at a node k to exchange data, the sequence of events will be as follows:

```
agent I arrives, ⟨ agent program of I executed⟩, agent I waits at k
```

Following this, the application program at node k resumes and continues until the other agent J arrives there. When J arrives at node k, the following sequence of events takes place:

```
agent J arrives, ⟨ data exchange with I occurs ⟩, agent I and J leave.
```

Then the application program at k resumes once again.

Agent-based stabilization can be viewed as a stabilizing extension of a distributed system as proposed in [24]. While Katz and Perry [24] emphasized the feasibility of designing stabilizing distributed systems, mobile agents have some interesting properties that make implementations straightforward. We first demonstrate the mechanism by presenting a stabilizing spanning tree construction [14] using a single mobile agent.

5.6.1 Stabilizing Spanning Tree Construction Using a Single Agent

To construct a spanning tree rooted at a given node of a graph $G = (V, E)$, we assume that the root is the home of the agent. Each node i has a parent $p(i)$ chosen from its immediate neighborhood. In addition, each node i has two other variables:

```
Child(i) ≡ {j:p{j}=i}
neighbor(i) ≡ {j:(i,j)∈e}
friend(i) ≡ {j:j∈N(i)∧j≠P(i)∧P(j)≠i}
```

The program of the agent consists of three types of actions: (i) actions that update the local variables of the process that it is visiting, (ii) actions that modify its briefcase variables, and (iii) actions that determine the next process that it will visit. The individual processes are passive.

A key issue in agent-based solution is graph traversal. To distinguish between consecutive rounds of traversal, we introduce a briefcase variable SEQ ($\in \{0, 1\}$) that keeps track of the most recent round of traversal. With every process i, define a Boolean $f(i)$ that is set to the value of SEQ whenever the

Program for the agent while visiting node i

agent variables NEXT, PRE, SEQ, C;
process variables f, child, p, neighbor, friend;

if $f(i) \neq$ SEQ $\rightarrow f(i) :=$ SEQ **fi**;
if PRE $\in$ friend$(i) \rightarrow p(i) :=$ PRE **fi**;
if

    Visit an unvisited child
    (DFS1) $\exists j \in$ child$(i) : f(j) \neq$ SEQ $\rightarrow$ NEXT $:= j; C := 0$
    When all neighbors have been visited, return to the parent
□   (DFS2) $\forall j \in$ neighbor$(i) : f(j) =$ SEQ $\wedge (C < N \vee$ friend$(i) = \emptyset) \rightarrow$
    NEXT $:= p(i); C := C + 1$
    Create a path to a node that is unreachable using DFS1
□   (DFS3) $\forall j \in$ child$(i) : f(j) =$ SEQ $\wedge \exists k \in$ neighbor$(i) : f(k) \neq$ SEQ $\rightarrow$
    NEXT $:= k; C := 0$
    Break a possible cycle
□   (DFS4) $(C > N) \wedge \exists k \in$ friend$(i) \rightarrow$ NEXT $:= k; p(i) := k; C := 0$
fi;

Program for the home process

Executed when the agent visits home
if $\exists j \in$ child: $f(j) \neq$ SEQ $\rightarrow$ NEXT $:= j; C := 0$
□ $\forall j \in$ neighbor: $f(j) =$ SEQ $\rightarrow$
            SEQ $:= 1$-SEQ;
            $C := 0$; NEXT $:= k : k \in$ child
fi

FIGURE 5.10   Spanning tree construction with a single agent.

process is visited by the agent. SEQ is complemented by the root before the next traversal begins. Thus, the condition $f(i) \neq$ SEQ is meant to represent that the node has not been visited in the present round.

The agent program has three basic rules, DFS1, DFS2, and DFS3, and is described[4] in Figure 5.10. The first rule DFS1 guides the agent to an unvisited child. The second rule DFS2 guarantees that when all children are visited the agent returns to the parent node.

Unfortunately, this traversal may be affected when the DFS spanning tree is corrupted. For example, if the parent links form a cycle, then no tree edge will connect this cycle to the rest of the graph. Accordingly, using DFS1, the nodes in the cycle will be unreachable for the agent, and the traversal will remain incomplete. As another possibility, if the agent reaches one of these nodes contained in a cycle before the cycle is formed, then the agent is trapped and cannot return to the root using DFS2.

---

[4]It disregards the details of how a process i maintains its *child* (i) and their flags.

To address the first problem, DFS3 will "force open" a path to the unreachable nodes. After reaching k, the agent will set $f(k)$ to true and $p(k)$ to i, that is, i will adopt k as a child. If the unreachable nodes form a cycle, then this rule will break it. This rule will also help restore the legal configuration when the spanning tree is acyclic, but not a DFS tree.

To address the second problem, the agent will keep track of the number of nodes visited while returning to the root via the parent link. For this purpose, the briefcase of the agent will include a nonnegative integer counter C. Whenever the agent moves from a parent to a child using DFS1 or DFS3, C is reset to 0, and and when the agent moves from a node to its parent using DFS2, C is incremented by 1. (Note: This will modify rules DFS1 and DFS2.) When C exceeds a predetermined value bigger than the size N of the network, a new parent has to be chosen, and the counter has to be reset. This is the essence of DFS4. A proof appears in [14].

Both the time complexity and the message complexity for stabilization are $O(n^2)$. Once stabilized, the agent needs $2(n-1)$ hops for subsequent traversals.

5.6.2 Agent Failure

If failures can hit the underlying distributed system, then they can hit the mobile agent too, resulting in its loss, or in the corruption of its state variables. This could do more damage than good and needs to be addressed. To deal with agent failure, we first introduce a *reliable agent*. Divide the agent variables into two classes: *privileged* and *nonprivileged*. Call an agent variable privileged when it can be modified only by its home process—all other variables will be called nonprivileged. Examples of privileged variables are SEQ or the id assigned to an agent by its home process. Then, an agent will be called reliable when it satisfies the following two criteria:

1. The agent completes its traversal of the network and returns home within a finite number of steps.
2. The values of all the privileged variables of the agent remain unchanged during the traversal.

An agent can be unreliable due either to the corruption of its privileged variables during a traversal or to routing problems. Note that, by simply being reliable, an agent cannot stabilize a distributed system. This leads to the adoption of a two-phased approach. In the first phase, we demonstrate how a reliable agent guarantees convergence and closure. In the second phase, we present methods by which unreliable agents eventually become reliable and remain reliable thereafter until the next failure occurs. This part will use some generic remedies independent of the problem under consideration. The generic remedies are as follows:

- **Loss of Agent** If the agent is killed, then the initiator discovers this using timeout and generates a new agent with a new sequence number. If the timeout is due to a delayed arrival of the original agent, then the original agent will eventually be killed[5] by the initiator.

- To avoid the risk of multiple agents with identical sequence numbers roaming in the system, the probabilistic technique of [25] can be used. It involves the use of a sequence number from a three-valued set $\sum = \{0, 1, 2\}$. If the sequence number of the incoming agent matches with the sequence number of the previous outgoing agent, then the initiator randomly chooses the next sequence number from $\sum$; otherwise the agent is killed.

- **Corruption of Agent Identifier** An agent is recognized by its home using the agent's id. If the id of the agent is corrupted, then the home process will not be able to recognize it, and the unreliable agent will roam the network forever. To prevent this, the supervisory program S of the agent counts the number of hops taken by the agent. As soon as this number exceeds a predefined limit $c \cdot R$ (c is a large constant and R is the round-trip traversal time), the agent kills itself. The same strategy works if, due to routing anomalies, the agent is unable to return home.

- **Corruption of Agent Variables** Here, the corruption of the nonprivileged variables is not a matter of concern because these are expected to be modified when the agent interacts with the underlying system. Our only concern is the possible corruption of privileged variables. To recover from such failures, we need to demonstrate that, despite the corruption of the privileged variables, eventually the agent reaches the global state to which it was initialized by its home process. For each agent-stabilizing system, as a part of the correctness proof, we need to prove the following theorem:

Theorem 28 *An unreliable agent is eventually substituted by a reliable agent.*

For the spanning tree generation algorithm using a single agent, all the generic fixes will hold. In addition, we will demonstrate how the system transparently handles an inadvertent corruption of the privileged variable SEQ.

5.6.3 Spanning Tree Construction Using Multiple Agents

We represent the agents by the uppercase letters $I, J, K, \ldots$ Our model is a synchronous one, where in unit time every agent takes a step. We assume that every agent visiting a process i leaves its "footprint" by writing its own id to a local variable $f(i)$, which is a set of agent identifiers. Each agent can find out which other agents visited process i by examining $f(i)$. The issue of bad data in $f(i)$ will be handled as follows: If an agent J discovers the footprint of another agent k that does not exist, then J will delete the K's footprint. Ordinarily all

---

[5]This is a shift from the static to the dynamic model but is necessary to keep the number of agents in control.

agents with a valid footprint are supposed to show up within a time period; otherwise they become fossils and are flushed out.

We now employ multiple agents to generate a spanning tree (not necessarily DFS) of g, with the hope of reducing the message or time complexity. Our static model uses a fixed number k of agents ($1 \leq k \leq n$). The proposed protocol is an adaptation of the Chen–Yu–Huang protocol [26] for spanning tree generation. The home of the agent with the smallest id is designated the root of the spanning tree—we call it the *root agent*. The spanning tree generation has two layers: In the first layer, the agents work independently and continue to build disjoint subtrees of the spanning tree until they meet other agents. In the second layer, the agents meet other agents to build appropriate bridges among the different subtrees; this results in a single spanning tree of the entire graph.

For the first layer, we will use the protocol of Figure 5.10. Therefore, we will only elaborate on the second layer, where a pair of agents K, L meet to make a decision about the bridge between them during an *unplanned meeting* at some process x. We will designate a bridge by the briefcase variable BB. When an agent K that is yet to form a bridge meets another agent L at a node x, it sets its briefcase variable BB to (L, x). Thereafter, node x will have two parents p_k and p_L from the two subtrees generated by K and L (see node j in Figure 5.11). Until a node x is visited by an agent K, $p_K = \phi$.

The maximum number of parents for any node is min (δ, k), where δ is the degree of the node and k is the number of agents. By definition, the root agent does not have a bridge (we use BB $= \perp$, $\perp$ to represent this).

In addition to BB, we add another nonnegative integer variable $Y (0 \leq Y \leq k)$ to the briefcase of every agent. By definition, $Y = 0$ for the root agent. Furthermore, during a meeting between two agents K, L, when K sets up its bridge to L, x, it also sets $Y(K)$ to $Y(L) + 1$. Thus, $Y(K)$ denotes "how many subtrees away" the subtree of K is from the root segment. In a consistent configuration, for every agent, $Y < k$. Therefore, if $Y = k$ for any agent, then the bridge for that subtree is invalidated.

Figure 5.12 describes the protocol for building a bridge between two subtrees. The home processes initialize each BB to $\perp$, $\perp$ once, but like other variables, these are also subject to corruption. The description of this protocol does not include the fossil removal actions.

Theorem 29 *For a given graph, if each agent independently generates disjoint subtrees, then the protocol in Figure 5.12 stabilizes to a spanning tree that consists of all the tree edges of the individual subtrees.*

Proof Outline As a consequence of the fossil removal mechanism, for every agent K, eventually BB $= \perp, \perp$ or L, i, where i is a process visited by both L and K. By definition, each subtree has *exactly one bridge* BB linking it with another subtree. Draw a graph g' in which the nodes are the subtrees (excluding the bridges) of g and the edges are the bridges linking these subtrees. Using the arguments in [26], we can show that g' will eventually be connected and acyclic. Therefore the set of edges (connecting a node with its parents) generated by the protocol of Figure 5.12 defines a spanning tree. $\square$

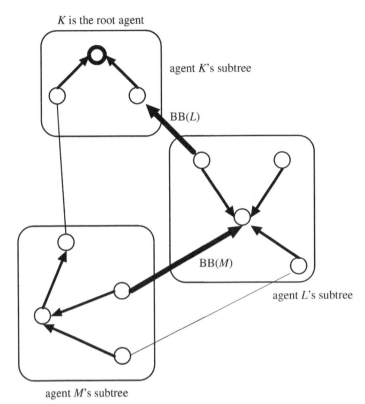

FIGURE 5.11 Spanning tree viewed as a graph with the nodes as subtrees and the edges as bridges.

Program for agent K while meeting agent L at node i

agent variables BB, NEXT, PRE;
process variables p {represents the parent of a node};
initially BB $= \perp, \perp$;

$$
\begin{aligned}
\textbf{do} \quad & BB(K) = L, i \wedge BB(L) = K, i \wedge K < L \rightarrow BB(K) := \perp, \perp; \\
\square \quad & BB(K) = \perp, \perp \wedge BB(L) \neq K, i \wedge K \neq \text{root agent} \wedge Y(L) \neq k \rightarrow \\
& \qquad BB(K) := L, i; \; Y(K) := Y(L) + 1 \\
\square \quad & BB(K) = L, i \wedge BB(L) \neq K, i \wedge Y(L) \neq k \wedge Y(K) \neq Y(L) + 1 \rightarrow \\
& \qquad Y(K) := Y(L) + 1 \\
\square \quad & BB(K) \neq L, i \wedge BB(L) = K, i \wedge p_K(i) \neq \text{PRE} \rightarrow p_K(i) := \text{PRE} \\
\square \quad & BB(K) = L, i \wedge Y(L) = k \wedge Y(K) \neq k \rightarrow Y(K) := k; \\
\square \quad & Y(K) = k \wedge BB(K) \neq L, i \wedge Y(L) < k - 1 \rightarrow \\
& \qquad BB(K) := L, i; \; Y(K) := Y(L) + 1 \\
\square \quad & BB(L) \neq K, i \wedge p_K(i) \neq \phi \rightarrow p_K(i) := \phi \\
\textbf{od}; & \\
\text{NEXT} & := \text{PRE}
\end{aligned}
$$

FIGURE 5.12 Program for building a bridge between adjacent subtrees.

Note that any existing spanning tree configuration is closed under the actions of the protocol.

To estimate the complexities, assume that each subtree is of equal size n/k. Let $M(s_K)$ be the number of messages required by a single agent K to build a subtree of size s_K starting from an *arbitrary initial state*. From [27], $M(s_K) = O(s_K^2)$ Also, once the subtree is stabilized, the number of messages required to traverse the subtree is $2 \cdot (s_K - 1)$. Since we assume $s_k = n/k$, the number of messages needed to build the k subtrees is $k \cdot M(n/k)$. To estimate the number of messages needed to detect a cycle in the graph g′ using the condition $Y \geq k$, consider a cycle $s_0 s_1 \cdots s_t s_0 (t \leq k)$ in g′, where each node is a subree. To correctly compute Y, each agent has to read the value of Y from the agent in its predecessor segment. This can take up to $1 + 2 + 3 + \cdots + (t - 1) = t(t - 1)/2$ traversals of the subtrees. Since the maximum value of t is k, for correctly detecting cycles in g′, at most $[k(k - 1)/2]/[2n/k]$ will be required. Also, each time a cycle is broken, the number of disjoint subtrees in g′ is reduced by 1 [26], so this step can be repeated no more than $k - 1$ times. Therefore the maximum number of messages needed for the construction of a spanning tree using a set of cooperating reliable agents will not exceed

$$k \cdot O\left(\frac{n^2}{k^2}\right) + \left(k \cdot \frac{k(k-1)}{2} \cdot \frac{2n}{k}\right) = O\left(\frac{n^2}{k} + n \cdot k^2\right)$$

To estimate the worst-case message complexity, we also need to take into account the overhead of fossil removal. This is determined by the number of hops taken by the agents to traverse the subtrees of size n/k, which is $O((n/k) \cdot k) = O(n)$. Note that this does not increase the order of the message complexity any further. The interesting result, at least with this particular protocol, is that as the number of agents increases, the message complexity first decreases and then increases. The minimum message complexity is $O(n^{5/3})$ when $k = O(n^{1/3})$.

To estimate the time complexity, assume that each of the k agents simultaneously builds subtrees of size n/k in time $O(n^2/k^2)$. The time required by the k agents to correctly establish their Y values is $k \cdot O(n/k) = O(n)$. At this time, the condition $Y \geq k$ can be correctly detected. The resulting actions reduce the number of disjoint subtrees by 1, so these actions can be repeated at most $k - 1$ times. The time complexity is thus $O(n^2/k^2 + n \cdot k)$. The overhead of fossil removal [which is $O(n/k)$] does not increase the time complexity any further. Therefore, the smallest value of the time complexity is $O(n^{4/3})$ when $k = O(n^{1/3})$.

5.7   CONCLUSION

Of the various possible roles that multiple mobile agents can play in a distributed system, this chapter picks four specific applications, illustrates how they work, and summarizes important results and open problems. For the proofs of these results, we encourage the readers to read the original articles.

Some of these problems can be solved using a single agent too. But a few problems (like the BHS problem) cannot be solved by a single agent and explicitly needs miltiple agents. The multiplicity of agents, on the one hand, accelerates certain applications, but on the other hand, synchronization becomes a tricky issue. Peleg [28] highlights several open problems related to coordination of multiple autonomous mobile robots (also known as robot swarms) for distributed computing application.

REFERENCES

1. P. Panaite and A. Pelc, Exploring unknown undirected graphs, *J. Algorithms*, 33:281–295, 1999.

2. A. Dessmark and A. Pelc, Optimal graph exploration without good maps, *Theor. Computer Sci.* 326: 343–362, 2004.

3. P. Fraigniaud, D. Ilcinkas, G. Peer, A. Pelc, and D. Peleg, Graph exploration by a finite automaton, in *Proceedings of the 29th International Symposium on Mathematical Foundations of Computer Science (MFCS'2004)*, Lecture Notes in Computers Science, Vol. 3153, Prague, Czech Republic, Aug. 22–27, 2004, pp. 451–462.

4. P. Fraigniaud, L. Gasieniec, D. Kowalski, and A. Pelc, Collective tree exploration, in *Proceedings of Latin American Theoretical Informatics (LATIN'2004)*, Lecture Notes in Computers Science, Vol. 2976, Springer, Buenos Aires, Argentina, Apr. 2004, pp. 141–151.

5. A. Dessmark, P. Fraigniaud, and A. Pelc, Deterministic rendezvous in graphs, in *Proceedings of the 11th Annual European Symposium on Algorithms (ESA'2003)*, Lecture Notes in Computers Science, Vol. 2832, Budapest, Hungary, Sep. 2003, pp. 184–195.

6. D. Kowalski and A. Pelc, Polynomial deterministic rendezvous in arbitrary graphs, in *Proceedings of the 15th Annual Symposium on Algorithms and Computation (ISAAC'2004)*, Hong Kong, Dec. 2004.

7. G. Marco, L. Gargano, E. Kranakis, D. Kriznac, A. Pelc, and U. Vaccaro, Asynchronous deterministic rendezvous in graphs, *Theor. Comput. Sci.,* 355(3): 315–326, 2006.

8. K. Chandy and L. Lamport, Distributed snapshots: determining global states of distributed systems, *ACM Trans. Computer Syst.,* 3: 63–75, 1985.

9. A. Arora and M. Gouda, Distributed reset, *IEEE Trans. Computers*, 43(9): 1026–1038, 1990.

10. N. Carriero and D. Gelernter, *How to Write Parallel Programs: A First Course*, MIT Press, Cambridge, Mass., 1990.

11. M. Herlihy, Wait-free synchronization, *ACM Trans. Program. Lang. Syst.* 11(1): 124–149, 1991.

12. L. Lamport, Concurrent reading and writing, *CACM Ser.*, 20(11): 806–811, 1977.

13. L. Lamport, On interprocess communication. Part II: Algorithms, *Distributed Comp.*, 1(2): 86–101, 1986.

14. S. Ghosh, Agents, distributed algorithms, and stabilization, Lecture Notes in Computers Science, Vol. 1858, pp. 242–251, 2000.

15. S. Ghosh and A. Bejan, Concurrent reading and writing with mobile agents, in *Proceedings of IWDC (International Workshop on Distributed Computing),* Lecture Notes in Computer Science, Vol. 2571, Springer, 2002, pp. 67–77.

16. S. Dobrev, P. Flocchini, G. Prencipe, and N. Santoro, Mobile agents searching for a black hole in an anonymous ring, in *Proceedings of the 15th International Symp. on Distributed Computing (DISC'01)*, Lecture Notes in Computer Science, Vol. 2180, Springer, 2001, pp. 166–179.

17. S. Dobrev, P. Flocchini, G. Prencipe, and N. Santoro, Searching for a black hole in arbitrary networks: Optimal mobile agent protocols, in *Proceedings of the 21st ACM Symposium on Principles of Distributed Computing (PODC'02)*, 2002, pp. 153–162.

18. S. Dobrev, P. Flocchini, R. Kralovic, G. Prencipe, P. Ruzicka, and N. Santoro, Searching for a black hole in hypercubes and related networks, in Bui and Fouchal (Eds.), *Proceedings of the 6th International Conference on Principles of Distributed Systems (OPODIS '02)*, Vol. 3, Suger, Saint-Dennis, France, 2002, pp. 171–182.

19. S. Dobrev, P. Flocchini, and N. Santoro, Improved bounds for optimal black hole search with a network map, in *SIROCCO 2004*, Lecture Notes in Computer Science, Vol. 3104, Springer, 2004, pp. 111–122.

20. J. Czyzowicz, D. Kowalski, E. Markou, and A. Pelc, Searching for a black hole in tree networks, in *Proceedings of the 8th International Conference on Principles of Distributed Systems (OPODIS '04)*, Lecture Notes in Computer Science, Vol. 3544, Grenoble, France, Dec. 15–17, 2005, pp. 67–80.

21. S. Dobrev, P. Flocchini, G. Prencipe, and N. Santoro, Multiple agents rendezvous in a ring in spite of a black hole, in *Proceedings of the 7th International Conference on Principles of Distributed Systems (OPODIS '03)*, Lecture Notes in Computer Science, Vol. 3144, La Martinique, French West Indies, Dec. 10–13, 2004, pp. 34–46.

22. S. Dobrev, R. Kralovic, N. Santoro, and W. Shi, Black hole search in asynchronous rings using tokens, in *CIAC*, Lecture Notes in Computer Science, Vol. 3998, Springer, 2006, pp. 139–150.

23. E. W. Dijkstra, Self stabilizing systems in spite of distributed control, *Commun. ACM*, 17(6):643–644, 1974.

24. S. Katz and K. Perry, Self-stabilizing extensions for message-passing systems, in *Proceedings of the 9th Annual Symposium on Principles of Distributed Computing (PODC'90)*, 1990, pp. 91–101.

25. T. Herman, Self-stabilization: Randomness to reduce space, *Distributed Comput.* 6:95–98, 1992.

26. N. S. Chen, H. P. Yu, and S. T. Huang, A self-stabilizing algorithm for constructing a spanning tree. *Inform. Process. Lett.*, 39:147–151, 1991.

27. S. Ghosh, Cooperating mobile agents and stabilization, Lecture Notes in Computers Science, Vol. 2194, 2001, pp. 1–18.

28. D. Peleg, Distributed coordination algorithms for mobile robot swarms: New directions and challenges, in *IWDC 2005*, keynote address, Lecture Notes in Computers Science, Vol. 3741, Springer, pp. 1–12.

29. T. Araragi, P. Attie, I. Keidar, K. Kogure, V. Luchanugo, N. Lynch, and K. Mano, On formal modeling agent computations, Lecture Notes in Artificial Intelligence, Vol. 1871, 2000, pp. 48–62.

30. F. Mattern, Virtual time and global states of distributed systems, in M. Cosnard, et al. (Ed.), *Proceedings of Parallel and Distributed Algorithms*, Chateau de Bonas, France, Elsevier, 1989, pp. 215–226.

31. G. Tel, *Distributed Algorithms*, Cambridge University Press, Cambridge, 2000.

PART III
Mobile Agent Based Techniques and Applications

6 Network Routing

ANDRE COSTA

ARC Centre of Excellence for Mathematics and Statistics of Complex
Systems, University of Melbourne, Melbourne, Australia

NIGEL BEAN

Applied Mathematics, University of Adelaide, Adelaide, Australia

6.1 INTRODUCTION

Since the early days of the ARPANET [1], adaptive decentralized routing
algorithms for communications networks have been in demand. This has driven
the development of a variety of autonomous systems, where routing decisions
are made by a collection of agents that are spatially distributed and where each
agent communicates only local information with is nearest neighbors. The first
such algorithm was the distributed Bellman–Ford algorithm [1] originally
developed for the ARPANET, whereby a routing "agent" resides at each network node and implements the local dynamic programming operations that are
required to solve the network shortest path problem.

Recent advances in the fields of multiagent systems, ant colony optimization,
and reinforcement learning have led to the proposal of mobile agent-based
algorithms for the task of network routing. The majority of mobile agent-
based routing algorithms extend the basic distance–vector framework [1],
whereby each node maintains a measure of the cost associated with reaching
each possible destination node, via each available outgoing link. However, while
traditional distance–vector algorithms (such as the distributed Bellman–Ford)
take administrator-assigned link weights, mobile agent-based algorithms
employ agents to actively gather route information, usually in the form of trip
time measurements. This information is then used to update the routing tables at
the network nodes in a manner that aims to reinforce good routes (e.g., those

Mobile Agents in Networking and Distributed Computing, First Edition.
Edited by Jiannong Cao and Sajal K. Das.
© 2012 John Wiley & Sons, Inc. Published 2012 by John Wiley & Sons, Inc.

with low delay) and divert data traffic away from unfavorable routes (those with high delay).

A characteristic of many (though not all) agent-based routing algorithms is that they employ stochastic routing policies for agents and data traffic. This provides an exploration mechanism which permits agents to discover and monitor alternative routes. It also provides a mechanism for multipath routing for data traffic, which is desirable for the purpose of load balancing. In particular, for algorithms such as AntNet [2, 3], which are inspired by the collective problem-solving behaviors of ant colonies, the stochastic routing tables play a role that is analogous to the chemical pheromone field which mediates the indirect communication between biological ants.

Agent-based routing algorithms possess a number of desirable attributes:

- They are designed to adapt in real time to sudden or gradual changes in traffic demands and network component failures.
- There is no centralized controller—The overall network routing policy is generated and updated by a population of mobile agents, none of which are critical (at the individual level) to the operation of the system.
- They are distributed, in the sense that agents make decisions using only local information, without having knowledge of the global state of the network, that is, a priori knowledge of all link delays and the state of the routing tables at every network node.

Together, these attributes confer a high level of *robustness*, making agent-based algorithms highly suited to the problem of adaptive routing in stochastic nonstationary environments.

This chapter is structured as follows. In Section 6.2, we give background information and a literature review. In Section 6.3, we describe the problem of network routing. A number of agent-based routing algorithms are described in Section 6.4 followed by a discussion and comparison in Section 6.5. Future research directions are given in Section 6.6, followed by conclusions in Section 6.7.

6.2 BACKGROUND AND LITERATURE REVIEW

Adaptive routing algorithms such as the distributed Bellman–Ford algorithm were first developed for the ARPANET in the 1970s. Although these algorithms did not explicitly employ mobile agents, they were an important precursor, because modern agent-based algorithms are based on data structures that are natural extensions of the basic distance–vector framework [1]. For the early ARPANET algorithms, it is useful to think of each node of the network as a routing agent, which communicates with its nearest neighbors to collectively solve a shortest path problem on the whole network. These algorithms typically performed single-path (as opposed to multipath) routing of data traffic, via the shortest path links. As a result, when they were augmented with real-time measurements of link delays and link queue lengths [1, 4], undesirable oscillations in the routing policy often resulted, requiring fine tuning of damping parameters in order to avoid instability. Faced with these difficulties, static routing schemes remained popular in practice (and are still widely used in modern computer networks).

In the late 1970s, distance-vector algorithms were further extended using ideas from nonlinear programming [5, 6]. These were more sophisticated than the early shortest path routing algorithms, in that they were designed to perform multipath routing and were able to attain system optimal routing policies. The trade-off is that they were also significantly more computationally demanding due to their requirement for the estimation of gradients. In particular, instead of maintaining estimates of the delay associated with reaching a destination node via each outgoing link, they maintained estimates of the marginal delay. This restricted their suitability to network environments in which traffic demands changed gradually or infrequently, that is, at the time scale of connections, rather than the much shorter time scale of packet trip times. As a result of these factors, they were never actually deployed in the ARPANET. However, we include them in our discussion because they serve as a useful point of reference, and they show how system optimal routing policies can be achieved by a distributed multiagent system. Furthermore, they may prove useful in future research on mobile agent systems, particularly in the light of recent advances in online gradient estimation techniques [7–9].

The idea of using mobile agents for routing first appeared in [10], although no actual algorithm was given. During the 1990s, the emergence of multiagent systems and reinforcement learning algorithms [11] led to renewed research interest in applications to adaptive network routing. In the context of control systems, reinforcement learning is concerned with algorithms which are able to learn optimal or near-optimal control policies using as input only a set of reward or penalty signals obtained directly from the system that is being controlled. Indeed, the approach of "embedding" a reinforcement learning algorithm within a multiagent system has received much attention in recent years (see, for example, [12]). In particular, the Q-learning algorithm [13] is one of the most well-known reinforcement learning algorithms and forms the basis for a number of agent-based routing algorithms for communications networks, such as Q-routing [14], predictive Q-routing [15], confidence-based Q-routing [16], and confidence-based dual-reinforcement Q-routing [17]. Much like the early ARPANET algorithms, these were designed to perform single-path shortest path routing.

More recently, actor–critic reinforcement learning algorithms [11] provided the motivation for the multipath system and traffic adaptive routing algorithm (STARA) [18, 19], which can attain a particular type of user-based optimization known as the Wardrop equilibrium. Other agent-based routing algorithms that use reinforcement learning include team-partitioned, opaque transition reinforcement learning (TPOT-RL) [20] and an online policy gradient algorithm for solving partially observed Markov decision problems (OLPOMDP) [21].

The field of research which has had perhaps the greatest impact on mobile agent-based routing algorithms is that of ant colony optimization (ACO) [22, 23], a well-established stochastic optimization method which is inspired by the emergent problem-solving capabilities of biological ant colonies. The ACO method has been applied to the problem of adaptive network routing, resulting in the well-known AntNet algorithm [2, 3, 24] for packet-switched networks. Extensive simulation experiments using AntNet [2] show that it performs exceptionally well in a range of realistic nonstationary network conditions. One of AntNet's strengths is that it was designed to perform a type of multipath routing known as *proportional routing*, with the specific aim of avoiding the instability and inefficiency that often arises with shortest path routing algorithms. Other routing algorithms that use antlike mobile agents include [25–29]. Indeed, there exist close connections between ant-based methods and reinforcement learning [30–32].

Mobile agent-based routing algorithms have also been applied successfully to wireless adhoc networks [33–35], where the network nodes are themselves mobile, resulting in a nonstationary network topology.

6.3   NETWORK ROUTING PROBLEM

Given a communications network with a set of origin–destination node pairs and a set of associated traffic demands, the routing problem involves determining which paths are to be used to carry the traffic. A path may comprise multiple hops passing through multiple routers and links, and typically, there are a number of alternative paths that are available for any given origin–destination node pair.

In principle, the task of routing can be viewed as an optimization problem whereby a routing algorithm attempts to optimize some performance measure subject to constraints. Examples of performance measures include the total average throughput (to be maximized) or the total average packet delay (to be minimized). Constraints arise from the limited transmission and processing capacity of network links and nodes. The algorithm designer may also introduce quality-of-service (QoS) constraints, such as maximum acceptable average delay jitter or packet loss rate [36]. Furthermore, a good adaptive routing algorithm is able to *track* an optimal or near-optimal solution in real time as network conditions change.

Some agent-based routing algorithms are designed to solve an explicit optimization problem using a "top-down" approach; that is, the agent behavior is programmed so that the collective outcome is the solution to a target optimization problem. In contrast, a number of popular agent-based routing algorithms are designed from the "bottom up"; that is, agent behavior is designed using heuristics that are inspired and informed by reinforcement learning and/or ACO. Thus, an answer to the question of *"what optimization problem is an agent-based routing algorithm attempting to solve"* is not always immediately apparent; indeed, in some cases, it may be that be that a complex multiobjective problem is being solved implicitly, without such a problem ever having been formulated explicitly by the algorithm designer. In this study, we survey agent-based routing algorithms within a unified framework, which provides some insights into their similarities and differences with respect to a number of key features, including their optimization goals.

Given that the majority of agent-based routing algorithms attempt to optimize delay-based performance measures, we focus in this chapter on delay as the primary measure of performance. Furthermore, we focus on algorithms for packet-switched networks, as these comprise the majority of mobile agent-based routing algorithms that have been proposed during the last decade.

We begin by surveying the decentralized data structures and functions that are common to the majority of agent-based routing algorithms that have been proposed in the literature to date. As we shall see, these data structures and their functions determine the nature of the optimization goals that can be achieved using a mobile agent system.

6.3.1 Data Structures and Function

The algorithms incorporate a delay estimation process and a routing update process as follows. For every possible destination node, each node maintains estimates of the travel time taken by a packet to reach the destination via each of the node's outgoing links. This is done via the active online measurement of network delays using mobile agents and constitutes a natural extension of static distance–vector routing algorithms which take administrator-assigned link weights as an exogenous input [1]. For every possible destination node, each node maintains a set of routing probabilities which are used to randomly route data packets arriving at the node via one of the outgoing links. The routing probabilities are updated periodically according to some function of the delay estimates that are generated by the agents and are modified in such a way as to reduce traffic flow on outgoing links which are estimated to have a high associated delay to the destination and vice versa.

Significantly, a node does not differentiate packets by their origin or history; all packets assigned to a given destination node are treated equally by the node, as they are routed according to the same set of routing probabilities. We use the term *stateless* (a term that we borrow from the language of networking [37]) to describe this type of routing. Thus, a characteristic property of agent-based algorithms is that link routing probabilities, rather than paths flows, are the primary control variables.

As a general network model, we consider a set of nodes $\mathcal{N}$ connected by a set of directed links $\mathcal{A}$, and we denote the set of neighbor nodes of i using $\mathcal{N}_i$. Consider a given destination node $d \in \mathcal{N}$. A stateless routing policy for packets with destination node d, represented by Ψ^d, is a decision rule for selecting an outgoing link from the set $\mathcal{A}$ at each node $i \in \mathcal{N} \setminus d$. A complete routing policy for all destination nodes is thus defined by the union $\bigcup_{d \in \mathcal{N}} \Psi^d$. For clarity of presentation and without loss of generality, we shall henceforth write all quantities and algorithmic update rules for a single destination node $d \in \mathcal{N}$. The reader should keep in mind that all of the quantities and update rules are replicated for all nodes that act as destination nodes for data traffic.

Specifically, a routing policy for a given destination d is represented by the probabilities $\psi_{ij}, i \in \mathcal{N} \setminus d, j \in \mathcal{N}$, where we set $\psi_{ij} = 0$ for $(i,j) \notin \mathcal{A}$. For a given node $i \in \mathcal{N} \setminus d$, the probabilities $\psi_{ij}, j \in \mathcal{N}_i$, define a probability distribution over the outgoing links. These probabilities are used for the stochastic routing of data packets, comprising the traffic demands that are placed on the network. We note that several agent-based routing algorithms employ shortest path routing, which corresponds to the "degenerate" case where all routing probabilities are either 0 or 1. For convenience, the routing policy for destination d can be represented by an $(N-1) \times (N-1)$ stochastic matrix Ψ, with entries $\psi_{ij}, i, j \in \mathcal{N} \setminus d$, where the remaining probabilities ψ_{id} are recovered as $1 - \sum_{j \in \mathcal{N}_i} \psi_{ij}$. Let s_i denote the traffic demand originating at node i, with destination node d, and let $s_i, i \in \mathcal{N} \setminus d$, denote the set of such demands.

The union $\bigcup_{d \in \mathcal{N}} \Psi^d$ uniquely determines the total expected rate of packet flow on link (i,j), denoted simply as $f_{ij}(\Psi)$ for clarity of notation. The reader is referred to [5, 31] for details of the calculation of these flows. The link flows induce queueing delays on each link, and their expected values, given by the functions $R_{ij}(f_{ij}(\Psi)), (i,j) \in \mathcal{A}$, can be described using a number of models [38]. Clearly, the functions $R_{ij}(x)$ must be increasing in x to reflect the fact that the expected delay incurred by a packet in traversing a link increases as the amount of traffic flow on the link increases.

In the following sections, we shall discuss a number of delay-based optimization problems that can be solved in a distributed manner by mobile agent systems, that is, by performing delay measurements and controlling only the routing probabilities $\psi_{ij}, (i,j) \in \mathcal{A}$. The objective functions that we consider are constructed using the delay functions $R_{ij}(x)$, and the routing probabilities are subject to the following constraints:

(i) The routing probabilities must belong to the set

$$\mathcal{P} = \left\{ \psi_{ij} \geq 0, (i,j) \in \mathcal{A}, \sum_{j \in \mathcal{N}_i} \psi_{ij} = 1, i \in \mathcal{N} \setminus d \right\}$$

(ii) In order for the link flows to be finite, it is necessary that the matrix Ψ specify a path of positive probability from all nodes $i \in \mathcal{N} \setminus d$ to d, which is equivalent to the condition that the matrix $I - \Psi$ be invertible, where I is the identity matrix [31].

(iii) Letting C_{ij} denote the (finite) capacity on link (i,j), Ψ must be such that

$$f_{ij}(\Psi) \leq C_{ij} \tag{6.1}$$

for all links $(i,j) \in \mathcal{A}$.

We note that constraint (iii) forces the link flows to be finite, which in turn implies that constraint (ii) must be satisfied; therefore it is sufficient to consider only constraints (i) and (iii). Furthermore, following [5, 6], the need to explicitly consider constraint (iii) can be eliminated by assuming that the link delay functions have the property

$$\lim_{x \to C_{ij}} R_{ij}(x) = \infty \tag{6.2}$$

for all $(i,j) \in \mathcal{A}$. This eliminates the need to introduce Lagrange multipliers corresponding to the capacity constraints because capacity violations are effectively "penalized" in the objective function via (6.2). Thus the only constraint which must be explicitly taken into account when modeling the routing algorithms is the normalization constraint (i).

6.3.2 User Equilibrium Routing

User equilibria arise when there exist multiple users of a finite set of shared resources. In the network routing context, a user usually represents a certain traffic demand which must be carried from an origin node to a destination node, while the finite resources correspond to the network link transmission capacities. Mathematically, this situation can be modeled as a multiobjective optimization problem. Let $\mathcal{Y}$ denote the set of users of the network, and let $f_{ij}^{y}(\Psi)$ denote user y's traffic flow on the link (i, j) under the routing policy Ψ. The expected flow-weighted delay incurred by user y on the link (i, j) is given by

$$J_{ij}^{y}(\Psi) = f_{ij}^{y}(\Psi) R_{ij}(f_{ij}(\Psi)) \tag{6.3}$$

The total expected flow-weighted delay incurred by user y over all links is then given by

$$J^{y}(\Psi) = \sum_{(i,j) \in A} J_{ij}^{y}(\Psi) \tag{6.4}$$

This suggests the constrained multiobjective program

$$\min_{\Psi \in \mathcal{P}} J^{y}(\Psi) \qquad y \in \mathcal{Y} \tag{6.5}$$

The program (6.5) on its own does not specify a way to balance the competing interests represented by each objective function. The fields of game theory and multiobjective optimization have given rise to a number of solution concepts that define what it means for a solution to (6.5) to be "optimal." These include *pareto optimality* as well as the celebrated *Nash equilibrium* [39]. However, given the stateless routing structure described in Section 6.3.1, agent-based routing algorithms are not able to systematically attain these operating points, except for a special case of the Nash equilibrium, known as the Wardrop equilibrium [40], which we describe shortly.

Nash equilibria arise when each user (origin–destination traffic flow) is "selfish," in that each user makes decisions in an effort to minimize its own incurred cost, without taking into consideration the effect that its own decisions may have on the costs incurred by other users. A routing policy Ψ represents a Nash equilibrium if no user $y \in \mathcal{Y}$ can unilaterally decrease its total expected delay $J^y(\Psi)$ by routing its traffic according to a routing policy that is different to Ψ. The reason why Nash equilibria cannot be attained in a systematic manner by agent-based algorithms is essentially due to their stateless property; all packets arriving at a node that have the same destination are routed according to a common set of routing probabilities, regardless of their origin or "history." Thus no node can be acting differently on behalf of any user. Indeed, algorithms that do attain Nash equilibria are not stateless in the sense defined above. For Nash routing, each user is allocated a finite set of paths, and Nash equilibria are attained by an iterative procedure, whereby users take turns in adjusting their own routing policy in response to the decisions made by other users (see, for example, the algorithms described in [41, 42]). In order to achieve the same degree of control, it would be necessary for agent-based routing algorithms to maintain a separate routing matrix Ψ *for each origin–destination node pair*, so that each user could have complete control of the routes used to carry its own traffic. However, this is beyond the scope of the data structures and functions that have been proposed to date for agent-based routing algorithms.

The Wardrop equilibrium arises from the Nash equilibrium in the limiting case where

1. the number of users becomes infinitely large,
2. the demand placed on the network by each individual user becomes negligible, so that routing decisions of individual users have no effect on the delays experienced by other users, and
3. the total traffic demand placed on the network remains constant.

This limiting case corresponds most closely to the physical interpretation of users being the *individual packets* that pass through the network. A Wardrop equilibrium routing policy Ψ is therefore characterized by the fact that no *packet* can unilaterally decrease its expected trip time from its origin to the destination by following a policy that is different from Ψ.

The derivation of the Wardrop equilibrium as a special case of the Nash equilibrium was originally given in [43]. A similar derivation that is specific to the context of stateless network routing is given in [31], yielding the following sufficient condition: Ψ corresponds to a Wardrop equilibrium if, for each $(i, j) \in \mathcal{A}$,

$$
\begin{aligned}
R_{ij}(f_{ij}(\Psi)) + D_j(\Psi) &\geq \min_{l \in \mathcal{N}_i} [R_{il}(f_{il}(\Psi)) + D_l(\Psi)] &&\text{if } \psi_{ij} = 0 \\
R_{ij}(f_{ij}(\Psi)) + D_j(\Psi) &= \min_{l \in \mathcal{N}_i} [R_{il}(f_{il}(\Psi)) + D_l(\Psi)] &&\text{if } \psi_{ij} > 0
\end{aligned}
\tag{6.6}
$$

where $D_l(\Psi)$ is the expected delay incurred by a packet currently at node l to reach the destination d under the routing policy Ψ. At the destination node, we impose the natural boundary condition $D_d(\Psi) = 0$. It is straightforward to show that the terms

$$
\min_{l \in \mathcal{N}_i} [R_{il}(f_{il}(\Psi)) + D_l(\Psi)]
$$

for $i \in \mathcal{N} \setminus d$, which appear on the right-hand side of (6.6), are Lagrange multipliers associated with the normalization constraints in $\mathcal{P}$. We also note that since the objective functions in the problem (6.5) are nonconvex in the routing probabilities, the ordinary Karush–Kuhn–Tucker (KKT) conditions [44] are not sufficient for a Wardrop equilibrium, and a slightly different method of proof is required (see [31] for details). The condition (6.6) can be understood as follows. Let

$$
D_{ij}(\Psi) = R_{ij}(f_{ij}(\Psi)) + D_j(\Psi)
\tag{6.7}
$$

and observe that this is the expected delay incurred by a packet at i to reach the destination node d via the outgoing link (i, j), under the routing policy Ψ. Then (6.6) implies that a Wardrop equilibrium has the property that for each node $i \in \mathcal{N} \setminus d$ all outgoing links at i which are used for sending traffic to the destination have the minimum expected delay and all outgoing links at i which are unused have an expected delay which is at least as large, thus coinciding with our earlier definition of a Wardrop equilibrium. Of course, given delay functions $R_{ij}(\cdot)$, the expectations $D_{ij}(\Psi), (i,j) \in \mathcal{A}$, can be calculated analytically. However, this requires global information, that is, knowledge of all entries of the matrix Ψ, which is not available to any individual node agent or mobile agent. These expectations must therefore be estimated by performing active trip time measurements using mobile agents. In particular, an agent-based routing algorithm that exploits the sufficient condition (6.6) in order to achieve approximate Wardrop routing policies is described in Section 6.4. Furthermore, we shall see that the equilibrium operating point of ant-based routing algorithms can be interpreted as an *approximate* Wardrop equilibrium.

6.3.3   System Optimal Routing

A common networkwide performance measure is the total expected flow-weighted packet delay in the network [1], given by

$$D^T(\Psi) = \sum_{(i,j) \in \mathcal{A}} f_{ij}(\Psi) R_{ij}(f_{ij}(\Psi)) \qquad (6.8)$$

Accordingly, a routing policy Ψ corresponds to a *system optimum* if it is an optimal solution to the constrained nonlinear program

$$\min_{\Psi \in \mathcal{P}} D^T(\Psi) \qquad (6.9)$$

We note that $D^T(\Psi)$ is nonconvex in the routing probabilities Ψ, and thus the corresponding KKT conditions for the routing probabilities are not sufficient for a minimum of (6.9). However, it is shown in [5] that a minor modification of the KKT conditions yields a first-order sufficient condition. In particular, letting

$$R'_{ij}(x) = \frac{dR_{ij}(x)}{dx}$$

then Ψ corresponds to a system optimal routing policy if, for all $(i,j) \in \mathcal{A}$,

$$
\begin{aligned}
R'_{ij}(f_{ij}(\Psi)) + \frac{\partial D^T(\Psi)}{\partial s_j} &\geq \min_{l \in \mathcal{N}_i}\left[R'_{il}(f_{il}(\Psi)) + \frac{\partial D^T(\Psi)}{\partial s_l} \right] \quad \text{if } \psi_{ij} = 0 \\
R'_{ij}(f_{ij}(\Psi)) + \frac{\partial D^T(\Psi)}{\partial s_j} &= \min_{l \in \mathcal{N}_i}\left[R'_{il}(f_{il}(\Psi)) + \frac{\partial D^T(\Psi)}{\partial s_l} \right] \quad \text{if } \psi_{ij} > 0
\end{aligned}
\qquad (6.10)
$$

where $R'_{ij}(f_{ij}(\Psi))$ gives the marginal packet delay associated with the link (i,j) and $\partial D^T(\Psi)/\partial s_j$ captures the marginal packet delay associated with the remaining journey from node j to d. At the destination node, we have the natural boundary condition $\partial D^T(\Psi)/\partial s_d = 0$. Again, the terms on the right-hand side of (6.10) for each $i \in \mathcal{N} \setminus d$ are Lagrange multipliers associated with the normalization constraints on the routing probabilities. Let

$$D'_{ij}(\Psi) = R'_{ij}(f_{ij}(\Psi)) + \frac{\partial D^T(\Psi)}{\partial s_j} \qquad (6.11)$$

and observe that this is the expected marginal delay incurred by a packet at i to reach the destination node d via the outgoing link (i, j) under the routing policy Ψ. Then (6.10) implies that a system optimal routing policy has the property that for each node $i \in \mathcal{N} \setminus d$ all outgoing links at i which are used for sending traffic to the destination have the minimum expected marginal delay, and all outgoing links at i which are unused have an expected marginal delay which is at least as large. A class of distributed algorithms which exploit the sufficient condition (6.10) in order to attain a system optimal routing policy is discussed in Section 6.4.

6.4 SURVEY OF AGENT-BASED ROUTING ALGORITHMS

Before giving a description of the algorithms, we introduce a classification system which will allow us to compare and contrast their salient features.

6.4.1 Classification System

Agent-based routing algorithms can be classified and differentiated by comparing the following key properties:

1. *Nature of Active Measurements That Are Performed by Mobile Agents*
   - The delay statistic that is sampled by a mobile agent is either *delay* or *marginal delay*.
   - The mechanism for generating samples of the delay statistic is either the *bootstrapping* approach or the *Monte Carlo* approach [11]. The bootstrapping approach involves using mobile agents that report the delay statistic associated with traveling from node i to d via node j by directly measuring the delay statistic on the link (i, j) and using node j's locally stored estimate of the delay statistic associated with the remainder of travel to the destination d. In contrast, the Monte Carlo approach involves the direct measurement of the delay statistic by explicitly measuring the delay statistic associated with a complete path from node i to d via node j. The Monte Carlo method therefore does not use any delay information that is "cached" at intermediate nodes between i and d.

2. *Exploration Mechanism* The exploration mechanisms employed by agent-based routing algorithms can be classified as either *on-policy* or *off-policy*—this terminology is borrowed from [11]. In the context of adaptive routing, exploration entails measurement of the delay characteristics of alternative routes. We shall say that an agent-based routing algorithm is on-policy if there exists a coupling between the agent exploration mechanism and the data routing policy, resulting in a trade-off between these tasks. In contrast, an off-policy learning algorithm is one where the exploration and decision-making mechanisms are decoupled in such a manner that there is no trade-off between these tasks.

3. *Single-Path or Multipath Routing* For a given origin–destination node pair, a *single-path* routing algorithm sends all data traffic over a single

route, so that $\psi_{ij} \in \{0, 1\}$ for all links (i, j). In contrast, a *multipath* routing algorithm has the ability to split the data traffic over more than one route, and we therefore have $\psi_{ij} \in [0, 1]$ for all (i, j).

4. *Characteristic Operating Point* The target operating points of the agent-based routing algorithms described in this chapter fall into one of four categories: *Wardrop routing, system optimal routing, shortest path routing,* and *proportional routing*. The first two were discussed earlier in this section. Shortest path routing entails sending all traffic on an outgoing link which is associated with the smallest estimated delay to the destination. The last category refers to the case where the proportion of data traffic routed on each outgoing link is a decreasing function of the associated delay estimate for the link.

We note that the characteristic operating point refers to the type of routing policy that is targeted by the algorithm and which would be reached sooner or later (depending on the rate of convergence) under stationary conditions. In a nonstationary environment, the target operating point changes with time and is then *tracked* by the algorithm.

In the following sections, we describe a collection of representative agent-based routing algorithms, and we classify them according to the features listed above. A summary of these classifications is provided in Table 6.1.

6.4.2   Algorithms Based on Q-Learning

The first adaptive routing algorithm for communications networks to be modeled explicitly on a reinforcement learning algorithm is known as Q-routing [14] and was inspired by the popular Q-learning algorithm [13]. In the Q-routing algorithm, a reinforcement learning agent is embedded into each node of the network. For every destination node d, each node i maintains estimates $Q_{ij}, j \in \mathcal{N}_i$, of the time taken for a packet to travel from i to d via the outgoing link (i, j). Consider a fixed destination node d and a given link (i, j). Every time that a packet with destination d traverses the link (i, j), the update

$$Q_{ij} := (1 - a)Q_{ij} + a(r_{ij} + Q_j^{\min}) \qquad (6.12)$$

is performed, where $a \in (0, 1]$ is a step size, or learning rate parameter, r_{ij} is the queueing plus transmission delay incurred by the packet in moving from node i to j, and

$$Q_j^{\min} = \min_{l \in \mathcal{N}_j} Q_{jl} \qquad (6.13)$$

which constitutes node j's estimate of the minimum possible trip time to reach d. In particular, the quantity $r_{ij} + Q_j^{\min}$ is transmitted back to node i by a special "signaling" packet, or mobile agent, which carries this information. Thus, Q-routing performs delay measurements via the bootstrapping approach.

TABLE 6.1 Classification of Agent-Based Routing Algorithms

	Routing	Operating Point	Measurement	Exploration
Distributed Bellman–Ford	Singlepath	Shortest path	Bootstrapping (fixed-link weights)	Off policy
Q-routing	Singlepath	Shortest path	Bootstrapping (delay)	Off policy
STARA	Multipath	Wardrop	Monte Carlo (delay)	On policy
AntNet and other ant-based algorithms	Multipath	Proportional	Monte Carlo (delay)	On policy
AntHocNet	Multipath	Proportional	Combination (delay)	On policy
Gallager and variants	Multipath	System optimal	Bootstrapping (marginal delay)	On policy
TPOT-RL	Singlepath	Shortest path	Monte Carlo (delay)	On policy
OLPOMDP	Multipath	System optimal	Monte Carlo (marginal delay)	On policy
SAMPLE	Multipath	Proportional	Bootstrapping (delay)	Off policy
MCWR	Multipath	Wardrop	Monte-Carlo (delay)	Off policy
Wardrop Q-routing	Multipath	Wardrop	Bootstrapping (delay)	Off policy

The policy used to route packets is to always select the outgoing link which has the minimum associated delay estimate, that is, single-path shortest path routing. This corresponds to the following update rule for the routing probabilities maintained at node i:

$$\psi_{ij} := \begin{cases} 1 & \text{if} \quad (i,j) = \arg\min_{(i,l):l \in \mathcal{N}_i} Q_{il} \\ 0 & \text{otherwise} \end{cases} \tag{6.14}$$

The inventors of Q-routing do not specify how to route packets in the event that two or more outgoing links have equal associated delay estimates. A natural way to resolve such situations would be to randomize uniformly over the set of outgoing links with equal minimum associated delay estimates.

The delay estimates Q_{ij} are analogous to the entries of a traditional distance–vector routing algorithm, such as the distributed Bellman-Ford [1]. The difference lies in the fact that actual trip time measurements, r_{ij}, form the basis for updating the estimates and these updates occur very frequently, whereas the entries of a distance–vector algorithm are typically static values that are assigned by a network administrator. The step size parameter a in (6.12) determines the weighting of new measurements versus the stored estimate.

As described above, the Q-routing algorithm routes all packets on the outgoing link which has the smallest associated Q-value. As a result, delay measurements for currently unused links are never obtained. This prevents the algorithm from being able to discover new paths which may be shorter than those currently in use. The inventors of Q-routing avoid this type of stagnation by introducing an exploration mechanism whereby every node periodically probes all of its outgoing links, including the ones which do not carry traffic. This task is performed by mobile agents, which periodically trigger the update (6.12) for all links $(i,j) \in \mathcal{A}$. This exploration mechanism ensures that delay information is obtained for all outgoing links, irrespective of which link is currently being used to route data traffic, and is therefore classed as off-policy. We also note that this exploration mechanism performs an analogous role to the broadcasting of distance estimates between neighboring nodes that occurs in most distance–vector routing algorithms, such as the distributed Bellman–Ford algorithm [1].

While the exploration mechanism greatly enhances the algorithm's ability to adapt to network changes, it was also found to result in oscillatory routing policies when traffic load on the network is high [14]. This is because shortest path routing can easily lead to the saturation of individual links and paths. Indeed, oscillations in the routing policy are a well-known phenomenon in adaptive routing algorithms that perform shortest path routing and were documented in the study of similar algorithms developed in the 1970s for the ARPANET [1, 4]. A number of heuristic modifications to the basic Q-routing scheme have been proposed, which aim to address this problem and improve efficiency in general. These include predictive Q-routing [15], confidence-based Q-routing [16], and confidence-based dual-reinforcement Q-routing [17].

For example, the confidence-based dual-reinforcement Q-routing algorithm associates a measure of confidence with each Q-value, which is used to vary the learning rate parameter so as to give greater importance to delay information which is estimated to be more up to date and thus a more accurate reflection of the true state of the network.

6.4.3 Algorithms Based on "Actor–Critic" Approach

In the field of reinforcement learning, an alternative to the Q-learning approach described in Section 6.4.2 is the "actor–critic" approach [11, 45]. While the former is derived from the value iteration algorithm for solving dynamic programming problems, the latter is derived from the policy iteration algorithm, and as such, the policy update plays a more prominent role. In particular, instead of switching between deterministic policies, actor–critic algorithms typically make incremental changes to a randomized policy in a "direction" which is estimated to lead to improvement.

The most common application of reinforcement learning is for solving unconstrained Markov decision problems, which are known to have deterministic optimal policies, and thus a randomized policy in an actor–critic algorithm typically converges to a deterministic one. However, the use of randomized policies is attractive in its own right in the context of network routing, as this allows the algorithm to balance traffic between any given origin–destination node pair over multiple paths and thus perform multipath routing. Indeed, the existence of optimal routing policies which are randomized reflects the presence of capacity constraints in a communications network.

These considerations motivated the development of STARA [18, 19], an agent-based multipath routing algorithm that was derived from the actor–critic reinforcement learning approach. Instead of shortest path routing, STARA has as its routing goal the Wardrop equilibrium, which, as discussed earlier, constitutes a form of user optimum.

STARA operates as follows: A packet entering the network at node i is routed to one of the neighboring nodes $j \in \mathcal{N}_i$, according to the probability mass function $\psi_{il}, l \in \mathcal{N}_i$. A similar selection process is repeated at node j and at each subsequent node until the packet reaches its destination d. Let the (random) sequence of nodes visited by the packet be denoted $i_1, i_2, \dots, i_n$, where n is the total number of nodes visited by the packet, and $i_1 = i, i_n = d$. We denote using $q_{i_k, i_{k+1}}, k = 1, \dots, n-1$, the packet's trip time measurement from node i_k to the destination d via the node i_{k+1}. Upon the packet's arrival at the destination, a mobile agent is sent back to every node $i_k, k = 1, \dots, n-1$, which triggers the following updates of the corresponding delay estimates,

$$Q_{i_k, i_{k+1}} := (1-a)Q_{i_k, i_{k+1}} + a q_{i_k, i_{k+1}} \qquad k = 1, \dots, n-1 \qquad (6.15)$$

where, as before, $a \in (0, 1]$ is a step size parameter. Thus, STARA performs delay measurements via the Monte Carlo approach (see Section 6.4.1) using data packets for the forward trip time measurement and mobile agents for backpropagation of the delay information.

Whenever the value Q_{ij} is updated, the set of routing probabilities maintained at i that pertain to the destination node d are updated according to the rule

$$\psi_{ij} := \psi_{ij} + b\psi_{ij}\left(\sum_{l \in \mathcal{N}_i} \psi_{il}Q_{il} - Q_{ij}\right) + \xi_{ij} \qquad (i,j) \in \mathcal{A} \qquad (6.16)$$

where $0 < b \ll 1$ is a step size parameter and ξ_{ij} is a randomized noise term which we discuss shortly. The set of routing probabilities $\psi_{ij}, j \in \mathcal{N}_i$, are then renormalized so as to remain in the set $\mathcal{P}$. Finally, the routing probabilities are perturbed according to the rule

$$\psi_{ij} := (1 - \varepsilon)\psi_{ij} + \varepsilon U_{ij} \qquad (i,j) \in \mathcal{A} \qquad (6.17)$$

where $0 < \varepsilon \ll 1$ and $U_{ij}, j \in \mathcal{N}_i$, is the probability mass function corresponding to the uniform distribution over the set of outgoing links at node i. Thus, the routing probabilities that are actually used for the routing of packets are a convex combination of the uniform distribution and the routing probabilities given by (6.16).

The update rule (6.16) has the following effect: The probability ψ_{ij} is increased by a small amount if the estimated delay, Q_{ij}, is less than the estimated average delay over all outgoing links, given by $\sum_{l \in \mathcal{N}_i} \psi_{il}Q_{il}$, and conversely, ψ_{ij} is decreased by a small amount if the estimated delay Q_{ij} is greater than the estimated average delay. If Q_{ij} is equal to the average estimated delay, then the routing probability ψ_{ij} does not change. It is shown in [19] that STARA converges to an approximate Wardrop equilibrium. Indeed, inspection of (6.16) shows that a fixed point of the algorithm occurs when

$$\sum_{l \in \mathcal{N}_i} \psi_{il}Q_{il} = Q_{ij} \qquad (6.18)$$

for all $(i,j) \in \mathcal{A}$. A connection between the fixed-point condition (6.18) and the sufficient condition for a Wardrop equilibrium given in (6.6) can be established by considering a fixed routing policy Ψ and observing that under this policy the zupdate rule (6.15) will result in convergence of the delay estimates Q_{ij} to the corresponding mathematical expectations $D_{ij}(\Psi)$ given by (6.7). Thus, with a little thought, we see that a routing matrix Ψ which satisfies the Wardrop condition (6.6) also satisfies the fixed-point condition (6.18). Furthermore, it is shown in [19] that fixed points of the algorithm are stable only if they correspond to Wardrop equilibria, and the presence of the noise term ξ_{ij} in (6.16) guarantees that unstable (non-Wardrop) equilibria which also satisfy the condition (6.18) are avoided.

A key feature of (6.16) is that for sufficiently small values of the parameter b a gradual deviation of traffic flow from high delay paths to lower delay paths is achieved. This contrasts with the "all-or-nothing" shortest path approach used by Q-routing and other adaptive shortest path algorithms and is the reason why STARA is able to converge to a multipath routing policy.

The update rule (6.17) constitutes an exploration mechanism parameterized by ε because it ensures that no routing probability can ever be set to zero, which in turn means that all links are eventually probed with probability 1 by sending small amounts of traffic on them. Thus, the delay characteristics of alternative paths can be measured, allowing the algorithm to detect changes in the network environment and reroute traffic accordingly. This exploration mechanism is classed as on-policy, because there exists a tight coupling between the data routing policy and the exploration mechanism such that there exists a direct trade-off between the tasks of exploration and decision making. In fact, STARA converges to an "ε-perturbed," or approximate, Wardrop equilibrium rather than a pure Wardrop equilibrium due to this exploration mechanism. Specifically, the decision-making mechanism steers the routing policy toward a Wardrop equilibrium, while the exploration mechanism perturbs the routing policy away from a Wardrop equilibrium. It follows that a closer approximation to a Wardrop equilibrium is achieved at the expense of reducing the probability (and hence frequency) of exploration, which in turn leads to a slower rate of adaptation when changes occur in a nonstationary environment. In Section 6.6, we describe how this perturbation can be eliminated without compromising the exploration of alternative paths.

6.4.4 Other Reinforcement Learning-Based Algorithms

TPOT-RL is a general reinforcement learning technique that is applied to network routing in [20]. In the network routing context, TPOT-RL reduces to a shortest path routing algorithm that is similar to Q-routing, but with the following key differences: The delay estimates are updated via the Monte Carlo approach instead of the bootstrapping approach, it performs on-policy exploration by making small perturbations of the shortest path data routing policy, and an additional state variable is maintained for each link, indicating whether the number of packets routed on a link over a recent time period of fixed length was "high" or "low." This additional state variable is taken into account when making routing decisions. A limited simulation study presented in [20] suggested that TPOT-RL has marginally superior performance to Q-routing.

OLPOMDP is presented in [21] in the context of network routing. It has an on-policy exploration mechanism in the form of a randomized data routing policy, which also allows for multipath routing. Mobile agents employ the Monte Carlo approach for the measurement of marginal delays, and the algorithm's routing goal is a system optimal routing policy. However, it employs a global rather than local method in order to generate marginal delay estimates, requiring that mobile agents measure the trip time from a given origin to destination and then subsequently broadcast *all* trip times to *every* node in the network, regardless of the packet's path. The latter feature therefore constitutes a significant departure from the completely decentralized framework of the other agent-based routing algorithms discussed in this chapter (including a system optimal distributed routing algorithm which we discuss in Section 6.4.6). The algorithm's potential was demonstrated via a small simulation example in [21].

6.4.5 Ant-Based Routing Algorithms

The first ant-based routing algorithm, proposed in [26], was designed for connection-oriented networks [46]. A more sophisticated version for routing in connectionless packet-switched networks, known as AntNet [2], was proposed by researchers working in the field of ACO [23], and similar algorithms were also proposed in [25–29]. In this survey, we focus exclusively on ant-based routing algorithms for connectionless packet-switched networks, as these have attracted the most research interest in recent years.

Ant-based routing algorithms employ "antlike" mobile agents, whose purpose it is to traverse the network and measure trip times between origin–destination node pairs. The trip time measurements made by the ants are then used to construct a data routing policy. Thus, ant-based algorithms operate separate ant and data traffic "layers" whereby the ant layer informs and drives the data routing update process. Specifically, for every destination node, ant-based routing algorithms maintain a set of ant routing probabilities $\phi_{ij}, (i,j) \in \mathcal{A}$, in addition to the data routing probabilities ψ_{ij}. Each node periodically creates ants, which are dispatched to each of the destination nodes. Consider a given distination node d. At node i, an ant is routed to one of the neighbor nodes $j \in \mathcal{N}_i$ according to the probability mass function $\phi_{il}, l \in \mathcal{N}_i$. The ant repeats this process at every subsequent node that it visits until the destination node is reached, experiencing the same queueing and transmission delays that are experienced by data packets at each of the links that it traverses. As before, we denote the (random) sequence of nodes visited by the ant using $i_1, i_2, \ldots, i_n$, where n is the total number of nodes visited by the ant, and let $q_{i_k, i_{k+1}}, k = 1, \ldots, n-1$, be the ant's trip time measurement from node i_k to the destination d (via the node i_{k+1}). The ant then retraces its path from d back to i, passing via every intermediate node $i_k, k = 1, \ldots, n-1$, thus performing on its return journey a similar role to the acknowledgment packet in STARA. Thus, like STARA, ant-based algorithms perform delay measurements via the Monte Carlo approach.

On its return path, the ant directly updates the routing probabilities, $\phi_{i_k, i_{k+1}}$, for $k = 1, \ldots, n-1$, using the corresponding trip time measurements $q_{i_k, i_{k+1}}$. To simplify presentation, we just give the update rule for the case of the origin node $i = i_1$ and first-hop node $j = i_2$. Similar updates are performed at the intermediate nodes. First, the routing probability ϕ_{ij} associated with the link (i, j) traversed by the ant is *incremented according to a rule of the form*

$$\phi_{ij} := \frac{\phi_{ij} + \delta(q_{ij})}{1 + \delta(q_{ij})} \tag{6.19}$$

where $\delta(x)$ is a function which determines the size of the increment. The specific functional form of $\delta(x)$ differs from one ant-based routing algorithm to another, but in all cases, $\delta(x)$ is a decreasing function of x. Second, the routing probabilities ϕ_{il}, $l \neq j$ corresponding to all other outgoing links at node i that were not traversed by the ant on its forward path, are updated according to a rule of the form

$$\phi_{il} := \frac{\phi_{il}}{1 + \delta(q_{ij})} \qquad l \in \mathcal{N}_i \setminus j \qquad (6.20)$$

Note that the update rules (6.19) and (6.20) ensure that the ant routing probabilities remain normalized, that is, that they remain in the set $\{\phi_{ij} : \phi_{ij} \geq 0, (i,j) \in \mathcal{A}, \sum_{l \in \mathcal{N}_i} \phi_{il} = 1 \text{ for all } i \in \mathcal{N} \setminus d\}$ provided they are initially in this set.

The update rules for the ant routing probabilities have the following effect. The routing probability ϕ_{ij} is increased by the update (6.19) each time that an ant traverses the link (i, j). Simultaneously, the remaining routing probabilities at node i are decreased by the update (6.20). With the passage of many ants across the network, the net effect is that routing probabilities associated with links that lie on low-delay paths experience a positive feedback effect leading to their increase, and conversely, routing probabilities associated with links that lie on high-delay paths are decreased. In particular, this positive and negative feedback occurs as a result of two distinct processes:

1. A transient process whereby in a given period of time a larger number of ants complete a round trip from origin to destination over a low-delay path compared with the number that complete a round trip over a high-delay path. This results in a larger *total* number of increments per unit of time to the routing probabilities associated with links that lie on low-delay paths.

2. A persistent process, resulting from the fact that the routing probability ϕ_{ij} is incremented by an amount $\delta(q_{ij})$, which decreases as the measured trip time Q_{ij} increases. Therefore, routing probabilities corresponding to links that are associated with high-delay paths receive a smaller reinforcement compared with links that are associated with low-delay paths.

The transient process described above is unique to ant-based routing algorithms and is not present in other agent-based routing algorithms. However, the differential reinforcement of paths that results from the persistent process is analogous to the delay-based updating of the routing policy that is carried out by the reinforcement learning-based algorithms and is crucial to an ant-based algorithm's ability to adapt to on-going network changes.

In addition, in some ant-based algorithms, the ant routing probabilities are perturbed according to the rule

$$\phi_{ij} := (1 - \varepsilon)\phi_{ij} + \varepsilon U_{ij} \qquad (i,j) \in \mathcal{A} \qquad (6.21)$$

This constitutes an exploration mechanism similar to the one employed by STARA, except that it is the *ants* which are routed according to the ε-perturbed policy rather than the data packets, the routing of which we discuss next.

The ant routing probabilities constitute an ant-based algorithm's representation of the current network state and are used in order to generate a data traffic routing policy. This is done via a proportional routing mechanism where the data routing probabilities ψ_{ij} are given by a transformation of the form

$$\psi_{ij} \propto (\phi_{ij})^{\beta} \qquad (i,j) \in \mathcal{A} \tag{6.22}$$

where β acts as a load-balancing parameter. For typical values of β (for example, $1 < \beta < 2$), the rule (6.22) usually results in traffic being routed over multiple paths between any given origin–destination node pair, rather than being routed on a single path. As a result, ant-based algorithms are able to perform a heuristic form of load balancing. We see that (6.22) generates a data packet routing probability mass function that is more strongly biased in favor of links which are most likely to be selected by ants, reflecting the idea that data should attempt to exploit the paths which are estimated by the ants to have the lowest delay. That is, exploration should be left to the ants, and the data traffic should exploit low-delay paths.

As noted above, ant-based routing algorithms are classed as on-policy, despite the fact that they employ a separate "layer" of ant traffic for exploration. This is because they achieve only a partial rather than complete decoupling of the tasks of exploration and data traffic routing. The reason for this is as follows. The role of ants is twofold: to explore alternative routes and to estimate the delays experienced by data packets. The first task is achieved by assigning to all links a nonzero probability of being selected by an ant. However, ants are not able to accurately perform the latter task, since ants and data packets are routed according to *different* policies [see expression (6.22)]. Specifically, the distribution of ant trip times is different from the distribution of data packet trip times, and the expected trip times will generally also be different. Therefore, the ants' measurement of network delays will in fact contain a systematic perturbation with respect to the actual delays experienced by data packets under the current data routing policy. These perturbations are a direct result of the exploratory actions of the ants and can be characterized as perturbations from a Wardrop equilibrium [31, 32].

A unique feature of ant-based algorithms is that the ants update the routing probabilities *directly*, according to (6.19) and (6.20), rather than first updating a set of delay estimates. However, some ant-based algorithms also employ an explicit delay estimation component. For example, AntNet maintains delay estimates that are analogous to the Q_{ij} maintained by STARA, and in addition, it maintains estimates of the variances of ant trip times. These statistics are used to determine a heuristic measure of "reliability" of trip time samples Q_{ij}, which is incorporated into the function $\delta(\cdot)$ used to determine the magnitude of increments to the routing probabilities in (6.19) and (6.20). This confers to

AntNet a greater degree of stability and counteracts undesirable oscillations in the routing policy [2]. Extensive simulation experiments presented in [2] demonstrate that AntNet performs very well in response to sudden changes in traffic demands or isolated network component failures. Further analysis and comments on the topic of ant-based routing algorithms can be found in [31, 32, 47–49].

6.4.6 System Optimal Routing Algorithms

A family of distributed algorithms that are able to achieve system optimal routing policies is described in [6]. Although these algorithms predate the era of multiagent systems and mobile agents, the data structures and functions at the network nodes fit within the distance–vector framework that is common to the other algorithms presented in this chapter, and the internode information transfer which is carried out by "signaling packets," as they were referred to in [6], can equivalently be cast in terms of modern mobile agent technology. In particular, they serve as a useful point of reference because they show how system optimal routing policies can be achieved by a distributed multiagent system. As mentioned in Section 6.2, they may prove useful in future research on mobile agent systems due to recent advances in online gradient estimation techniques [7–9].

We give a brief overview of the simplest of these routing algorithms, known as Gallagher's algorithm, which was first proposed in [5]. For every destination node d, each node $i \in \mathcal{N} \setminus d$ maintains estimates $Q'_{ij}, j \in \mathcal{N}_i$, where Q'_{ij} is the marginal delay associated with traveling from i to d via the outgoing link (i, j). These estimates are periodically updated according to the rule

$$Q'_{ij} := R'_{ij} + Q'_{j} \tag{6.23}$$

where R'_{ij} is an estimate of the marginal delay associated with the link (i, j) and Q'_{j} is an estimate of $\partial D^T(\Psi)/\partial s_j$ (see Section 6.3). Techniques for measuring these quantities using mobile agents are discussed at the end of this section. In particular, Q'_{j} is calculated by each neighbor node j according to

$$Q'_{j} = \sum_{l \in \mathcal{N}_j} \psi_{jl} Q'_{jl} \tag{6.24}$$

and transmitted back to node i by a mobile agent. Similarly all nodes $i \in \mathcal{N} \setminus d$ calculate Q'_{i} and broadcast this value to all of their immediate neighbors. Of course, $Q'_{d} = 0$ due to the boundary condition $\partial D^T(\Psi)/\partial s_d = 0$. In terms of our classification scheme (see Section 6.4.1), we see that Gallagher's algorithm performs marginal delay measurements via the bootstrapping approach. We also note that the values Q'_{ij} constitute estimates of the *average* "first derivative length" [1] of used paths from i to d via the link (i, j) under the current routing policy Ψ.

The routing probabilities are updated as follows. Each node i calculates

$$A_{ij} := Q'_{ij} - \min_{l \in \mathcal{N}_i} [Q'_{il}] \qquad (6.25)$$

and

$$\Delta_{ij} := \min \left[\psi_{ij}, \frac{bA_{ij}}{f_i} \right] \qquad (6.26)$$

for each $j \in \mathcal{N}_i$, where $b > 0$ is a step size parameter and f_i is an estimate of $\sum_{j \in \mathcal{N}_i} f_{ij}(\Psi)$, the total traffic flow passing through node i, with destination d. Let M_{ij} be a node $l \in \mathcal{N}_i$ which achieves the minimization in (6.25). Then

$$\psi_{ij} := \begin{cases} \psi_{ij} - \Delta_{ij} & \text{if } j \neq M_{ij} \\ \psi_{ij} + \sum_{l \in \mathcal{N}_i, l \neq M_{ij}} \Delta_{il} & \text{if } j = M_{ij} \end{cases} \qquad (6.27)$$

Note that the update rule given by (6.25)–(6.27) preserves the normalization of the routing probabilities. In [25], the case where multiple nodes achieve the minimum in (6.25) is not addressed; however, a natural extension whereby the increase in traffic is divided evenly over such links is given in [31].

We see from (6.27) that Gallagher's algorithm increases the proportion of traffic sent on the link with the minimum estimated marginal delay to reach d and reduces the proportion of traffic sent on the remaining links (which have larger estimated marginal delay).

If a routing policy is a fixed point of Gallagher's algorithm, then it must satisfy the sufficient condition for a system optimal routing policy given in (6.10). To see this, let Ψ be a fixed point of Gallagher's algorithm and suppose that all marginal link delay estimates R'_{ij} are equal to their exact mathematical expectations $R'_{ij}(\Psi)$. Then, recalling (6.11), we have $Q'_{ij} = D'_{ij}(\Psi)$ for all links (i, j). We observe that the fixed-point condition for Gallagher's algorithm implied by (6.25) and (6.26) implies the sufficient condition (6.10).

In addition, in order to ensure that the routing policy Ψ is loop free, Gallagher's algorithm maintains at each node i a list of outgoing links which are "blocked." Blocked links are excluded from all of the calculations in (6.25)–(6.27), and the routing probabilities associated with blocked links are automatically set to zero. The reader is referred to [5, 6] for details.

Gallagher's algorithm and those in [6] differ significantly from the other algorithms described in this chapter, in that they require the online estimation of *marginal* link delays rather than link delays. There are a number of approaches to performing this estimation. The simplest is to assume a specific functional form for $R_{ij}(f_{ij})$ from which $R'_{ij}(f_{ij})$ can readily be calculated given an estimate of the link flow f_{ij}. The drawback of this approach is that any formula for R involves assumptions which may not be warranted. Another approach is to perform nonparametric estimation of the marginal link delays. This can be done via finite-difference methods, as in [50], or via infinitesimal perturbation analysis, as described in [7–9, 51]. In either case, these methods require trip time measurements, which can be performed by mobile agents.

The finite-difference method requires an *actual perturbation of the data routing policy* and produces biased estimates, whereas the method of infinitesimal perturbation analysis does *not* require an actual perturbation of the policy and has been shown to produce asymptotically unbiased estimates. The latter (more modern) approach is therefore preferable, although in both cases the computational demand is significantly greater than that required to estimate (simple) delays, or trip times. In addition, the amount of time required to accurately estimate marginal delays is large compared with that required to estimate delays. These are perhaps the primary reasons for which distributed system optimal routing algorithms have not been deployed in real networks, despite the fact that they are well researched and can be implemented within the distance–vector framework.

Finally, we note that the exploration mechanism employed by Gallagher's algorithm is, in principle, off-policy, because there is a decoupling between the mechanism for probing the marginal delays of alternative routes (via the nearest-neighbour broadcasting of marginal delay estimates) and the mechanism for the routing of data traffic. However, Gallagher's algorithm effectively becomes on-policy whenever perturbations of the data routing policy are performed in order to estimate marginal delays. This is always the case when the method of finite differences is employed. Furthermore, the method of infinitesimal perturbation analysis cannot be applied to estimate the marginal delay of a link which carries no data traffic, and thus some small amount of data traffic would need to be carried on all links in order to generate estimates. Thus, the exploration mechanism of Gallagher's (and other related system optimal algorithms [6]) tend to be on-policy in practice.

6.4.7 Algorithms for Mobile Ad Hoc Networks

Agent routing algorithms for mobile ad hoc networks extend the fundamental ideas already described for wired networks. In particular, for ad hoc networks, agent-based routing algorithms are augmented with a broadcast (rather than unicast) mechanism for monitoring the network topology and connectivity, which itself is nonstationary due to node mobility. Thus, at each node, the possible destination entries and nearest-neighbor entries are added and removed as the topology and connectivity change with time.

The idea of using mobile agents for routing in mobile ad hoc networks first appeared in [33]. More recently, a reinforcement learning-based algorithm called SAMPLE was proposed in [35]. In SAMPLE, each node maintains a local model of the reliability of outgoing links, which is used to drive a model-based version of the Q-learning algorithm (see Section 6.4.2). Another promising approach is given by the algorithm AntHocNet [34], which is derived from the original AntNet algorithm for wired networks. AntHocNet employs a combination of bootstrapping between nearest neighbors for topology discovery and Monte Carlo measurement of route delays using mobile agents. Both SAMPLE and AntHocNet were found to perform favorably compared with ad hoc on-demand distance–vector (AODV) routing [52], a popular algorithm for mobile ad hoc networks.

6.5   COMPARISON OF ROUTING ALGORITHMS

Table 6.1 shows a classification of agent-based routing algorithms in terms of properties 1–4 presented in Section 6.5. This classification system provides a useful basis from which to compare the algorithms and to highlight their relative strengths and weaknesses.

Regarding property 1, Q-routing and Gallagher's algorithm employ the bootstrapping approach to delay and marginal delay measurement, respectively. This is because in both algorithms each node uses mobile agents to perform direct measurements of the delay (or marginal delay) on its immediate outgoing links and then uses information that is cached at neighboring nodes in order to construct an estimate of the remaining delay (or marginal delay) associated with reaching the destination node. On the other hand, STARA, ant-based algorithms, TPOT-RL, and OLPOMPD employ the Monte Carlo approach, since mobile agents report direct measurements of packet trip times over entire paths from origin to destination, rather than using delay information cached at neighboring or intermediate nodes. With all other properties 2–4 (compared below) being equal, it is not immediately clear which of these approaches is superior in the context of adaptive network routing, and this remains an open question for future research.

Regarding property 2, only Q-routing is an off-policy algorithm; all of the other algorithms are on-policy, due to some form of coupling between the exploration mechanism and the data routing mechanism. In principle, an off-policy exploration mechanism is preferable, as this ensures that the exploration of alternative routes does not interfere with or adversely affect the routing of data traffic. We note that STARA and ant-based routing algorithms, which are both on-policy, can be easily modified to perform off-policy exploration, as will be described in Section 6.6.

Regarding property 3, single-path routing is attractive because of its simplicity, ease of implementation, and the fact that all packets traveling between a given origin–destination node pair are routed on the same path (provided the routing policy is not oscillatory), which is in turn desirable as this leads to the minimum possible resequencing delays for packets arriving at the destination. On the other hand, it is well known that adaptive shortest path routing is prone to instability, in the form of oscillations of the routing policy, especially when the traffic load on the network is high [1]. Indeed, when the total traffic load placed on the network is large, it may be impossible to route all traffic over any given single path. These considerations motivate the need for multipath routing algorithms, which are less likely to be unstable if properly designed and can reduce network delays and increase throughput by splitting traffic flows over multiple paths. In this regard, multipath routing algorithms are superior to single-path algorithms.

Regarding property 4, of the agent-based algorithms that have the capacity to perform multipath routing, and hence have the ability to perform load balancing of traffic, there is no immediate and straightforward answer to the question *"which type of operating point, or optimisation goal, is best?"* It may be tempting to suggest that the system optimal routing goal of Gallagher's and related algorithms is superior, but it is important to consider that an optimal routing policy can result in some traffic streams experiencing low delays while subjecting others to very high delays. In other words, there is no guarantee of "social welfare" for all traffic streams in this case. Also, even though all used paths between a given origin–destination node pair have the same marginal delay under a system optimal routing policy, they do not in general have the same delay, and this can lead to significant packet resequencing delays at the destination for packets belonging to the same traffic stream [1]. Perhaps the most significant drawback of achieving a system optimum is the added computational requirement associated with the estimation of marginal delays, which necessarily slows down the rate of convergence and adaptation of the algorithm.

The consideration of the actual delays incurred by users of the network leads to the notions of user optimality and user equilibria that were discussed in Section 6.3. We observed that the Wardrop equilibrium is the only type of user equilibrium that is systematically attainable within the distributed distance–vector framework of agent-based routing algorithms. Significantly, no estimation of derivatives is necessary in order to achieve it. The Wardrop equilibrium also has the desirable property that all packets traveling between a given origin–destination node pair have the same expected delay. As a result, packet resequencing delays at the destination are, on average, the smallest possible among the algorithms that perform multipath routing. Indeed, the authors of [19] are of the opinion that "The Wardrop equilibrium is a curious animal, certainly deserving more attention from the telecommunications community than what it seems to have hitherto mustered (p. 383)."

Of the algorithms that perform multipath routing, STARA, Gallagher's and OLPOMDP were designed using a "top-down" approach (as discussed in Section 6.3), with the goal of Wardrop or system optimal routing. In contrast, ant-based routing algorithms were not designed with a particular optimization concept as their goal. The primary motivation for their development was to harness the emergent problem-solving behaviors of ant colonies and to design routing algorithms which could adapt rapidly in response to network changes and perform a degree of adaptive load balancing (via proportional routing) instead of shortest path routing. For example, in simulation experiments presented in [2], AntNet is shown to have superior performance in a nonstationary environment compared with algorithms that are currently deployed in the

Internet, such as distributed Bellman–Ford [1] and OSPF [53] (which both employ shortest path routing), and was also shown to have superior performance to the Q-routing algorithm described in Section 6.4. The fact that AntNet performs adaptive multipath routing instead of shortest path routing is one of the primary reasons for its superior performance. An analysis of the equilibrium properties of ant-based routing algorithms in [31, 32] suggests that, similarly to STARA, AntNet effectively achieves "perturbed" Wardrop equilibria. This is not surprising given that the ant-based routing algorithms measure path delays and deviate data traffic flow from high-delay paths to low-delay paths, much like STARA.

6.6 NEW DIRECTIONS: HYBRID AGENT-BASED ALGORITHMS

In this section, we propose two alternative approaches for the future development of mobile agent-based routing algorithms, incorporating the strengths of the algorithms surveyed in this chapter, as identified in Section 6.5. Assuming that marginal delay estimates are unavailable, we propose that a desirable hybrid of the algorithms surveyed in this chapter should

1. have the ability to perform multipath routing,
2. target the Wardrop equilibrium as its routing goal, and
3. have an off-policy exploration mechanism.

The first approach that we propose, which we call Monte Carlo Wardrop routing (MCWR), is a hybrid of STARA and ant-based routing algorithms and employs the Monte Carlo method for the measurement of path delays. An implementation of MCWR first appeared in [31, 32]. The second, which we call Wardrop Q-routing, is a hybrid of Q-routing and STARA and employs the bootstrapping method for the measurement of path delays. A comparison of these approaches is shown in Table 1.1.

The proposed MCWR approach uses the same update rules as STARA, but the updates are driven by measurements performed by a separate layer of mobile agents. In particular, the mobile agents are routed *according to the same routing policy as data packets*, Ψ, *except for the agent's first hop*, which is selected uniformly at random from the set of outgoing links as the agent's node of origin. Upon reaching the destination, the mobile agent retraces its path and updates the appropriate delay estimates at each node that it visited on its forward path, according to the rule (6.15). Each update of a delay estimate triggers an update of the set of routing probabilities, according to the rule (6.16).

The MCWR approach differs from STARA in the following ways. Exploration of all possible routes is carried out by a separate layer of mobile agents, rather than the data packets, thus borrowing from the ant-based approach (in STARA, the forward trip time is measured using data packets, and a mobile agent is created at the destination node for the purpose of backpropagating the information to the origin node). This ensures that every link continues to be selected by agents, even if the current data routing probability for the link is set to zero. In particular, there is no need for the ε-perturbation that is employed by STARA [see (6.17)]. This is an important factor in the decoupling of the exploration mechanism from the data routing mechanism and also permits MCWR to converge to a pure rather than ε-perturbed Wardrop equilibrium.

The MCWR approach differs from ant-based routing algorithms in that the mobile agents perform exploration exclusively when selecting their first hop and follow the current data routing policy for all subsequent routing decisions until the destination node is reached. As a result, unlike in ant-based algorithms, the trip times reported by agents have the same distribution as the trip times of data packets, and thus agent exploration does not introduce a bias in the measurement of network delays. A simulation study comparing an implementation of MCWR with ant-based routing is presented in [31, 32], which suggests that the MCWR approach is more efficient than the ant-based approach with respect to a number of equilibrium delay-based performance measures.

An alternative to the MCWR approach is to perform delay measurements using the bootstrapping method, and we refer to this as Wardrop Q-routing. In particular, instead of sending mobile agents from their node of origin all of the way to the destination, an agent created at node i selects an outgoing link uniformly at random, (i, j), say, and requests information from node j about the remaining trip time to the destination node d (under the current routing policy), and then returns to node i. Upon returning to node i, the agent performs the update

$$Q_{ij} := (1 - a)Q_{ij} + a(r_{ij} + Q_j^{\text{ave}}) \qquad (6.28)$$

where $a \in (0, 1]$ is a step size parameter, r_{ij} is the random delay incurred by the agent in traversing the link (i, j), and

$$Q_j^{\text{ave}} := \sum_{l \in \mathcal{N}_j} \psi_{jl} Q_{jl} \qquad (6.29)$$

Note that Q_j^{ave} constitutes node j's local estimate of $D_j(\Psi)$, the average trip time of a data packet from node j to d, under the current routing policy [see (6.7)]. As before, each update of a delay estimate $Q_{ij}, j \in \mathcal{N}_i$, triggers an update of the set of routing probabilities $\psi_{ij}, j \in \mathcal{N}_i$, according to the rule (6.16).

The update rules (6.28) and (6.29) should be compared with the Q-routing update rule (6.12). We see that Q-routing and Wardrop Q-routing share the bootstrapping approach, which involves the use of local estimates of trip times maintained at the neighbor node. The difference lies in the fact that an agent in the Q-routing algorithm requests the local minimum estimate, $Q_j^{\min}$, given by (6.13), reflecting the fact that all data traffic is routed on links with the minimum associated delay estimates, that is, single-path shortest path routing. In contrast, an agent in the Wardrop Q-routing algorithm requests the average trip time estimate, Q_j^{ave}. The "minimization" step is then performed in small increments by (6.16), thus allowing the algorithm to converge to a multipath rather than single-path routing policy.

Finally, if every node periodically probes every outgoing link in the manner described above, then this guarantees that every alternative path is explored, and there is no need for the ε-perturbation introduced by the exploration mechanism (6.17) that is employed by STARA. Importantly, the use of a separate agent layer, where agents perform exploration only on their first hop, guarantees that the exploration mechanism is decoupled from the data traffic routing mechanism and is thus off-policy.

As we shall discuss in Section 6.7, it remains an open problem to compare the performance of the bootstrapping and Monte Carlo methods for delay measurement. We believe that the MCWR and Wardrop Q-routing approaches would constitute appropriate vehicles for such a study.

6.7   CONCLUSIONS

We have surveyed and compared a number of agent-based routing algorithms within a unified framework. In particular, we have brought together ideas from "traditional" nonlinear optimization and game theory and more recent ideas from the fields of multiagent systems, reinforcement learning, and ant-based routing algorithms, and we have shown how these ideas overlap and interrelate within the specific setting of network routing in packet-switched networks.

One of our primary aims has been to provide a "roadmap" and platform for future research and comparison of mobile agent-based routing algorithms. In particular, we have identified the following aspects as key defining features: the mechanism for delay measurement and information-passing between nodes, the exploration mechanism, and the type of routing goal sought. These considerations led to our proposal of two alternative hybrid mobile agent algorithms, which combine several of their strengths, as outlined in Section 6.6.

A fertile area for further investigation is the investigation of bootstrapping versus Monte Carlo delay estimation. The question of which of these approaches results in the best performance, when all other factors such as load balancing and exploration are equal, remains open. A number of empirical studies have compared these approaches in the context of solving Markov decision problems (see [54] for an overview and references), but neither has been shown to possess a universal advantage for all situations. An interesting avenue for further research would be to establish if and when one of these approaches yields superior performance in the context of adaptive network routing. Indeed, the MCWR and Wardrop Q-routing approaches presented in Section 6.6 provide a vehicle for such a study, although we note that care must be taken in the selection of a performance measure which leads to a fair comparison. Following the approach employed in [25, 34], a natural measure would be the rate of convergence (speed of adaptation) to a desired operating point achieved with a given "overhead" cost, measured in total mobile agent hops.

We conclude by noting that the trade-off between adaptive behavior and stability, first identified in early adaptive algorithms for the ARPANET, remains a pertinent issue for mobile agent routing systems and is notoriously difficult to study within a theoretical domain. While reliance on ad hoc methods for tuning algorithm parameters via simulation remains a common and useful practice, there exists a need for improved theoretical support for analysis of the trade-offs between timely adaptive behavior and stability.

ACKNOWLEDGMENTS

Andre Costa would like to acknowledge the support of the Australian Research Council Centre of Excellence for Mathematics and Statistics of Complex Systems. Nigel Bean would like to acknowledge the support of the Australian Research Council through Discovery Project DP0557066.

REFERENCES

1. D. Bertsekas and J. Gallager, *Data Networks*, Prentice-Hall, Englewood Cliffs, NJ, 1992.
2. G. Di Caro and M. Dorigo, AntNet: Distributed stigmergetic control for communications networks, *J. Res.* 9:317–365, 1998.
3. G. A. Di Caro and M. Dorigo, Two ant colony algorithms for best-effort routing in datagram networks, in *Proceedings of PDCS'98—10th International Conference on Parallel and Distributed Computing and Systems*, Las Vegas, Nevada, NV, Oct. 28–31, 1998.
4. D. Bertsekas, Dynamic behavior of shortest path routing algorithms for communication networks, *IEEE Trans. Automatic Control*, 27:60–74, 1982.
5. R. Gallagher, A minimum delay routing algorithm using distributed computation, *IEEE Trans. Commun.*, 25:73–85, 1979.
6. D. Bertsekas, E. Gafni, and R. Gallagher, Second derivative algorithms for minimum delay distributed routing in networks, *IEEE Trans. Commun.*, 32:911–919, 1984.
7. J. Spall, *Introduction to Stochastic Search and Optimization: Estimation, Simulation, and Control*, Wiley, Hoboken, NJ, 2003.
8. H. Kushner, *Stochastic Approximation Algorithms and Applications*, Springer, New York, 1997.
9. F. J. Vazquez-Abad, Strong points of weak convergence: A study using RPA gradient estimation for automatic learning, *Automatica*, 35:1255–1274, 1999.
10. S. Appleby and S. Steward, Mobile software agents for control in telecommunications networks, *BT Technol. J.*, 12(2):104–113, 1994.
11. R. S. Sutton and A. G. Barto, *Reinforcement Learning: An Introduction*, MIT Press, Cambridge, MA, 1998.

12. D. Kudenko, D. Kazakov, and E. Alonso, *Adaptive Agents and Multi-Agent Systems II*, Lecture Notes in Artificial Intelligence, Vol. 3394, Springer, Berlin, Heidelberg, 2005.

13. C. Watkins and P. Dayan, Q-learning, *Machine Learning*, 8:279–292, 1992.

14. J. A. Boyan and M. L. Littman, Packet routing in dynamically changing networks: A reinforcement learning approach, in J.D. Cowan, G. Tesauro, and J. Alspector (Eds.), *Advances in Neural Information Processing Systems*, Vol. 6, Morgan Kaufmann, San Francisco, CA, 1993, pp. 671–678.

15. S. P. M. Choi and D.-Y. Yeung, Predictive Q-routing: A memory-based reinforcement learning approach to adaptive traffic control, in D. S. Touretzky, M. C. Mozer, and M. E. Hasselmo (Eds.), *Advances in Neural Information Processing Systems*, Vol. 8, MIT Press, Cambridge, MA, 1996, pp. 945–951.

16. S. Kumar and R. Miikkulainen, Confidence-based Q-routing: An on-line adaptive network routing algorithm, in C. H. Dagli, M. Akay, O. Ersoy, B. R. Fernandez, and A. Smith, (Eds.), *Proceedings of Artificial Neural Networks in Engineering, Smart Engineering Systems: Neural Networks, Fuzzy Logic, Data Mining, and Evolutionary Programming*, 8, 1998.

17. S. Kumar, Confidence based dual reinforcement Q-routing: An on-line adaptive network routing algorithm, Technical Report AI98–267, 1998, p. 1. Available at http://citeseerx.ist.psu.edu/viewdoc/summary?doi = 10.1.1.33.1555.

18. P. Gupta and P. R. Kumar, A system and traffic dependent adaptive routing algorithm for ad hoc networks, in *Proceedings of the 36th IEEE Conference on Decision and Control*, San Diego, Dec. 1997, pp. 2375–2380.

19. V. S. Borkar and P. R. Kumar, Dynamic Cesaro-Wardrop equilibration in networks, *IEEE Trans. Automatic Control*, 48(3):382–396, 2003.

20. P. Stone, TPOT-RL applied to network routing, in *Proceedings of the 17th International Conference on Machine Learning*, 2000, pp. 935–942.

21. N. Tao, J. Baxter, and L. Weaver, A multi-agent policy-gradient approach to network routing, in *Proceedings of the 18th International Conference on Machine Learning*, 2001, pp. 553–560.

22. M. Dorigo, V. Maniezzo, and A. Colorni, The Ant System: Optimization by a colony of cooperating agents, *IEEE Trans. Syst. Man Cybernet.*, 26:29–41, 1996.

23. M. Dorigo and T. Stutzle, *Ant Colony Optimization*, MIT Press, Cambridge, MA, 2004.

24. G. Di Caro, Ant colony optimisation and its application to adaptive routing in telecommunications networks, Ph.D. thesis, Faculté des Sciences Appliquées, Université Libre de Bruxelles, Belgium, 2004.

25. K. A. Amin and A. R. Mikler, Agent-based distance vector routing: A resource efficient and scalable approach to routing in large communications networks, *J. Syst. Software*, 71:215–227, 2004.

26. R. Schoonderwoerd, O. Holland, J. Bruten, and L. Rothkrantz, Ant-based load balancing in telecommunications networks, *Adaptive Behav.*, 5:169–207, 1996.

27. E. Bonebeau, F. Henaux, S. Guerin, D. Snyers, P. Kuntz, and G. Theraulaz, Routing in telecommunications networks with "smart" ant-like agents, in *Proceedings of IATA '98, Second Int. Workshop on Intelligent Agents for Telecommunication Applications*, Lecture Notes in Artificial Intelligence, Vol. 1437, Springer, Berlin, Heidelberg, 1998.

28. M. Huesse, D. Snyers, S. Guerin, and P. Kuntz, Adaptive agent-driven load balancing in communication networks, *Adv. Complex Syst.*, 1:237–254, 1998.

29. T. White, B. Pagurek, and F. Oppacher, Connection management using adaptive mobile agents, in *Proceedings of International Conference on Parallel and Distributed Processing Techniques and Applications (PDPTA'98)*, CSREA Press, 12–16 July, 1998, pp. 802–809.

30. D. Subramanian, P. Druschel, and J. Chen, Ants and reinforcement learning: A case study in routing in dynamic networks, in *Proceedings of IJCAI-97, International Joint Conference on Artificial Intelligence*, Morgan Kaufmann, 1997, pp. 832–839.

31. A. Costa, Analytic modelling of agent-based network routing algorithms, Ph.D. thesis, School of Applied Mathematics, University of Adelaide, Adelaide, 2003.

32. N. Bean and A. Costa, An analytic modelling approach for network routing algorithms that use "ant-like" mobile agents, *Comput. Networks*, 49:243–268, 2005.

33. N. Minar, K. H. Kramer, and P. Maes, Cooperating mobile agents for dynamic network routing, in A. L. G. Hayzelden and J. Bigham (Eds.), *Software Agents for Future Communication Systems*, Springer, Heidelberg, Germany, 1999, pp. 287–304.

34. F. Ducatelle, G. Di Caro, and L. M. Gambardella, Using ant agents to combine reactive and proactive strategies for routing in mobile ad-hoc networks, *Int. J. Computa. Intell. Appl.*, 5(2):169–184, 2005.

35. J. Dowling, E. Curran, R. Cunningham, and V. Cahill, Using feedback in collaborative reinforcement learning to adaptively optimize MANET routing, *IEEE Trans. Syst. Man Cybernet. Part A: Syst. Humans*, 35:360–372, 2005.

36. C. Aurrecoechea, A. Campbell, and L. Hauw, A survey of QoS architectures, *Multimedia Syst.*, 6:138–151, 1998.

37. L. Leon-Garcia and I. Widjaja, *Communication Networks: Fundamental Concepts and Key Architectures*, McGraw-Hill, New York, 2004.

38. L. Kleinrock, *Queueing Systems, Vol. 1*, Wiley, New York, 1975.

39. M. Osborne and A. Rubenstein, *A Course in Game Theory*, MIT Press, Cambridge, MA, 1994.

40. J. G. Wardrop, Some theoretical aspects of road traffic research, *Proc. Inst. Civil Eng. Part II*, 1:325–362, 1952.

41. Y. Korilis and A. Lazar, On the existence of equilibria in noncooperative optimal flow control, *J. ACM*, 42(3):584–613, 1995.

42. A. Orda, R. Rom, and N. Shimkin, Competitive routing in multi-user communications networks, *IEEE/ACM Trans. Networking*, 1:510–521, 1993.

43. A. Haurie and P. Marcotte, On the relationship between Nash-Cournot and Wardrop equilibria, *Networks*, 15:295–308, 1985.

44. D. Bertsekas, *Nonlinear Programming*, Athena Scientific, Belmont, MA, 1999.

45. V. R. Konda and V. S. Borkar, Actor-critic-type learning algorithms for Markov decision processes, *SIAM J. Control Optimization*, 38:94–123, 1999.

46. A. Tanenbaum, *Computer Networks*, Prentice-Hall International, Hemel Hempstead, 1988.

47. J. Sum, H. Shen, C. Leung, and G. Young, Analysis on a mobile agent-based algorithm for network routing and management, *IEEE Trans. Parallel Distrib. Syst.*, 14(3):193–202, 2003.

48. G. Takahara and C. Leith, A control framework for ant-based routing algorithms, in *Proceedings of IEEE Congress on Evolutionary Computation*, Vol. 3, CEC'03, 2003, pp. 1788–1795.

49. K. M. Sim and W. H. Sun, Ant colony optimisation for routing and load-balancing: Survey and new directions, *IEEE Trans. Syst. Man Cybernet. Part A: Syst. Humans*, 33:560–573, 2003.

50. A. Segall, The modeling of adaptive routing in data-communication networks, *IEEE Trans. Commun.*, 25:85–98, 1977.

51. C. Cassandras, V. Abidi, and D. Towsley, Distributed routing with on-line marginal delay estimation, *IEEE Trans. Commun.*, 38:348–359, 1990.

52. C. E. Perkins and E. M. Royer, Ad hoc on-demand distance vector routing, in *Proceedings of the 2nd IEEE Workshop on Mobile Computing Systems and Applications*, 1999, pp. 90–100.

53. S. Thomas, *IP Switching and Routing Essentials*, Wiley, Hoboken, NJ, 2002.

54. D. Bertsekas and J. Tsitsiklis, *Neuro-Dynamic Programming*, Athena Scientific, Belmont, MA, 1996.

7   Resource and Service Discovery

PAOLO BELLAVISTA, ANTONIO CORRADI,
and CARLO GIANNELLI

DEIS—University of Bologna, Bologna, Italy

7.1   INTRODUCTION

The dynamic discovery of resource and service components at run time is a crucial functionality in any execution environment where it is not feasible to assume that (i) all interacting entities have full static knowledge about each other and (ii) this mutual knowledge does not change over time. Therefore, given that almost all modern execution environments permit some forms of dynamicity (change in the set/characteristics of available resources and service components, client mobility, resource/service mobility, etc.), discovery has become a necessary element of any distributed systems nowadays.

The discovery process consists of several substeps, as extensively described in Section 7.2: Resources/service components willing to be discovered should advertise their availability; clients should be able to perform queries on the set of resources/service components available in the client discovery scope; once retrieved, that discovered entity should be bound to the client for the whole duration of the interaction by possibly maintaining the session state in the case of multiple successive client requests.

The support of all these steps, available in any traditional discovery solution, is significantly complicated when dealing with mobile computing environments. Mobile clients can change their discovery scope due to their movements while either involved in a discovery query or accessing discovered resources/service components. Discoverable entities can even migrate due to server host mobility, more and more common in peer-to-peer networks. In response to client mobility and changed patterns of request traffic, it could be interesting to have the possibility to either duplicate or move discoverable entities as well as maintain colocality with their currently served clients [1]. The code and state mobility features typical of the mobile agent (MA) programming paradigm can

Mobile Agents in Networking and Distributed Computing, First Edition.
Edited by Jiannong Cao and Sajal K. Das.
© 2012 John Wiley & Sons, Inc. Published 2012 by John Wiley & Sons, Inc.

well fit and further improve these mobility scenarios: On the one hand, mobile clients can exploit MA-based mediators that follow them at run time to transparently maintain/requalify the bindings to discovered entities; on the other hand, MAs could represent the primary enabling technology to develop and deploy discoverable entities that can even move during service provisioning by preserving their reached execution state.

After having introduced the main concepts related to discovery in mobile computing scenarios, the chapter aims at overviewing the most important discovery solutions in the literature by sketching their main characteristics according to an original taxonomy. The proposed classification is organized in three different and growing degrees of mobility:

    i. *Low-mobility* solutions, where only discovery clients can sporadically move at run time, even after having retrieved a discoverable entity and before having terminated using it

    ii. *Medium-mobility* solutions, where the frequency of client movements significantly increases and discoverable resources (files, cached data, devices, etc.) can change their location at run time

    iii. *High-mobility* solutions, where discoverable service components can migrate at run time together with their code and their reached execution state because they are implemented as MAs.

The state of the art will also permit us to identify some trends that are recently emerging as solution guidelines for resource/service discovery in highly mobile computing scenarios. In particular, we claim the suitability of middleware-level solutions based on the exploitation of possibly mobile mediators, fully aware of current location and context information about served clients and discoverable entities. These mediators are in charge of transparently managing the bindings to discovered entities independently of the various forms of mobility that clients, resources, and service components may exhibit.

The remainder of the chapter is organized as follows. Section 7.2 defines the terminology and primary subprocesses involved in resource/service discovery. Section 7.3 presents our original taxonomy and uses it to classify the main discovery research activities in the literature. Section 7.4 identifies the most relevant solution guidelines recently emerging, while hot topics for current discovery research and concluding remarks end the chapter.

7.2   MOBILITY AND RESOURCE/SERVICE DISCOVERY

Over the past few years resource/service discovery has been the subject of active research and development efforts in both academia and industry; the primary goal was facilitating dynamic interworking among clients and dynamically retrieved entities with minimal administration intervention. Nowadays, several discovery suites are available, targeted at different computing environments and exhibiting different features [2]. The aim of this section is first to define the characteristics and terminology of the different tasks involved in the discovery process and then to show how mobile code programming paradigms, especially MAs, can affect the design and implementation of these tasks.

7.2.1 Definitions and Concepts

The discovery process involves several entities. *Discovery clients* are the active entities that usually start the process by requesting the list of available discoverable items; in several discovery solutions, this list only includes the items that respect the criteria specified at request time. Discoverable entities are usually distinguished into *resources*, such as printers, storage devices, network access points, and data files, and *service components*, which offer logical services by allowing clients to invoke code execution and returning service results. In contrast to resources, discoverable service components can maintain their reached execution state (session), possibly to exhibit differentiated behavior over time, also in response to the same service requests.

Any discovery solution has to define how discoverable entities should register and provide the information needed to be retrieved (*naming* and *description*). Typically, discovery systems impose unique identifiers for registered discovered entities. In open and dynamic scenarios, it is usually unfeasible to assume that clients know the identifiers of needed resource/service components. Therefore, client retrieval requests are typically based on some forms of description of the searched entity. Most discovery solutions (see the following section) exploit very simple pattern-matching mechanisms based on entity names. Some state-of-the-art proposals have recently started to extend this approach by permitting the description of discoverable entities via their interfaces and/or name–attribute pairs.

Once it is decided how to identify and describe discoverable entities, any discovery solution has to provide a registration/query service. To support service registration, different options are available. Nonscalable solutions suitable for small-scale deployment environments with limited dynamicity exploit broadcast-based announcement of available entities; announcements can be either clientdriven (in response to a new client broadcast message with a discovery query) or resourcedriven (any time a discoverable entity enters a new group of potential clients). Most interesting registration solutions adopt centralized or distributed registries to maintain names/interfaces/attributes of discoverable entities. Distributed registries can work in an isolated way by maintaining only their nonreplicated partition of registrations or can coordinate to maintain a globally consistent, partially replicated, and partially partitioned registry space. Coordinated registries can be organized in a flat way, hierarchically, or in a hybrid intermediate way, as more generally in many distributed naming servers.

To support resource/service component retrieval, some discovery solutions require clients to specify the name, interface, or attribute values for searched discoverable entities. Most recent proposals also permit us to specify search patterns: Discoverable entity selection is based on a syntactic comparison for compatibility between the specified pattern and the registered names/interfaces/attribute values.

A more relevant aspect of discovery solutions is how they determine the search space scope for client discovery queries, that is, the set of discoverable entities that should be searched in response to a discovery query from one client. First we need to consider the difference between local and global scopes. The first case includes discovery solutions that decide that any client could be interested only in discoverable entities which are currently located within the client locality (in the same wired network locality, at a single-hop distance in wireless ad hoc networks, in the same administrative domain, etc.). In the second case, in contrast, any registered entity can be discovered by any client, thus complicating scalability and coordination over geographical-wide deployments. A few recent and very interesting research efforts are providing more advanced forms of discovery scope where the set of searchable resources depends not only on client position but also on its preferences, application-level requirements, and history of previous interactions: The ultimate goal is to transparently provide clients with personalized and context-dependent discovery [3].

Finally, a very relevant aspect of advanced mobility-enabled discovery solutions is how they address the issue of managing the bindings between clients and discovered entities in the case of mutual mobility after discovery-based retrieval and before termination of resource/service access. Traditional discovery solutions do not confront this issue: If a binding is broken at service provisioning time, the client is notified about it and has the full duty of managing the event, possibly by redoing the query and restarting the resource/service request on the newly retrieved entity instance. Changing an entity instance with an equivalent new one is easily possible only if the offered service is stateless. Recently, some discovery supports have started to investigate how to provide mechanisms and tools for automatically requalifying broken bindings, with no impact on the client application implementation. Those solutions tend to provide differentiated binding strategies that automatically rebuild links to moved entities, or requalify bindings to novel equivalent discoverable instances, or even move/duplicate resources and service components to maintain the search scope invariant. As better detailed in Section 7.3, these solutions can extensively benefit from mobile code technologies and especially from MA-based support implementation.

7.2.2 Suitability of Mobile Code Technologies for Discovery Solutions

As observed for the above case of rebinding support and stated in the introduction, mobility affects the discovery process and forces us to consider novel mechanisms, tools, and strategies to dynamically adapt several discovery steps to react to different forms of mobility exhibited by different discovery entities.

In all these cases, mobile code programming paradigms, in particular the MA technology, play the twofold role of exacerbating the dynamicity of clients, resources, and service components on the one hand and representing a very suitable implementation solution for mobility-enabled discovery on the other hand.

By exploiting the terms defined in the introduction, in low-mobility deployment scenarios, clients can move at run time (typically in an unpredictable way), that is, during a discovery query or after binding to discovered entities but before ending to use them. In most discovery solutions, client movements may produce changes in the client discovery scope and bound entities could become unreachable. Traditional discovery approaches do not address those issues: Mobile clients have visibility of failures in their discovery queries or are unable to access the retrieved entities; they are in charge of explicitly restarting the discovery process; and the management of discovery-related client mobility should be directly embedded into the client application logic code in a way that is not transparent. As a consequence, in addition to affecting the development of client application logic, they tend to support only execution environments where node movements are relatively rare events: If mobility frequency increases too much, these traditional discovery solutions are not able to correctly propagate updates and the client-visible discovery scopes may become inconsistent.

Novel discovery solutions tend to provide some forms of middleware support to low-mobility problems: The primary solution guideline is of assisting mobile clients with middleware-level mediators in charge of transparently refreshing discovery scopes and of requalifying broken bindings in response to client mobility. Those mediators sometimes execute directly on the same hosts of assisted clients, thus ensuring mediator–client colocality for the whole discovery sessions notwithstanding client movements.

However, it is more frequent the case of discovery clients running on resource-limited devices with strict constraints on available memory, battery power, connectivity bandwidth, and connectivity time (often simply because of high costs of wireless connectivity). In this case, it can be extremely useful to deploy mediators on fixed network hosts in the same current locality of the assisted clients: Those mediators could autonomously work on behalf of their clients and without consuming their limited resources (by also permitting temporary disconnections to reduce connectivity time). However, mediators running in the fixed network can lose colocality with their clients depending on run time client movements. For this reason, mobile code programming paradigms are extremely suitable: They are necessary to enable the dynamic movement of discovery mediators at run time to follow client mobility. If mediators have to preserve the reached execution state when migrating, the MA technology has to be exploited.

Medium-mobility deployment scenarios further complicate the case above by allowing higher frequency in client movements and resource mobility at run time. Therefore, in this situation, not only can the set of resources in the discovery scope rapidly vary, for example, due to resource frequently entering/ exiting an administrative domain, but also bindings can break because of resource change of allocation. These problems are counterbalanced by the significant advantage of having the possibility to move resources at run time depending on dynamic factors, such as change of traffic and/or of traffic distribution in the served localities. Let us observe that resource replication and deployment could be considered a particular case of resource movement, where the old resource instance maintains its original location while the new copy moves.

All the problems of refreshing discovery scopes and requalifying already established bindings to resources remain the same as described for low mobility. Therefore, code mobility technologies have the same crucial role of enabling implementation solution. In addition, mobile code can be extremely suitable for moving/replicating resources that are associated with server-side code to access there. For instance, when moving a printer from one network locality to another, the printer spooler demon should be moved too and thus requires migrating its code at run time. The spooler acts as a server-side mediator to access the printer resource and benefits from continuously maintaining colocation with the managed resource, similarly to the dual case of discovery client mediators.

In the high-mobility scenario, the complexity of managing dynamic changes in both discovery scope and established bindings is further exacerbated. In particular, here also service components (together with their code and reached execution state) could be interested in moving during service provisioning, basically for the same reasons resources benefit from mobility, for example, to dynamically balance workload among localities, to follow client group movements, and to increase/decrease replication degree depending on request traffic. Note that this is a logical form of mobility which could be present also when physical nodes are not mobile and service components change their execution environments by migrating from one host to another one, as in the case of MA-based service components.

On the one hand, the additional complexity of managing discovery in high-mobility scenarios stems from the fact that service components usually are interested in migrating by preserving their session state. That implies that discovery supports should not only provide mechanisms to move the reached execution state but also distinguish between stateless interactions (for instance, to requalify client bindings to new, just started, instances of service components) and stateful interactions (where clients should maintain bindings to exactly the same service instances independently of mutual movements). On the other hand, service components are often clients of other service components: Migrating a component could affect the network of bindings established not only with its clients but also with other service components of which it is currently a client. Advanced discovery proposals can benefit from MA solutions to support high-mobility scenarios simply because exactly the same problem had to be faced in the MA research and in the development of state-of-the-art MA systems: A mobile agent is a service component, can move at run time by carrying its session state, and has the problem of reestablishing its bindings to clients and resources/service components when resuming its execution after migration.

In short, mobility-enabled discovery solutions could significantly profit from both mobile code enabling technologies and the results obtained from research about code mobility and MAs, as pointed out by the examples of the following section.

7.3 RELATED WORK

This section tries to classify all the relevant discovery protocols/systems proposed in the literature by taking into consideration their capability to react to discovery entity movements by either adapting service discovery scope or performing automatic service rebinding or both. The presentation of existing solutions is organized according to our original taxonomy that distinguishes low-, medium-, and high-mobility deployment environments for discovery solutions.

7.3.1 Low-Mobility Discovery Solutions

Traditional and widely accepted service discovery protocols, such as Jini [4], Service Location Protocol (SLP) [5], Universal Plug and Play (UPnP) Simple Service Discovery Protocol (SSDP) [6], Bluetooth Service Discovery Protocol (SDP) [7], and Salutation [8], address the basic issues related to the dynamic discovery of resources and service components. The main objective is to advertise the availability of discoverable entities to mobile devices approaching a new location in order to increase client capabilities with resources/services usually deployed on fixed nearby nodes. These solutions have been designed with a nomadic mobility deployment scenario in mind: Mobile clients are expected to request that locally available entities not to move while accessing them.

Jini, SLP, and SSDP mainly focus on discovery in infrastructure-based and administrator-controlled networks, where services are typically deployed on the fixed-network infrastructure and wired/wireless nodes play the role of clients. All these solutions model the different entity roles defined in the previous section and provide repositories for resource/service registration and lookup. In particular, Jini exploits Lookup servers to store information about the available set of discoverable entities and provides client-sided Java proxies to invoke services. SLP clients (called user agents) look up available services (called service agents) either by multicasting requests and waiting for unicast responses from matching services or by contacting a service repository (called directory agent) to get the currently available service set. SSDP discoverable entities announce their presence with an "alive" message at startup; SSDP clients look up available services by broadcasting User Datagram Protocol (UDP) messages in their local network.

Bluetooth SDP and Salutation do not strongly differentiate roles among discovery entities. Each node is always able to both request and offer resources/services in a peer-to-peer fashion. In particular, Bluetooth SDP well fits peer-based ad hoc networks as its primary deployment scenario. The Salutation protocol is mainly based on Salutation managers, perceived from clients as local service brokers connected with other remote Salutation managers. Clients request for and offer resources/services to their local managers, which eventually contact other managers to widen the discovery search. Managers exploit underlying Salutation transport managers to retrieve nearby managers via multicasting, static configuration, or interrogation of a centralized directory.

Regardless of their specific characteristics, the aforementioned widespread protocols do not address the specific issues of performing discovery in a highly dynamic mobile environment where new resources/services become available and old ones disappear frequently.

On the one hand, discovery protocols based on entity description repositories (e.g., Jini and, optionally, Salutation and SLP) assume rather static discovery management operations and do not fit well highly mobile scenarios with frequent update procedures, which tend to be expensive and resource consuming. In addition, let us observe that Jini is the only discovery solution that supports a form of lease, that is, any Jini lookup server entry has an expiration time after which the corresponding information is automatically discarded. The other discovery solutions do not exploit any lease mechanism: Resources/services should be explicitly removed from registries, thus increasing the possibility of inconsistent registry entries whenever a discoverable entity become abruptly unreachable due to, for example, the crash of the hosting node or its movement outside the wireless client coverage range in single-hop communication environments.

On the other hand, other service discovery protocols (Bluetooth SDP, SSDP, and, optionally, SLP and Salutation) perform discoverable entity advertisement in a client-driven pull fashion by multicasting or broadcasting client requests. This generates heavy-bandwidth consumption and long discovery delay times, which are not appropriate for a highly dynamic mobile environment. Note that SLP and Salutation can be optionally based on a centralized registry for discoverable entity descriptions.

It is important to stress that, although Bluetooth SDP and Salutation have also been designed for ad hoc deployment scenarios, they do not fit well highly dynamic execution environments. In fact, both Bluetooth devices and Salutation managers retrieve nearby nodes and hosted resources/services either in a costly manner by multi-/broadcasting client requests or in a very static way by exploiting configuration settings or a single centralized directory.

Most important, all these low-mobility discovery protocols do not provide any rebinding mechanism for clients to automatically and transparently switch among equivalent resources/service components. Rebinding mechanisms are crucial in dynamic environments where the available resource/service set continuously changes and it is necessary to support medium- and high-mobility deployment scenarios.

For a more detailed and general-purpose description of widely diffused low-mobility discovery solutions, refer to [2, 9], where they are classified according to an alternative, more traditional taxonomy not based on mobility characteristics.

7.3.2 Medium-Mobility Discovery Solutions

The widespread diffusion of mobile devices has provided new challenging scenarios where devices freely move and exploit wireless technologies for mutual interaction either with the traditional network infrastructure via wireless access points (infrastructure-based wireless Internet) or in a completely peer-to-peer collaboration mode with other wireless nodes (mobile ad hoc networks). Mobile devices move while taking advantage of dynamically discovered entities and while offering their resources. Therefore, possibly frequent device movements generate resource movements too, and the traditional discovery solutions sketched in the previous section cannot apply to this highly dynamic scenario.

First, medium mobility may cause abrupt variations in discovery scopes, even if mobile clients do not move. On the one hand, it is impossible to support discovery in a static manner merely exploiting resource/service registries since they become inconsistent quickly. On the other hand, the needed frequent broadcasting requests would waste the often limited wireless bandwidth available. Thus, the main guideline is to provide discovery protocols able to provide a suitable trade-off between discovery scope update costs and consistency. For example, a commonly adopted solution principle is to maintain the set of currently discoverable resources/services as a distributed soft state located at discovery clients: Announcements are exploited to specify not only a resource/service description but also an announcement lease interval; clients autonomously consider their locally stored discovery entries valid until their lease expires.

In addition, clients and resource movements claim for frequent rebinding management. Discovery solutions should provide mechanisms for the automatic update of resource/service references, for instance, based on service type to rebind to a functionally equivalent laser printer in the current client locality when the client moves outside its usual office locality, including its usual own printer. Moreover, automatic rebinding must be triggered not only by resource/service availability but also by considering the currently applicable context (see Section 7.4.3), for example, by preferring a discoverable entity because it better fits user preferences and/or application-level quality requirements.

In contrast to the solutions in the previous section, the discovery approaches sketched in the following propose mechanisms to (i) efficiently announce, discover, and request for discoverable entities when both clients and resources move and (ii) automatically rebind to new equivalent entities in response to broken binding.

DEAPspace aims at providing a discovery protocol for wireless, mobile, adhoc, single-hop networks able to maximize the diffusion of information about available resources/services among close mobile clients [10]. There is no centralized repository; the efficient dissemination of discovery registries is obtained by forcing any participating node to periodically broadcast not only the list of local resources/services it offers but also the lists of its recently nearby devices (diffusion of recent worldview). To maximize knowledge sharing, one node broadcasts its lists more frequently if it detects that the worldview of one nearby device lacks at least one of the entities it offers or when the leases of some entries are going to expire.

To increase the rapidity of discoverable entity information dissemination, it could be interesting to forward resource/service announcements/queries not only to neighbors but also along multihop paths. Konark is similar to DEAPspace but, instead of locally broadcasting the entire worldview each time, it sends only information missing at the neighbors [11]. Moreover, Konark spreads discovery registries by exploiting multihop ad hoc communications. In addition, it provides discovery clients implementing the tiny SDP and resources running a Hyper Text Transfer Protocol (HTTP) server to answer to binding requests.

Rubi has the specific primary goal of addressing registry dissemination in a mobility-enabled execution environment [12]. Rubi distinguishes different classes of mobility frequencies. In the case of frequent client/resource mobility, Rubi exploits a reactive algorithm to reduce updated message traffic: Each node performs lookup in a peer-to-peer fashion by individually caching the received registry information, similarly to DEAPspace and Konark. In the case of rare client/resource movements, Rubi exploits a proactive algorithm to keep disseminated information probabilistically up to date by delegating to relays the duty of storing the most updated discoverable entity descriptions; Rubi relays work similarly to directory servers.

Allia [13] is a multihop discovery framework that aims at performing dissemination and caching of discoverable entity descriptions, like Konark and Rubi. Any Allia mobile device advertises its offered services to neighbors. The main original aspect is that Allia is policybased; that is, it exploits high-level rules, modifiable at run time, to manage service information. In particular, when a node receives a new entity description, it exploits its locally enforced policies to decide to either forward, stop, or discard it.

In the case of medium-mobility deployment scenarios, it is crucial to increase discovery efficiency not only via effective mechanisms for registry dissemination but also with efficient solutions for propagating discovery queries and for establishing client-to-resource bindings. Lanes proposes a two-dimensional overlay structure that exploits an axis for registry dissemination and another different axis for distributing search requests [14]. The main idea is to differentiate paths and strategies for announcement and request messages: Since resource/service announcements have usually long-term effects, they should benefit from proactive propagation; on the contrary, given that a discovery query is typically a one-time action, a reactive anycast communication should be more appropriate. By exploiting its overlay network, Lanes proactively forwards announcements to nodes along the same line, while it reactively propagates discovery requests by anycasting nearby lines and limiting the search scope by exploiting resource/service description in the query.

Group-based service discovery (GSD) performs group management in ad hoc networks and exploits groups to efficiently perform discovery requests [15]. In particular, (i) GSD operates a limited advertising of resources/service components by propagating announcements for a given number of hops; (ii) it dynamically caches advertisements and information about current group compositions; and (iii) it efficiently performs discovery requests by exploiting a selective group-based request forwarding. Moreover, GSD resource/service description is based on the Ontology Web Language (OWL), an extensible and semantic-based description language that can help in matching discovery requests/offers based on semantics and not only on syntactic comparison between names, interfaces, and attributes.

The other crucial point in medium-mobility deployment scenarios is the support to efficient rebinding in response to unpredictable and frequent changes in the discovery search scope. The computer-aided design package Q-CAD proposes a rebinding mechanism specifically designed for pervasive environments [16]. Q-CAD first looks for resources that satisfy client requirements included in the discovery query; if several resources/service components are in the discovery scope and are compatible with the requirements, Q-CAD automatically selects the one maximizing a utility function that determines the applicable quality metrics. Q-CAD can answer discovery requests either reactively or proactively: In the first case, Q-CAD reacts to explicit client requests for (re)binding; in the second mode, Q-CAD permits us to specify which context changes trigger (re)binding operations.

While other discovery proposals are explicitly targeted to ad hoc networks, service-oriented network sockets (SoNSs) transparently support connection-oriented semantics in infrastructure-based wireless networks [17]. The main idea is to maintain client-to-resource bindings independently of entity movements during a service session by automatically requalifying bindings according to the strategies specified in high-level policies.

Tripathi et al. [18] focuse on adapting rebinding decisions to different user preferences, service characteristics and system settings. In particular, when a client moves among different network localities, to perform rebinding Tripathi et al. [18] take into account user role, user privacy concerns, and access authorizations based on context. Bindings could be either permanent or context based. In the case of permanent bindings, the discovery support tries to maintain them via multihop paths even if client and discovered entities are no longer in the same locality (remote reference); context-based ones use implicit and automatic rebinding triggered by context changes. The work of Tripathi et al. [18] is implemented in terms of MAs: MAs, running in the fixed network infrastructure, perform context monitoring and resource/service rebinding to preserve the limited resources of mobile client devices. Let us stress that [18] exploits MAs to assist client discovery operations and not to actively migrate resources and service components, as the advanced discovery proposals presented in the next section do.

Finally, the Atlas discovery solution is relevant because it introduces the policy-based exploitation of location awareness in rebinding mechanisms [19]. Atlas adapters wrap each discoverable entity and exploit failure determination mechanisms to notify relevant context modifications, for example, a client or resource location change. Triggered by failure detection, the Atlas Service Binder replaces the discoverable entity no longer available and tries to automatically rebind the involved clients with another equivalent instance of it.

7.3.3 High-Mobility Discovery Solutions

In this mobility case, not only do clients requesting for services and discoverable resources move freely, but also service components can migrate at provisioning time between different network nodes, also independently of actual physical node movements. In this highly dynamic scenario, service advertisement, discovery, and rebinding become more and more challenging, calling for state-of-the-art discovery solutions.

Hermann et al. [20] propose a discovery protocol that considers the possibility that service components can autonomously replicate and migrate to optimize resource usage and message latency in ad hoc networks. A self-repairing lookup service is able to efficiently cope with the inconsistencies caused by service migrations by applying a lazy, request-driven update protocol. The implemented lookup opportunistically exploits discovery responses to update local cached registration data with the new network locations of migrated/replicated services. The exploitation of discovery responses, instead of active announcements of migration/duplication updates, is a growingly adopted approach which can achieve a suitable trade-off between imposed network load and registration data consistency. Note that Hermann et al. [20] do not consider explicitly the possibility of triggering service component migration; they only focus on the efficient discovery of service components capable of mobility.

According to Riva et al. [21], service components can migrate among nodes of an ad hoc network. Both clients and discoverable entities continuously monitor their contexts to promptly trigger rebinding and migration, respectively. The main goal is to dynamically adapt the association with discovered resources/service components depending on quality requirements. Similarly to Hermann et al. [20], Riva et al. [21] completely hide clients from the fact that service components can move at run time; service migration is completely transparent from the point of view of the implementation of the client application logic.

By following the principle that different rebinding strategies after service migration can be suitable for different application requirements, the following discovery proposals allow developers to explicitly specify their preferred rebinding rules. Several solutions provide rebinding supports inspired by the seminal work on binding requalification in MA systems by Fuggetta et al. [22]: Depending on resource/service characteristics and client requirements, they tend to allow looking for new resources of the same type, migrating/copying discoverable entities, or remote referencing already bound entities independently of mobility.

The Dynamic Adjustment of Component InterActions (DACIA) framework supports the design and implementation of reconfigurable mobility-enabled applications by providing developers with mechanisms to dynamically alter mobility-triggered reconfiguration rules [23]. For instance, DACIA applications can explicitly request the migration, duplication, connection, and disconnection of specific middleware components that encapsulate discoverable entities. However, DACIA-based applications must explicitly specify how to perform mobility-triggered service adaptation within the same application code, thus not permitting clear separation of resource/service mobility logic from service application logic.

Instead, the Mobility Attributes Guide Execution (MAGE) project introduces mobility attribute programming abstraction to describe the mobility semantics of discoverable service components [24]. Developers can associate service components with different mobility attributes to control their dynamic allocation: For instance, the remote procedure call (RPC) attribute determines that the discoverable entity should be invoked remotely independently of its mobility, while the MA attribute specifies that the entity should migrate in response to run time event notifications. However, MAGE leaves to developers the burden of manually implementing the proper binding between discovery entities by exploiting the visibility of the specified mobility attributes.

Another interesting approach is FarGo, which supports the specification of high-level declarative policies influencing the run time allocation of mobile service components [25]. Explicit relationships between related components, namely complets, specify if the migration of a shared component should trigger, for example, the migration of another component (pull), the duplication of another component in the new locality (duplicate), or the discovery of an equivalent component in the new location (stamp). Migration policies are expressed in a high-level script language that permits us to specify the operations to be performed in response to events generated by a monitoring facility that controls discoverable resources/service components and the overall performance of the execution environment.

To further improve flexibility, Tanter and Piquer [26] use reflection to define customizable binding strategies implemented as basic reusable meta-objects attached to any mobility-enabled discoverable service component. For instance, it is possible to specify a rebinding strategy invoked whenever an MA arrives to a new host in order to discover and bind to a suitable instance of required resources. However, the linking between MA-based service components and binding strategies is performed at the beginning of the execution and cannot change at provision time without an execution restart.

With a greater support of dynamicity, the Context-Aware Resource Management Environment (CARMEN) permits us to specify migration/binding strategies in terms of high-level obligation policies and to modify them even during service provisioning, without any impact on service implementation [1]. CARMEN exploits profiles to describe discoverable entity characteristics and policies to manage migration, binding, and access control. Given that CARMEN is a good example of some solution guidelines recently emerging in the discovery research area, a description of its main characteristics can be found in Section 7.4.4.

To summarize some crucial points analyzed in Section 7.3, Figure 7.1 depicts a possible classification of registration dissemination and rebinding mechanisms. About announcement/search, *When* and *Where* respectively represent the events triggering and the scope of announcement/search procedures. The

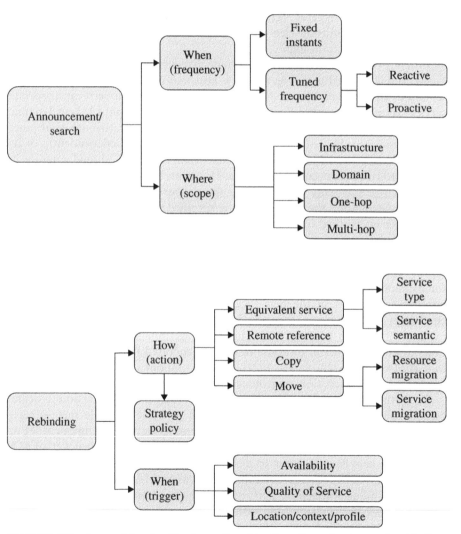

FIGURE 7.1 A possible classification of registration dissemination and rebinding mechanisms.

Infrastructure scope is the announcement/search scope coinciding with the underlying network locality, while *Domain* delegates to upper management layers the definition of scope boundaries. About rebinding, *How* is the actual action a triggered rebinding operation should perform, possibly depending on what is specified in the *Strategy Policy*. *When* represents the monitored indicators which can trigger binding management operations.

Table 7.1 organizes the discovery solutions according to the classification of Figure 7.1 (which covers some relevant aspects among the ones pointed out in

TABLE 7.1   Related Work Main Characteristics

Mobility Degree	Related Research	Announcement/Search		Rebinding	
		Frequency	Scope	Action	Trigger
Low	Jini [4]	Proactive (lease based)	Infrastructure	n.a.	n.a.
	SLP [5]	Fixed	Infrastructure	n.a.	n.a.
	SSDP [6]	Fixed	Infrastructure	n.a.	n.a.
	SDP [7]	Fixed	One-hop	n.a.	n.a.
	Salutation [8]	Fixed	Domain	n.a.	n.a.
Medium	DEAPspace [10]	Proactive	Onehop	n.a.	n.a.
	Konark [11]	Proactive	Multihop	n.a.	n.a.
	Rubi [12]	Reactive/ proactive	Multihop	n.a.	n.a.
	Allia [13]	Proactive (policy based)	Multihop	n.a.	n.a.
	Lanes [14]	Reactive/ proactive	Multihop	n.a.	n.a.
	GSD [15]	Proactive	Multihop	Semantic search	n.a.
	Q-CAD [16]	Reactive	Domain	Service type	QoS
	SoNS [17]	Reactive	Domain	Service type, policy	QoS
	[18]	Reactive	Domain	Policy	Context
	Atlas [19]	Reactive	Domain	Policy	Location
High	[20]	Proactive	Multihop	n.a.	n.a.
	[21]	Proactive	Multihop	n.a.	n.a.
	DACIA [23]	n.a.	Domain	Static policy	Aapplication dependent
	MAGE [24]	n.a.	Domain	Static policy	Resource invocation
	FarGo [25]	n.a.	Domain	Policy	QoS, context
	[26]	n.a.	Domain	Static policy	Resource invocation
	CARMEN [1]	n.a.	Domain	Policy	Context

Note: n.a = not applicable.

Section 7.3). Static policies do not allow us to change the rebinding strategy at run time, while more powerful regular policies provide this capability. Note that only a few proposals perform discovered entity binding management, and only [1, 25] do it with a policy-based approach suitable for highly mobile execution environments.

7.4   EMERGING SOLUTION GUIDELINES

The wide set of discovery-related research activities presented in the previous section certainly exhibit different characteristics and different levels of suitability depending on the mobility level of deployment scenarios. Despite this differentiation, here we try to point out some common solution guidelines that are emerging in the variegated field of discovery solutions, especially when dealing with highly dynamic execution environments (high density of highly mobile discovery entities). We claim that most recent discovery approaches tend to adopt three main directions of solution, of increasing complexity for discovery supports: (i) the exploitation of mediators to decouple clients from discoverable entities, (ii) the full visibility of location information to properly guide rebinding decisions, and (iii) the full awareness of context data to suitably inform rebinding management. All the guidelines can benefit from implementation solutions that exploit state/code mobility features.

7.4.1   Mediator-Based Discovery Infrastructures

In the last years, distributed service provisioning to mobile clients has often motivated infrastructure-based middleware solutions, for example, to dynamically downscale service contents, in order to suit the specific characteristics and limits of portable access terminals. In addition to dynamic content negotiation and tailoring, the crucial challenge is to properly handle mobility at service provisioning time, which requires several other support operations that may be too expensive to be done by severely limited devices on their own.

Discovery is the primary example of such a support operation, given that it calls for local/global finding of discoverable entities, binding to needed ones, and binding adjustments depending on varying conditions at provisioning time. Environment exploration, negotiation with discoverable entities, global identification and retrieval of applicable user profiles and resource/service descriptions, and location-/context-dependent binding management operations can be too expensive and time consuming to be directly managed by terminals with severe resource constraints (see also Section 7.4.4).

For this reason, recent discovery solutions propose distributed infrastructures of mediators (often called discovery brokers, or interceptors, or stubs, or proxies) that run in the fixed network on behalf of and close to mobile clients. We claim that the adoption of these mediators, which we will call discovery proxies in the following, is a crucial design solution for state-of-the-art and future discovery solutions in mobile computing scenarios.

The idea behind discovery proxies is quite simple (and common to other fields of distributed systems): to partially decouple, possibly in space and in time, discovery clients from discovered entities in order to hide application developers from the complexity of discovery management operations and to reduce the client-side discovery costs. For instance, proxies can cache discovery registration data by exploiting the typically large storage capabilities of fixed network hosts; they can determine the applicable discovery scopes without affecting the limited computing power and bandwidth of wireless clients, also proactively with regard to client discovery queries; they can allow temporary disconnection of their clients by returning discovery results at client reconnection; and they can transparently requalify bindings to discovered entities, for instance in response to context variations, with no impact on the implementation of the client/service application logic.

In particular, as pointed out in Section 7.3, several discovery management approaches are recently pushing toward the exploitation of mobile discovery proxies to assist mobile clients. Mobile proxies can be dynamically deployed, where and when needed, to work on behalf of mobile clients in their current network localities and can follow client movements during service sessions. Given the relevance of local access in mobile computing, proxy mobility is considered a crucial feature and is motivating the adoption of code mobility programming paradigms to implement discovery infrastructures of proxies. Among the different mobile code technologies, MAs tend to be preferred because of their additional capability of migrating the reached execution state, particularly relevant to maintain client session state notwithstanding client/proxy mobility.

In addition, the MA adoption facilitates the achievement of crucial properties, such as dynamicity, autonomy, and full visibility of the underlying execution environment. Discovery supports in open deployment scenarios should be highly dynamic also in the sense that it should be possible to migrate, modify, and extend the infrastructure of distributed discovery proxies with new components and protocols; dynamic installation and code distribution typical of MAs are decisive when dealing with mobile clients with highly heterogeneous hardware/software characteristics. MAs can autonomously work, either proactively or in response to explicit discovery queries, even temporarily disconnected from their served clients: In the case of wireless mobile clients with strict constraints on available bandwidth and communication reliability, MAs can minimize the requested client connection time by requiring client connectivity only at the moment of discovery query injection and of result forwarding. Finally, the MA programming paradigm typically permits us to achieve full visibility of underlying system implementation details and execution environment conditions because those forms of awareness are required to drive MA mobility decisions [22]: In particular, the full visibility of location and context information is crucial to enable efficient discovery management supports, as illustrated in the following sections.

7.4.2 Location Awareness

Location can be defined as physical position of any entity involved in discovery processes, that is, clients, resources, and service components. In general, traditional distributed middleware solutions tend to hide the location information to facilitate the high-level transparent design of services for fixed-network environments. However, in the mobile computing scenario, the explicit visibility of location is necessary both at the middleware level to perform efficient location-dependent management operations (as in the case of discovery) and at the application level to enable the realization of location-dependent services.

Notwithstanding the growing diffusion of positioning systems and localization techniques, the management of location information is still a complex and challenging issue in open and dynamic deployment scenarios. On the one hand, a wide set of heterogeneous positioning solutions are currently available, with no widely accepted standard specifications to uniformly access location information and to control positioning systems. On the other hand, an open problem remains of how to effectively handle possibly frequent modifications in location data without introducing excessive overhead in terms of both network traffic and data processing.

Notwithstanding the presence of these issues, novel discovery approaches should exploit location information as a crucial element to decide which discoverable entities belong to a client discovery scope. Discovery scopes should primarily include the discoverable resources/service components currently located a limited distance from their requesting clients (in direct wireless visibility or at a few wireless hops to reduce expensive multihop traffic, especially in ad hoc networks). There is the need to favor local access in order to reduce the complexity and the costs associated with discovery-related management operations: Reducing the size of the discovery scope not only accelerates search and selection processes but also significantly contributes to decrease the overhead of binding management operations. In fact, discovery supports typically have to monitor any modification in the availability of entities in the discovery scope and any change in context conditions of interest for them (see the following section); these modifications can trigger expensive middleware operations of binding management. In addition, local access to discovered resources/service components generally contributes to decrease the overhead in the phases of resource/service request/result delivery.

In addition to efficiency, let us observe that automatic rebinding according to the locality principle is also a very simple way to implement location-dependent services: Service providers can deploy different equivalent instances of service components (e.g., with the same interface) in any locality where they are willing to provide differentiated contents; anytime a client changes its access locality, it is transparently rebound to a local instance of the currently accessed service; that service instance provides results depending on its location to all currently served clients.

Let us finally note that the adoption of MA-based technologies can significantly simplify the realization of location-aware discovery services. In fact, the MA programming paradigm is location aware by nature because MAs should have full visibility of location information to make proper decisions in order to guide their autonomous migrations. Therefore, MAs can easily propagate their visibility of the underlying execution environment to support location-aware discovery when exploited both as mediators and as services themselves. For instance, in the former case, location-aware MA mediators usually execute in fixed hosts close to their associated mobile nodes, thus having full knowledge of client positions and favoring interaction with resources and services colocated either in the same network domain or in close proximity. In the latter case, MA-based services can dynamically migrate depending on location visibility of traffic requests, hence increasing the possibilities of local location-aware access to discovered services.

7.4.3 Context Awareness

Location visibility is not the only crucial property of which discovery solutions should be aware. We claim that there is the need for full context visibility to enable effective discovery management solutions for mobile computing environments. By following a widespread definition, the context of a client at a specified time is the whole set of user preferences, discoverable entity descriptions, and environmental conditions that can influence service provisioning to that client. A notable example of context is the user personal profile of preferences: For instance, a user can express his or her interest in audio streaming service components with specified minimum quality requirements and in Web page contents in the Italian language.

Similarly to what was pointed out for location awareness in the previous section, novel discovery supports should have full visibility of context-related information and should exploit this awareness to increase the effectiveness of discovery-related operations. First, context should drive the determination of reduced discovery scopes with only the discoverable entities that are compliant with current context requirements. For instance, from the point of view of resource consumption it is uselessly expensive to include in the discovery scope service components that provide results in a format not supported at the current user access terminal. In addition, context should also be exploited to prioritize items in the list of discoverable entities, thus simplifying the selection by the final client/user: That facilitation is crucial especially when using portable client devices with small display size and limited browsing capabilities.

However, the monitoring, control, and effective management of context information is still a hard technical challenge, in particular if coupled with the goal of minimizing network traffic overhead. Middleware solutions that continuously monitor all possible execution environment indicators and immediately notify all discovery entities about the occurred modifications are obviously unfeasible and not scalable in highly mobile computing scenarios. The current research in context-aware discovery supports is concentrating on dynamically choosing the most suitable trade-off between responsiveness of context monitoring and promptness of requalifications of bindings to subsets of discoverable entities (see also the following section). There is the need for solutions capable to perform effective context management operations by permitting us to consider only the context indicators of interest for resources/ service components in the discovery scope and, hopefully, only for a minimum subset, for example, the ones with currently ongoing sessions to discovery clients.

Also context-aware discovery solutions can significantly benefit from the adoption of MA-based technologies. In fact, context monitoring, control, and management are power-consuming and computationally heavy activities. Moreover, they usually require full visibility of both application requirements and execution environment characteristics in order to transform potentially huge amounts of raw monitoring data in high-level useful information for context-aware discovery. On the one hand, MA-based mediators deployed on the fixed-network infrastructure may autonomously monitor relevant context indicators (especially related to resources/services hosted in the wired network) with no need of continuous connectivity with the associated clients. The result is an important reduction in client node overhead, in terms of both power consumption and computing power. On the other hand, the migration of new MA-based mediators when and where required at provisioning time provide the needed flexibility and dynamicity, for instance to manage newly installed context sources or to inject new monitoring data-processing behavior when context-related policies are updated (see also the following section). In that way, MAs hide most complexity associated with context awareness from mobile nodes and service providers, thus leveraging and accelerating the development of context-aware discovery solutions.

Finally, notice that location can be modeled as a particular case of context, where the position (possibly mutual) of the different discovery entities is the only information about the execution environment taken into account to influence the discovery process. Therefore, all the technical issues of hetero-geneity and dynamicity put into evidence in the previous section also apply to context awareness, with the further relevant complications deriving from the multiplicity of potential context information sources.

7.4.4 Mobile Discovery Proxies with Location and Context Awareness: CARMEN Case Study

One interesting middleware for mobility-enabled discovery that exploits all three solution guidelines described above is CARMEN, which uses MA-based proxies acting as discovery mediators to transparently requalify bindings to discovered entities depending on both location and context visibility.

CARMEN allows service providers, system administrators, and final users to specify different kinds of metadata in a declarative way at a high level of abstraction. CARMEN metadata influence the dynamic determination of client discovery scope and the adaptation of client bindings to discovered entities without any intervention on the application logic, according to the design principle of separation of concerns. In particular, CARMEN exploits two types of metadata: *profiles* to describe the characteristics of any discovery entity modeled in the system and *policies* to manage migration, binding, and access control (see Figure 7.2).

CARMEN profiles describe users, resources, service components, and sites. In particular, *user profiles* maintain information about personal preferences, interests, and security requirements for any CARMEN registered user. *Resource profiles* report the hardware/software characteristics of client access terminals and discoverable resources. *Service component profiles* describe the interface of discoverable service components as well as their properties relevant for binding management decisions, for example, whether a service component can be copied and migrated over the network. *Site profiles* provide a group abstraction by listing all the resources/service components currently available at one CARMEN host. CARMEN adopts eXtensible Markup Language (XML)–based standard formats for profile representation to deal with the Internet openness and heterogeneity: the World Wide Web Consortium Composite Capability/ Preference Profiles (CC/PP) for user/device profiles [27], Web Service Definition Language (WSDL) for the service component interface description [28], and the resource description framework (RDF) for the site collections of resources [29]. For instance, Figure 7.2 shows the CC/PP-compliant profile for a PalmOS device resource hosting the K Virtual Machine (KVM)/Connected Limited Device Configuration (CLDC)/Mobile Information Device Profile (MIDP) software suite [30]. CARMEN profiles are opportunistically disseminated among discovery entities with a protocol similar to Lanes [1, 14].

In addition to profiles, CARMEN expresses discovery management policies as high-level declarative directives. CARMEN distinguishes two types of policies: *access control* policies to properly rule the determination of discovery scopes depending on security permissions and *mobility-handling* policies to guide discovery decisions in response to run time variations, mainly due to mobility. Mobility-handling policies, in their turn, include two different types of discovery-related policies: *migration policies* specify under which circumstances, where, and which service components have to migrate triggered by user/resource movements; *binding policies* define when and which binding management strategy to apply to update the set of currently accessed service components after any change of client context.

CARMEN supports four different binding management strategies among which service administrators can dynamically choose by specifying a binding policy:

- *Resource movement* states that currently accessed service components are transferred along with their client when it moves.
- *Copy movement* specifies to copy bound service components and to migrate them to follow client movements.

182

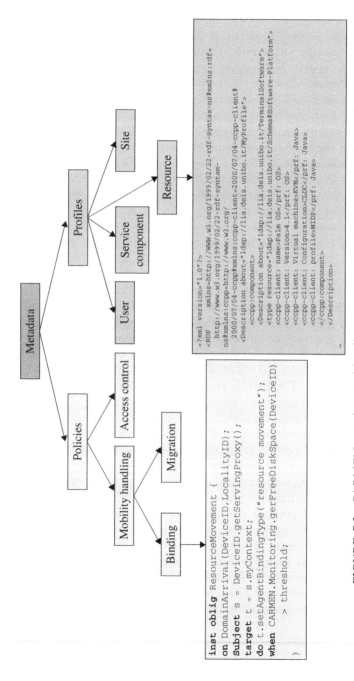

FIGURE 7.2 CARMEN metadata to drive discovery management: taxonomy and examples.

- *Remote reference* automatically modifies client bindings after migration to refer remotely its bound resources/service components.
- *Rebinding* specifies that the CARMEN discovery support, triggered by client movements, has to rebind the client to equivalent resources/service components available in the new locality.

For instance, depending on the chosen binding policy and the current applicable context, a user movement can either trigger the copy of a service component to his or her new access locality or request the reconnection to an equivalent local one, for example, to automatically provide location-dependent service contents. CARMEN adopts the Ponder language for policy specification [31]. In particular, mobility-handling policies are expressed as declarative event−action−condition rules defining the actions that policy subjects must perform on target objects when specific events occur (see Figure 7.2).

The CARMEN discovery solution is centered on the distributed and dynamic deployment of mobile MA-based proxies over the fixed network to play the role of discovery mediators on behalf of usually resource-limited wireless client terminals. CARMEN provides any user, at the start of his on her service session, with a personal mobile proxy, called a shadow proxy, that is responsible for metadata-dependent management of bindings to discovered entities. Shadow proxies usually follow their associated users in their movements among different Secure and Open Mobile Agent (SOMA) domains, carry the reached service state, and make it possible to migrate service sessions dynamically. Other different mobility policies for shadow proxies can be specified and enforced, for instance, to support disconnected asynchronous operations.

Shadow proxies retrieve the profile of characteristics of their companion devices and the profile of preferences of their users at their instantiation. Always at instantiation time, they determine the applicable discovery scopes by including only the discoverable entities of interest for their clients depending on profile metadata, current location, and execution context. Discovery scopes list resource/service component identifiers, tagged as active or passive: A discoverable entity becomes active when the client asks the proxy to access and use it. For any active entity, the shadow proxy also maintains the corresponding resource descriptor, the identifier of the binding strategy to apply, and a reference object that encapsulates the specific mechanism for implementing the associated binding strategy.

Let us finally observe that discovery-related information at clients, such as discovery scopes, is not continuously reevaluated in CARMEN to reduce the overhead introduced by the discovery support: The discovery scope of a client and the binding strategy decisions for its active resources are redetermined by the CARMEN support anytime the proxy changes its allocation, that is, typically anytime the client changes its wireless access point to the Internet; context monitoring is performed only for the indicators that can trigger obligation policies that involve currently active entities. Again, there is the need to partially sacrifice global consistency and freshness of discovery-related information in order to limit the overhead of complex location- and context-dependent discovery management.

7.5   CONCLUDING REMARKS AND "HOT TOPICS" IN CURRENT RESEARCH

The widespread diffusion of mobile wireless devices has resulted in service provisioning scenarios of market interest, where mobile nodes freely move while accessing/offering services from/to other peers and the Internet infrastructure. No assumption about the static knowledge of available resources and service components is realistic. Therefore, mobile computing stresses the need for novel discovery solutions capable of flexibly providing access to currently discoverable entities while preserving possibly limited client capabilities, for example, in terms of power consumption and available bandwidth.

As pointed out in the chapter, a first direction of solution is the adoption of opportunistic discovery announcements/queries, thus sacrificing consistency and completeness of discovery registration data but limiting computational/network overhead. A second guideline is that discovery solutions should also support transparent rebinding in order to avoid the growth of complexity of developing mobility-enabled distributed applications. To trigger transparent requalification operations, rebinding solutions should not only consider the availability of currently accessed discoverable entities. Also in the case of still-available entities, context modifications should stimulate rebinding management in order to maximize application-specific quality metrics, such as local access and load balancing.

Rebinding operations can be long and expensive for resource-limited clients in terms of both computational and traffic overhead. MAs represent a very suitable technology to implement middleware components that assist and follow mobile nodes by performing rebinding on their behalf. Recent MA-based discovery solutions also suggest the exploitation of high-level declarative policies to drive entity rebinding in order to achieve the required flexibility not impacting the application implementation. Most important, the MA adoption in discovery solutions enables a crucial management function that significantly widens the possibilities for systems/service administrators: MA-based service components can be migrated at run time to achieve several possible goals of different nature, for example, to continuously maintain colocality between mobile clients and their accessed service components.

The relevant research results obtained in the presented challenging scenarios, even with high node mobility, are stimulating the development of discovery solutions that fit well other emerging deployment environments, with different and specific characteristics, such as the wide research area of pervasive ubiquitous environments and the most peculiar case of mobile grids.

Future pervasive scenarios will exhibit an enormous density of mobile nodes, resources, and service components, which may continuously and abruptly move/disconnect, by provoking temporary interruptions in connectivity with extremely high frequency. For instance, a user with a personal digital assistant could gather a huge amount of context information (from temperature and humidity data to shop sales and cinema show-on lists) while driving a car on a motorway by interacting with sensors hosted in encountered cars and with context data repositories located on the way, that is, at gas stations. This context information should be collected and processed very rapidly, for example, when the user moves rapidly, in order to be useful to guide the discovery process (also the encountered cars may host discoverable resources and service components of interest, but they are transient too). On the one hand, those scenarios require extremely lightweight discovery protocols to effectively exploit the usually short time intervals of mutual client−resource reachability. On the other hand, they push to the extreme the necessity of flexible solutions to transparently and asynchronously carry on discovery requests and to hide clients from communication unreliability. The MA properties of asynchronicity and full visibility of their execution environment make MAs a promising implementation technology for pervasive scenarios.

A completely different kind of deployment scenarios of recent interest are grids, that is, wide-scale networks composed of nodes typically without strict limitations on hosted resources and willing to share them collaboratively. In particular, in mobile grids, service components can dynamically migrate toward their needed resources driven either by availability reasons (the resources are available on different hosts in different time intervals) or by load-balancing purposes. In addition, these collaborative sharing scenarios may exhibit relevant dynamicity since the set of current participants (and associated resources and service components) may frequently vary. MA-based mediators can represent an efficient way to dynamically rebind to discovered shared entities in response to relevant context changes, transparently and with no impact on client/resource implementations for mobile grids.

The adoption of MA-based solutions in both pervasive and mobile grid environments is still in its infancy (for this reason, we only mention it in this section) and will certainly deserve more attention in future research. The open challenges are primarily two: (i) to support prompt and energy-efficient resource/ service management in order to take into account the extreme dynamicity of pervasive environments and (ii) to dynamically distribute the required know-how only when and where really required (and deinstall it when no longer needed) guided by dynamic resource availability by relying on context-based service migration in mobile grid environments. Therefore, MA-based migration mechanisms and discovery solutions with automatic rebinding support are expected to find future application in those emerging execution environments.

ACKNOWLEDGMENTS

This work was partly funded by the MIUR FIRB WEB-MINDS, the MIUR PRIN MOMA, and the CNR Strategic IS-MANET Projects.

REFERENCES

1. P. Bellavista, A. Corradi, R. Montanari, and C. Stefanelli, *Context-aware middleware for resource management in the wireless internet, IEEE Trans. Software Eng.,* 29(12):1086–1099, 2003.
2. F. Zhu, M. W. Mutka, and L. M. Ni, *Service discovery in pervasive computing environments, IEEE Pervasive Comput.,* 4(4):81–90, 2005.
3. P. Bellavista, A. Corradi, R. Montanari, and A. Toninelli, *Context-aware semantic discovery for next generation mobile systems, IEEE Communi. Mag.,* 44(9):62–71, 2006.
4. K. Arnold, R. W. Scheifler, J. Waldo, A. Wollrath, R. Scheifler, and B. O'Sullivan, The Jini (TM) Specification, Addison-Wesley, June 1999.
5. E. Guttman, C. Perkins, J. Veizades, and M. Day, *Service Location Protocol,* Vol. 2, Internet Engineering Task Force, Request for Comment 2608, June 1999.
6. Y. Y. Goland, T. Cai, P. Leach, Y. Gu, and S. Albright, *Simple Service Discovery Protocol (SSDP),* available: http://www.upnp.org/download/draft_cai_ssdp_v1_03.txt.
7. Bluetooth Specification v. 2.0, Specification Vol. 3, Part B, Service Discovery Protocol (SDP), available: http://bluetooth.com/Bluetooth/Learn/Technology/Specifications/.
8. *Salutation architecture specification,* Salutation Consortium, 1999 (the reference website http://www.salutation.org is no more available; the Salutation Consortium dissolved on June 2005).
9. G. G. Richard III, *Service advertisement and discovery: Enabling universal device cooperation, IEEE Internet Comput.,* 4(5):18–26, 2000.
10. M. Nidd, *Service discovery in DEAPspace, IEEE Personal Commun.,* 8(4):39–45, 2001.
11. L. C. Lee, A. Helal, N. Desai, V. Verma, and B. Arslan, *Konark: A system and protocols for device independent, peer-to-peer discovery and delivery of mobile services, IEEE Trans. Sys Man Cybernet, Part A,* 33(6):682–696, 2003.
12. R. Harbird, S. Hailes, and C. Mascolo, *Adaptive resource discovery for ubiquitous computing,* Paper presented at the Workshop on Middleware for Pervasive and Ad-hoc Computing (Middleware), Toronto, Canada, Oct. 2004.
13. O. Ratsimor, D. Chakraborty, A. Joshi, T. Finin, and Y. Yesha, *Service discovery in agent-based pervasive computing environments, Mobile Networks Appl.,* 9(6): 679–692, 2004.
14. M. Klein, B. König-Ries, and P. Obreiter, *Lanes—A lightweight overlay for service discovery in mobile ad hoc networks,* Paper presented at the 3rd Workshop on Applications and Services in Wireless Networks, Berne, Switzerland, July 2003.
15. D. Chakraborty, A. Joshi, T. Finin, and Y. Yesha, *Toward distributed service discovery in pervasive computing environments, IEEE Trans. Mobile Comput.,* 5(2): 97–112, 2006.
16. L. Capra, S. Zachariadis, and C. Mascolo, *Q-CAD: QoS and context aware discovery framework for adaptive mobile systems,* Paper presented at the IEEE International Conference on Pervasive Services, Santorini, Greece, July 2005.
17. U. Saif and J. M. Paluska, *Service-oriented network sockets,* Paper presented at the USENIX International Conference on Mobile Systems, Applications and Services, San Francisco, CA, May 2003.

18. A. R. Tripathi, T. Ahmed, D. Kulkarni, R. Kumar, and K. Kashiramka, *Context-based secure resource access in pervasive computing environments*, Paper presented at the 2nd IEEE Conference on Pervasive Computing and Communications Workshops, PerCom Workshops 2004, Orlando, FL, Mar. 2004.

19. A. Cole, S. Duri, J. Munson, J. Murdock, and D. Wood, *Adaptive service binding middleware to support mobility*, Paper presented at the 1st International ICDCS Workshop on Mobile Computing Middleware, Providence, RI, May 2003.

20. K. Herrmann, G. Muhl, and M. Jaeger, *A self-organizing lookup service for dynamic ambient services*, Paper presented at the 25th IEEE International Conference on Distributed Computing Systems, Columbus, OH, June 2005.

21. O. Riva, T. Nadeem, C. Borcea, and L. Iftode, *Mobile services: Context-aware service migration in ad hoc networks*, available: http://www.cs.rutgers.edu/pub/technical-reports/technical report dcs-tr-564.

22. A. Fuggetta, G. P. Picco, and G. Vigna, *Understanding code mobility, IEEE Trans. Software Eng.*, 24(5):342–361, 1998.

23. R. Litiu and A. Prakash, DACIA: A mobile component framework for building adaptive distributed applications, Paper presented at the International Middleware Symp. Principles of Distributed Computing, Portland, OR, July 2000.

24. E. Barr, M. Huangs, and R. Pandey, *MAGE: A distributed programming model*, paper presented at the 21st IEEE International Conference on Distributed Computing Systems, Phoenix, AZ, Apr. 2001.

25. O. Holder, I. Ben-Shaul, and H. Gazit, *Dynamic layout of distributed applications in FarGo*, paper presented at the 21st International Conference on Software Engineering, Los Angeles, CA, May 1999.

26. E. Tanter and J. Piquer, *Managing references upon object migration: Applying separation of concerns*, paper presented at the 21st International Conference Chilean Computer Science Society, Punta Arenas, Chile, Nov. 2001.

27. W3 Consortium, *Composite capability/preference profiles (CC/PP)*, available: http://www.w3.org/Mobile/CCPP.

28. F. Curbera, M. Duftler, R. Khalaf, N. Mukhi, W. Nagy, and S. Weerawarana, *Unraveling the Web services: An introduction to SOAP, WSDL, and UDDI, IEEE Internet Comput.*, 6(2):86–93, Mar.–Apr. 2002.

29. S. Decker, P. Mitra, and S. Melnik, *Framework for the semantic Web: An RDF tutorial, IEEE Internet Comput.*, 4(6):68–73, Nov.–Dec. 2000.

30. C. E. Ortiz and E. Giguere, *Mobile Information Device Profile for Java 2 Micro Edition (J2ME): Professional Developer's Guide*, Wiley, Hoboken, NJ, 2001.

31. Imperial College, *Ponder*, available: http://www-dse.doc.ic.ac.uk/Research/policies/ponder.shtml.

8 Distributed Control

JIANNONG CAO

Internet Computing and Mobile Computing Lab, Department of Computing, Hong Kong Polytechnic University, Hung Hom, Kowloon, Hong Kong

SAJAL K. DAS

Center for Research in Wireless Mobility and Networking, Department of Computer Science and Engineering, University of Texas at Arlington, Arlington, Texas

YUDONG SUN

Computing Laboratory, Oxford University, Oxford, England

XIANBING WANG

School of Computers, Wuhan University, Wuhan, China

8.1 INTRODUCTION

Recent years have seen an explosion of interest in wide-area distributed applications running on a global network environment like the Internet. Due to its fast expansion, the Internet has become the platform of choice for many important applications that led to commercial and social activities being available through various services provided on the Internet. On the other hand, people have also seen the potential of the Internet on forming a super-computing resource out of networked computers and research efforts have been made to develop wide-area parallel computing infrastructure that provide access to high-end computational capabilities on the Internet.

Another prevailing technology is mobile computing: ubiquitous access to information, data, and applications. Ubiquitous access refers to the ability for users to access these computing resources from almost any terminal, whether personal or public. The Internet established solid foundations for wide-area ubiquitous computing systems. Further evolution of Internet technologies will

Mobile Agents in Networking and Distributed Computing, First Edition.
Edited by Jiannong Cao and Sajal K. Das.
© 2012 John Wiley & Sons, Inc. Published 2012 by John Wiley & Sons, Inc.

yield a wide-area network based on component-oriented, dynamic applications which will support efficient, scalable resource sharing for a large number of mobile and nomadic users. While mobile users access the Internet from a portable computer, nomadic users may move from terminal to terminal. In either case, ideally a user would be able to accomplish the same tasks with same effort from any location on either his portable computer or any Internet connected terminal. This requires management of distributed data and application resources over a wide area, including automated replication and consistency management.

To deploy a large-scale ubiquitous computing system, a mechanism to manage shared distributed resources over a wide-area, fault-prone network is required. To the end user, the entire network and any terminal attached to it is one large virtual host. The end user does not care how or where the data and applications are stored, because distributed hosts on the fixed network collaborate to provide data and application hosting for individual users. With data scattered over a wide-area network, redundancy of data for high availability, fault tolerance, and security are very important. The system should integrate replication and consistency management of data as a fundamental feature. A wide-area network such as the Internet, especially with mobile hosts, is subject to frequent shifts in topology and network conditions. Such volatility in topology is attributed to (1) frequent changes in the availability of various intermediate network hosts, (2) mobility of mobile hosts such as laptops, and (3) general shifts in network usage patterns which may affect bandwidth and host availability.

The classical architecture for the development of distributed applications involves two parts: a lower layer implementing the communication protocol and a higher layer implementing the actual algorithms of the application while using the functionality of the lower layer. There are two fundamental drawbacks with this traditional architecture: (1) The complexity of the lower layer, that is, the protocol layer increases dramatically as the needs of the higher level application layer increase. (2) The separation between these layers may not be clear, and as a result, distributed applications require different methodologies than those required by centralized applications and are harder to implement and verify.

The mobile agent (MA) paradigm [1] can be used to provide a solution to these problems in distributed applications. In the agent paradigm, a distributed application is broken down into components (usually on a task basis), which are then implemented as agents. Agents contain both computational logic to perform their task and state information needed to complete the task or information gathered while completing the task. These agents are then used to carry out the algorithms of the system and act on behalf of either human users or other agents (as subagents).

An agent does not need to be concerned with lower layer communication protocols. It instead behaves more like a traditional, self-contained, centralized application. All the responsibility of transporting agents, that is their code and state information, and maintaining communication between agent hosts is delegated to an agent system. The tasks an agent performs are within a local context, hence simplifying interfaces and protocols to a level of functions and procedures within an application. This brings about a clear line of separation between the communication protocol layer and the application layer. Furthermore, changes in the needs of applications are localized to the application layer. Additionally, changes in the communication protocol do not have an effect on the application itself.

In this chapter, we present a MA-enabled framework for structuring and building distributed systems and applications over the Internet. The framework is called MAWSG (MA-enabled, Web sever group) and has following characteristics:

1. MAWSG provides an approach to overcome the difficulties that hamper tight interaction between the servers. Because an MA can package a conversation and dispatches itself to a destination host, taking advantages of being in the same site as the peer server, interacting with the peer locally and autonomously making decisions, it allows us to design algorithms that make use of up-to-date system state information for decision making and eliminates unnecessary remote communications [2].

2. MAWSG provides a proactive, adaptive and scalable distributed control scheme, taking advantages of MA in the sense that an MA can encapsulate the distributed control protocol and can be dynamically dispatched according to the current system configuration and state. As such, a mobile agent can be programmed to automatically tolerate transit faults and accommodate dynamic changes of the network.

3. MAWSG can support disconnected operations by letting MAs carry out tasks for a server temporarily disconnected from the network. After being dispatched, the MAs become independent of the creating server and can operate asynchronously and autonomously.

Our approach focuses on realizing fundamental functions in structuring and building distributed systems and applications, including cooperation, coordination, and other distributed control functions using a collection of autonomous, cooperating MAs. Cooperating MAs are a collection of MAs that come together for the purpose of exchanging information or in order to engage in cooperative task-oriented behaviors [3]. In MAWSG, an individual server has a well-defined interface with primitives that can be invoked by mobile agents to access the information about the server, such as central processing unit (CPU) statistics, the memory statistics, and the number of Web requests served and to perform some local operations. Cooperating MAs encapsulate policies and algorithms for their interaction and coordination in order to implement various distributed control functions. In addition to the advantages described above, using cooperating MAs allows us to provide clear and useful abstractions in building network services through the separation of different concerns.

MAWSG demonstrates a new paradigm of distributed system design. For distributed control functions which require tight interactions between peer entities and quick response to system events, cooperative mobile agents are especially useful because they can move closer to the local servers, conduct local interactions, and achieve the coordination as required. The advantage of the MAWSG is that it clearly separates cooperation among servers and the local operations. Our goals of this chapter in designing MAWSG are as follows:

1. Study under which circumstances MA will be more beneficial than message passing or traditional methods in distributed control systems, such as load balancing, checkpointing, and fault tolerance.
2. Evaluate the performance of MA-based methods compared to traditional methods.
3. Design a uniform interface for realizing distributed control.

In addition to exploiting generic techniques, some protocols have been developed for performing load balancing [4, 5], replication consistency management [6], distributed mutual exclusion [7], checkpointing of Web server computations [8], distributed deadlock detection [9], and distributed consensus [10].

The rest of this chapter is organized as follows. Section 8.2 provides the MAWSG framework and its simulator. In Section 8.3, we describe how to use MAs to achieve consensus. Section 8.4 describes how to achieve dynamic load balancing of network services while Section 8.5 presents a MA-based checkpointing and rollback scheme. Finally, Section 8.6 concludes the chapter.

8.2   MAWSG FRAMEWORK

8.2.1   System Model

Figure 8.1 shows the general system model of a network service provided by a collection of servers for various purposes such as information distribution, trading, network management, and computation. Servers can be (partially) replicated and grouped in two cases:

a. *Server Cluster* Servers reside at a single geographical site and are created for performing the same task; this is mainly for reliability and computation speedup (parallel processing) purposes.
b. *Server Group* Servers reside at multiple sites and perform their tasks autonomously but, from time to time, cooperate to help each other; this is mainly for balancing of information/services, reducing service latency, and improving system availability and reliability.

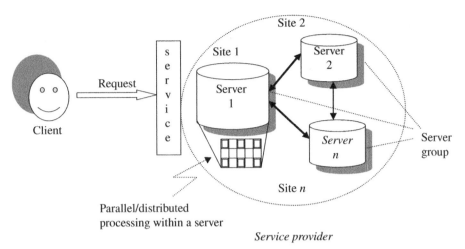

Site 2

Site 1

Server 2

Server 1

Server *n*

Server group

Site *n*

Request

Client

s
e
r
v
i
c
e

Parallel/distributed
processing within a server

Service provider

FIGURE 8.1   Network service and server group.

This chapter focuses on the server group system model, where the network service is provided by a group of servers distributed over a wide-area network, thus providing various kinds of services such as information distribution, network management, trading, and software download.

8.2.2   MAWSG Framework

The MAWSG framework is aimed at designing various distributed control functions for wide-area network server groups. All servers in the group may have the same functionality and carry the same data, but this is not a requirement. In general, we only require that each of the member servers is capable of handling individual client requests; they have partial overlap in their information and functionality and need to cooperate with each other from time to time in order to fulfill clients' requests. Each request arriving at the server group can be executed at a subset of the member servers. The servers are often heterogeneous in terms of the hardware configuration, operating systems used, and storage and processing capabilities. During execution, the capacity of a server changes over time according to the number of requests the server is processing.

We consider building network server groups with the support of MAs. In traditional distributed client/server systems using a message-passing-based approach, the coordination code has to be integrated into the server service model itself. This mixes the functionality of providing services with the service-independent operation for maintaining relationships among the servers and ensuring desirable performance. Whenever a new coordination protocol is used or a new feature is introduced, the server code must be reimplemented. Using MAs allows us to develop a flexible system architecture for Web server groups by having MAs carry out coordination tasks for the cooperating Web servers in the group. This provides us clear and useful abstractions in providing network services through the separation of different concerns. The server site functionality can be separated from the operations for maintaining the logical relationship between group members and providing the desired level of performance, which are realized by a collection of autonomous, cooperating MAs.

Figure 8.2 illustrates the system architecture of MAWSG. Mobile agents have their own identity and behavior and are capable of navigating through the underlying network, performing various tasks at the sites they visit, and communicating with other MAs. Mobile agents are sent by or on behalf of the

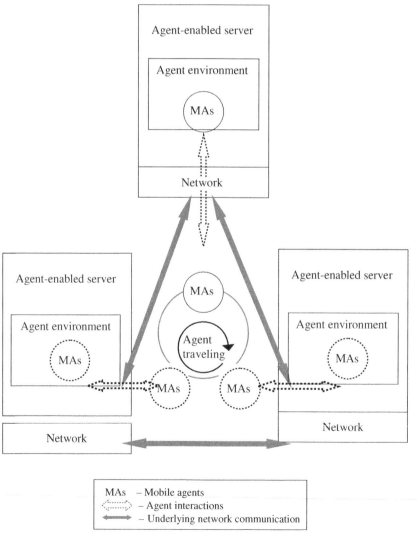

FIGURE 8.2   System architecture for MAWSG.

servers for coordinating their activities. The servers can use both message passing and MAs in their computation.

Mobile agents are capable of interacting with the stationary servers they visited for reading and writing specific data on that server. The individual server has a well-defined interface with primitives that can be invoked by mobile agents to access the information about the server, such as the CPU statistics, memory statistics, and number of Web requests served, and to perform some local operations. Cooperating MAs encapsulate policies and algorithms for their interaction with the servers and coordination with other agents. There will be a network connection between each pair of server hosts, which can be set up when the whole framework is initialized. The agents can be arranged to travel on the shortest path across the sites to collect information and dispatch jobs or dynamically decide their routes according to the current network traffic status.

Figure 8.3 shows the structure of the MA server, where the MA is used as an aid to the servers for achieving their coordination. Distributed control

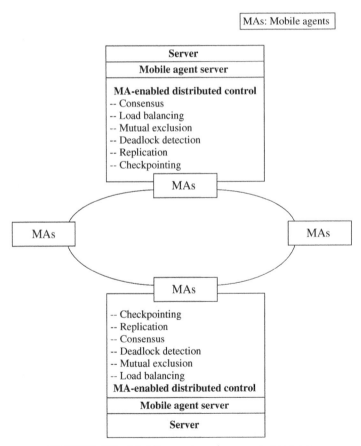

FIGURE 8.3   Mobile agent environment structure.

functions, which maintain logic relationship between group members and provide the desired level of performance, such as checkpointing, load sharing, distributed mutual exclusion, and consensus, are realized by a collection of autonomously cooperating MAs. Since the MA carries the program it executes in its entirety as it propagates through the network, utilizing the information provided by the local servers, a flexible and adaptive coordination scheme could be developed. There are several common issues that need to be addressed in designing an MA-enabled distributed control algorithm:

- *Who owns the MAs and what are the roles of the MAs?* Mobile agents can be created and dispatched by individual servers in the group for their specific purposes. Mobile agents can also be owned by the system and shared by all servers. In the former case, the MAs act as information collectors and/or negotiators on behalf of their owner servers, while in the latter case, MAs can travel around the servers to update their knowledge about the state of other servers and to coordinate their activities.

- *What will an MA do at a visited server site?* An MA can request the local state information from the visited host and collect shared information left by other MAs who has previously visited the host. An MA can also update the local host's knowledge about the other hosts along the traveling path of the MA.

- *What will be carried by an MA?* What kind of data or job information will be carried by an MA in different distributed control algorithms?

- *How can MAs communicate?* Mobile agents can interact with each other and with the server hosts either directly through message passing or indirectly using the *stigmergy* technique—mobile agents interact with the traces left in the environment by one another [3]. By sharing each other's information, unnecessary traversal to remote server sites may be eliminated and thus network traffic is reduced. Also, faster decision making can be achieved.

8.2.3   Simulation Environment

Based on the MAWSG model, we have designed a software environment to simulate and evaluate MA algorithms on the Aglet platform [11, 12]. As shown in Figure 8.4, the architecture of the software environment consists of four modules:

1. The *user interface module* has the functionality to enable the user to configure the simulation session and then start it. By sending the commands through the Session Control component, the user can set the parameters, including the number of simulated hosts, network topology, and network bandwidth. User Interface also provides support for the user to monitor the execution condition of the whole system (by

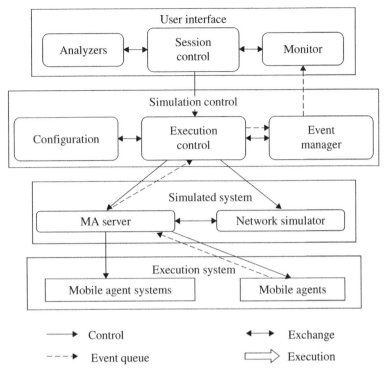

FIGURE 8.4   Architecture of MA group simulation environment.

Monitor) and to show the information about the performance evaluation of the mobile agents (by *Analyzer).*

2. The *simulation control module* is composed of Configuration, Execution Control, and Event Manager. Configuration can retrieve and save the information of simulation configuration. Simulation configuration consists of MA server configuration and network configuration. The former is about the system environment configuration mainly on the conditions of a simulated host such as CPU capacity, memory constraint, and workload. The latter focuses on the simulation of the underlying network, including network topology, bandwidth of the links, and network traffic. After finishing the environment configuration, Execution Control will place the agents to each logical host for execution and then launch the simulation. During the execution of the simulation, it will receive the events from the Simulated System module and forward the events to Event Manager.

Event Manager gathers the events sent by Simulated System during the simulation and forwards them to Monitor. The events from the simulators sent to Event Manager are usually interleaved and thus disordered. It is Event Manager's task to sort the incoming events and send a set of ordered event queues to the monitor and analyzer, which would extract the information about the performance of the MAs in the events.

3. The *simulation system module* largely implements the MA server model and Network Simulator involved with the simulation of network topology and communication cost. The network topology and the bandwidth are configured by the user, and the real-time communication cost is generated by using the specified network simulation algorithm given in the configuration.

4. In the *execution system module*, the MA algorithm implemented by Mobile Agents is the target of the evaluation. In the simulation environment, each MA is associated with a control object whose tasks are to generate the event when the agent is going to take a *behavior* action, send the event to the MA server's Controller, and resume the execution of the MA once it receives the commands from the Controller. As a part of the MA being transparent to the user, the control object is created within its owner agent and accompanies the agent during the agent's lifetime. In this way, the behavior of the MA can be recorded and the Event Manager module can obtain the details of the simulation execution of the agents in the MA algorithm to evaluate the performance.

In the following sections, we describe the design and simulation of distributed algorithms for various dynamic control functions in MAWSG.

8.3 MOBILE AGENT–ASSISTED SCHEME FOR DISTRIBUTED CONSENSUS

Consensus is a fundamental problem in distributed systems. It states that there are a set of n processes $\{p_1, p_2, \ldots, p_n\}$, each process p_i initially proposes a value v_i, and all nonfaulty processes have to decide on one common value v which is equal to one of the proposed values. A process *fails* if it behaves abnormally, for example, by crashing. A process is *correct* if it never fails. More precisely, the consensus problem is defined by the three following properties [13]:

- *Termination* Every correct host eventually decides on some value.
- *Validity* If a host decides v, then v was proposed by some host.
- *Agreement* No two correct hosts decide differently.

The agreement property applies to only correct hosts. Thus, it is possible that a host decides on a distinct value just before crashing. Uniform consensus prevents such a possibility. It has the same termination and validity properties but the following agreement property:

- *Uniform Agreement* No two hosts (correct or not) decide differently.

Consensus has been extensively studied over the last two decades both in synchronous and asynchronous distributed systems [14–16]. In a synchronous system, message delays and speeds of relative processes are bounded and these bounds are known. In contrast, none of these bounds exist in the asynchronous system. Fischer et al. [15] proved that consensus cannot be solved deterministically in an asynchronous system that is subject to even a single crash failure. In this chapter, we consider the uniform consensus problem for synchronous distributed systems.

8.3.1   Mobile Agent–Based Consensus Algorithm

To evaluate the MAWSG framework, we consider the consensus problem in synchronous systems with crash failures [10]. Here, the hosts can be the web servers. The design of our proposed algorithm uses the *pigeon hole principle*. Initially, $t+1$ hosts are randomly chosen, called *coordinating host* (*CH*), where $t < n$ is the maximum number of hosts that can crash and n is the number of all processes. We assume that every host has a unique identity over the network and $t+1$ hosts are randomly chosen as CHs. For simplicity, we choose first $t+1$ hosts, whose identities range from 1 to $t+1$, as CHs. Therefore, there is at least one CH which never crashes. We call such a CH a *correct* CH. Because there is at least one correct CH and correct CHs never destroy MAs, the information obtained from them is the most complete and reliable.

Our algorithm is round based. In each round, an MA is dispatched from each host to every live CH. On each live CH, the MAs exchange information with each other and then return home. On returning home, a stationary agent, called the *master agent*, performs calculations on the value set and a flag of every MA to obtain the information that indicates whether the algorithm can terminate or not. If execution of the algorithm is not yet complete, another round starts. Otherwise, the algorithm terminates, thus making decisions based on the current value set, and also all MAs stop their executions. The decision on termination is made as follows. Every MA carries a flag, Altered_Flag, which indicates whether some new values are added into its value set during last traveling. The flags of all agents created by a host will be combined after they return home. If there exists at least one agent with its flag set to true, the host will start the next round. Note that setting a flag to true can also reflect that some values were deleted by the intersection calculation at the previous round.

Algorithm 1 Executed by a Mobile Agent	Algorithm 2 Executed by a Master Agent
1. *Initialization*	1. *Initialization*
2. define $f_{i,j}$ as Altered_Flag of $A_{i,j}$	2. $V_i := \{v_i\}$
	3. $A_i := \emptyset$
3. *Algorithm*	4. for each CH_j, create_agent($a_{i,j}$), init($a_{i,j}$), $A_i := A_i \cup \{a_{i,j}\}$
4. while (alive)	
5. wait_signal() //wait signal from its creating host	5. *Loop*
	6. repeat
6. $V_{i,j} := V_i;$	7. for each $a_{i,j} \in A_i$, signal($a_{i,j}$)
7. migrate_to(CH_j)	8. if itself is a Coordinating Host then
8. $V_{CH_j} = V_{CH_j} \cup V_{ij}$	9. wait_timeout(Ts) // all agents migrate to it can arrive
9. wait_signal() //wait signal from the CH	
	10. signal all arrived agents
10. $f_{i,j} = (V_{CH_j} \neq V_{ij})$? true : false	11. wait_timeout(Ts) // its all agents can return
11. $V_{i,j} = V_{CH_j}$	12. else wait_timeout(2Ts) // its all agents can return
12. migrate_to(h_i) // return home	
	13. end if
13. end while	14. $A_i := \{$all agents that came back$\}$
	15. if Ai = $\emptyset$ then $f_i :=$ false
	16. else
	17. $V_i = \bigcap_{a_{i,j} \in A_i} V_{i,j}$
	18. $f_i = OR_{a_{i,j} \in A_i} f_{i,j}$
	19. end if
	20. until $f_i =$ false
	21. $A_i := \emptyset$ // delete all agents
	22. $d_i := \min(V_i)$

Algorithm 1 is executed by an MA, $Agent_{i,j}$, on behalf of host h_i traveling to CH_j; $V_{i,j}$ denotes the value set of $Agent_{i,j}$ and V_h denotes the value set of a host h; T_s denotes the maximum timeout that an agent migrates between any two hosts; $f_{i,j}$ denotes the Altered_Flag of $Agent_{i,j}$. This algorithm terminates as soon as the agent is destroyed by its home host (see explanation for Algorithm 2).

In each iteration of the loop, the MA $Agent_{i,j}$ waits for the signal from its home host. After receiving the signal from its home host, it copies the value set maintained by the master agent and migrates to CH_j. After arrival at CH_j, the agent gives its value set to CH_j. Then it waits for a signal from CH_j. When it has received the signal, every agent which is dispatched to CH_j must have arrived at CH_j, and V_{CH} will contain complete values of all arriving agents. Agent$_{i,j}$ then compares V_{CH} and the value set $V_{i,j}$ and sets the flag $f_{i,j}$ to indicate whether they are different. Finally, the agent copies the value set of CH_j and returns to its home host.

Algorithm 2 is executed by the master agent at each host h_i. The master agent first initializes the variables. There are a total of $t+1$ CHs, namely CH_0, $CH_1, \ldots CH_t$. Initially V_i contains its proposed value, v_i. The master agent maintains an agent set, A_i, for holding a mobile agent created by itself. Initially it creates and initializes $t+1$ MAs, namely, $a_{i,0}, a_{i,1}, \ldots a_{i,t}$. All these MAs are added to the set A_i.

Then the master agent sends a signal to every MA in the set A_i. The MAs will migrate out of host h_i after receiving the signal (Algorithm 1, lines 6–7). If the master agent is on a coordinating host, it waits for timeouts for all agents migrating to it to arrive, then signals these agents to collect information and return home. Finally it waits for another timeout for all its agents returning. Otherwise, the master agent just waits for two timeouts for all its agents returning.

After returning, the agent will stay at host h_i until it receives another signal from the master agent of its home host. If no agents return, which means all live CHs have made a decision and terminated, the master agent will set f_i to false to exit the loop and make a decision. Otherwise, it calculates the intersection of value sets of all agents in the set A_i and store the resulting set into its own value set V_i. It applies the OR operation on Altered_Flag variables of all agents in A_i and stores the result into the local variable f_i.

If f_i is true, implying that some agent in A_i reported that the values collected from its CH are not consistent, the master agent will go back to line 7 to start a new round. If not, the loop exits and all MAs will be destroyed, and the decision is made according to the current value set of host h_i.

8.3.2   Performance and Simulation

The proposed algorithm achieves uniform consensus and its upper bound is $\lceil (t+1)/2 \rceil$; the proof can be found in [10]. In the presence of up to t crash failures, uniform consensus can be solved by our algorithm within $\lceil (t+1)/2 \rceil$ rounds. The meaning of a *round* notion in this chapter is different from traditional message-passing consensus algorithms in both synchronous and asynchronous systems, in which each process executes sequentially the following steps in each round r [13]: (1) It sends a round r message to the other processes, (2) it waits for a round r message from the other processes, and (3) it executes local computations. The time cost of a round in our proposed algorithm is nearly double the time cost of a round in traditional synchronous consensus algorithms [13]. Our proposed algorithm can stop early as in [14], and uniform consensus can be solved by our algorithm within $\lceil (f+1)/2 \rceil + 1$ rounds according to the number of failures *f* that actually occur, where $f < t$. Unlike traditional early-stopping consensus algorithms in synchronous systems, such as the scheme in [14], where every process needs to broadcast a *decide* message after it has made a decision, our proposed algorithm does not need to do this.

In the simulation, we consider two types of costs: time and number of dispatched mobile agents. In the previous section, we discussed these costs for the best and worst cases. The experiments aim to find average costs. We consider an even distributed random model for the number of hosts that can fail. In other words, supposing *t* is the maximum number of hosts that can fail, the actual number of hosts that fail in a series of executions of the algorithm is evenly distributed in the closed interval $[0, t]$.

In simulation, MAs are simulated as Java threads. The main thread of the simulation application is in charge of simulating crashes. After initialization, it sleeps for a while and randomly chooses a host object and calls it "down" by invoking its crash() method.

The following properties may affect the performance of the algorithm: (1) the *actual* number of hosts that crash and (2) the frequency of crash events. We analyze two aspects of performance, namely, execution time and total number of agent migrations.

8.3.2.1 Number of Migrations to Actual Crashes
The simulation environment is comprised of 20 hosts, up to 19 of which can crash. In other words, the number of actual crashes is in the range of $[0, 19]$. The simulation is executed 5000 times. This is enough to generate multiple instances for each possible number of actual crashes. Figure 8.5 shows the results obtained from the executions.

We plot the graph of number of migrations (*Y* axis) against actual crashes (*X* axis) in Figure 8.5.

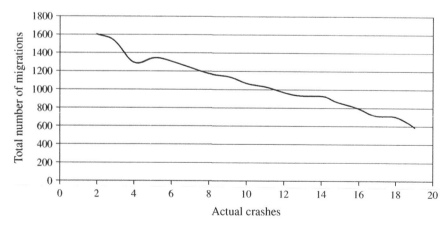

FIGURE 8.5 Number of migrations versus actual crashes.

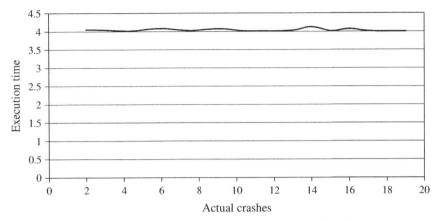

FIGURE 8.6   Execution time (ms) versus actual crashes.

We found that the more hosts crash in the network, the less migrations occur. This is reasonable because the crashes can destroy MAs. More crashes will destroy more MAs. This leads to a decrease in total migrations.

A more important implication of the results illustrated in the graph is that the network loading (reflected by the total number of migrations) decreases if more hosts crashed. That is, crash does not cause unwanted effects in network loading.

8.3.2.2   *Execution Time to Actual Crashes*

Figure 8.6 shows average execution time as a function of the number of crashes. From the figure, we found that the execution time is not affected very much by the actual number of crashes, which is different from our expectation; just two rounds are needed for all hosts to make a decision. The reason for this is that, during a round, the effect on the algorithm's execution time by a single crash or multiple crashes within a partition is the same. The probability of the occurrence of the worst case is too small, and we cannot reach this case by executing the simulation thousands of times.

The experiments illustrate how performance is affected by the properties of the network. These properties are determined by factors such as the number of actual failed hosts and frequency of the crashes.

8.4   MOBILE AGENT–ASSISTED DISTRIBUTED DYNAMIC LOAD BALANCING

Load-balancing protocols are designed for servers in a distributed system to balance their workload. The purpose is to avoid performance degradation caused by a high load imbalance. This is achieved by letting the servers cooperatively monitor the global system load information and distribute/redistribute the load among the servers whenever necessary. However, most of these algorithms are designed for closely coupled distributed systems [7].

Load-balancing algorithms can be *static* or *dynamic* and can use either *centralized* or *distributed* control [18, 19]. In static load-balancing algorithms, decisions about the allocation of requested jobs to the sites are made before execution. This is not practical for some system and application environments if the information about the requests to be executed (execution time) and the execution requirements (current load, transfer times, and communication costs) cannot be known in advance. Also, the static approach suffers from the weakness that it does not take the fluctuation of the system load into account. In dynamic load balancing, the system attempts to find the best server for an incoming request at run time. Decisions are made using the information on the current system states. However, this approach requires gathering and maintaining system state information because in order to achieve load balancing a snapshot of the global loading information in a specific time frame must be acquired.

A load-balancing algorithm can be adaptive or nonadaptive depending on whether only information on the current system states is used or both the previous behaviors of the system together with the current state are used to make decisions [20]. An adaptive algorithm can modify its policies and parameters used in the algorithm according to the previous and current system behaviors. Furthermore, load-balancing algorithms can be classified according to who will initiate the load distribution process (source initiative or destination initiative) and the way jobs will be allocated (e.g., preemptive or nonpreemptive). If the overloaded server is responsible for finding other servers to balance its workload, the strategy is called a source initiative. On the other hand, if a lightly loaded server looks for requests to process from overloaded servers, the strategy is called destination initiated [21]. In preemptive allocation, a job can be reallocated even after its execution starts, while in nonpreemptive allocation, once a job starts execution, it cannot be selected for reallocation [22].

Our framework MAWSG provides a novel MA-enabled load-balancing mechanism on distributed Web servers [5]. A server can dispatch MAs to the system when required. After leaving the source server, a mobile agent becomes an independent object that is autonomous in action. Mobile agents can embed optional policies of load balancing and travel from one server to another to interact with the servers on the site. The on-site interaction can acquire the latest load information. The optional policies can deal with the load-balancing requirement according to the states of the servers and the client requests. The MA-based approach can minimize the network traffic and enrich the flexibility of the load-balancing mechanism. Multiple MAs can exist simultaneously in the system and perform asynchronous operations. They can strengthen the scalability and availability of distributed Web servers. The MAWSG framework can specify the behaviors of mobile agents and the policies of load information gathering and load distribution to implement efficient load balancing on varied Web server systems from a local-area cluster and a wide-area system.

8.4.1 Mobile Agent–Enabled Load-Balancing Scheme

The design of a load-balancing mechanism for network services should include the following policies:

- *Information-Gathering Policy* Maintains the information about the workload at the servers. The policy is made up of two components: frequency of information exchange and method for dissemination of the information. There is a trade-off between having accurate information and minimizing the overhead. It also includes the estimation and specification of workload, for example, processor load, length of queue, and storage utility.
- *Initiation Policy* Determines who initiates the process of load balancing. The initiator can be the source server, the destination server, or both (symmetric initiations).
- *Job Transfer Policy* Decides when the initiator should consider reallocating some requests to other servers. The decision can be made based on only the local state or by exchanging global processor load information.
- *Selection Policy* Determines which particular job to reallocate. Nonpreemptive policies select tasks from the set of jobs which are yet to begin execution. Preemptive policies expand this set to include all jobs located at the processor.
- *Location Policy* Determines to which servers the jobs should be reallocated. The simplest location policy is to choose a server at random. More complicated policies use negotiation, where the initiator negotiates with each member in a subset of servers.

These policies must be represented and implemented in appropriate system components. A key issue common to all the components is how to define the workload at a server. Estimation of workload should consider factors including job characteristics and processor and storage utilization at the site.

Figure 8.7 illustrates the structure of the load-balancing mechanism at a server of the network server group. It contains three major component agents, namely the server module agent (SMA), the load information agent (LIA), and the job-dispatching agent (JDA). These agents cooperate to perform load balancing for the local server.

8.4.1.1 Server Module Agent
This is a stationary agent that interacts with the server's functionality. It accepts and serves the incoming job requests, and when the server becomes overloaded, it will create a JDA to trigger the job selection/migration process. The SMA is the largest component in the system and can be decomposed into four independent submodules:

- **Job Generator (G)** Receives incoming requests and inserts them into the job queue.

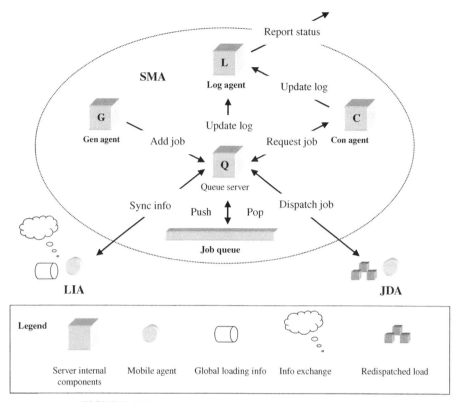

FIGURE 8.7   System components and their interactions.

- **Job Processor (J)** Processes the jobs located in the server job queue by submiting them to the server.
- **Log Agent (L)** Maintains the log of the server activities and provide an interface between the server module and the external system.
- **Queue Sever (Q)** This is the most critical submodule in the SMA. Its functions include job queue management, message queue management, storing of the global loading information, and triggering of the job-dispatching agent to redispatch a request to some other server.

In our design, we use a source initiative initiation policy, that is, the load-balancing process is initiated by a heavily loaded server who needs help. The SMA at a server site implements the job transfer policy and the selection policy. It needs to monitor and measure the local workload to decide when to reallocate a job to some other servers. Once it decides that the condition for job transfer becomes satisfied, the SMA needs to select a suitable job for reallocation from the local job queue.

The transfer policy used in the SMA is a threshold policy: Job reallocation is triggered if and only if the local workload exceeds some threshold T.

The threshold can be either static or dynamic. In the former case, it is set with a predefined value, while in the latter case, it is defined according to the current global load status, for example, the average of the global workload calculated from the local copy of the global load information stored at a site. Once the threshold is exceeded, an attempt is made to select an appropriate job from the job queue and transfer it to another server. Several strategies can be used for the selection policy. In the simplest way, the SMA continuously removes the last job in the local job queue and insterts it into a job reallocation list until either the local workload becomes less than the threshold or there is no destination server available for reallocation. A more complex strategy considers the time spent on transferring a job to a remote server. A job from the job queue of the local site A is reallocated to a remote server B if $t_{w(A)} > t_{m(AB)} + t_{v(B)}$, where $t_{w(A)}$ and $t_{w(B)}$ are the expected waiting times spent by the job at server A and server B, respectively, and $t_{m(AB)}$ is the time for the JDA to migrate from server A to server B. As a special case, we can consider the distance between the client and the server where the client's request is to be processed, so that access to the nearest server can be implemented whenever appropriate.

The job reallocation list will then be passed to a newly created job trasfer agent for relocating the jobs.

8.4.1.2   Load Information Agent

This is an MA that implements the information-gathering policy component of the load-balancing mechanism. It continuously travels through the servers to collect the global information about the workload and resource utilization at the participated servers. Each server site stores its own copy of the global workload information, which will be used for the SMA to make decisions on job reallocation. The global load information will be maintained by the LIAs shared by all the servers in the system. An LIA presents itself at a server site and obtains, in real time, the accurate, current load information at that site.

Figure 8.8a depicts a single LIA traveling from one site to another continuously, carrying updated load information of previously visited server sites. At each site it visits, an LIA synchronizes itself with the global load information stored at the server. When it arrives at a server site, it will first update its knowledge about the local workload at the site. Then it will retrieve the local copy of the global load information stored at the site, which is in the format of a table containing one entry for each server in the group, and synchronize it with the information collected from the sites it previously visited. As shown by the example in Figure 8.9, the synchronization is done by using the timestamp associated with each entry of the load information table, which represents the local time of a server when its workload information is last measured and collected.

Using only one LIA to move among all server sites to collect and maintain global workload information may result in delay in updating the load information tables at individual server sites. This delay will have impact on the effectiveness of load balancing because the timeliness and information coverage

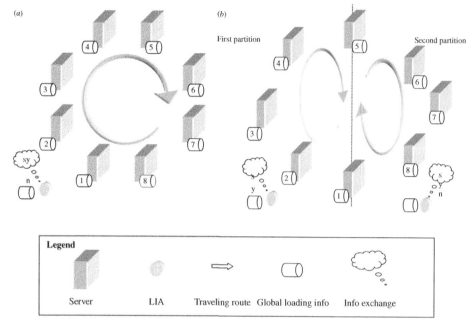

FIGURE 8.8   LIA traveling through network to maintain global load information.

Sever	Agent		Server	Agent
S1 info: t1	S1 info: t3		S1 info: t3	S1 info: t3
S2 info: t3	S2 info: t2	→	S2 info: t3	S2 info: t3
S3 info: t4	S3 info: t9		S3 info: t9	S3 info: t9

FIGURE 8.9   Synchronization of global load information at a server.

of the global workload snapshot are the most critical factors in determining system performance. Without up-to-date global load information, it is unlikely that an accurate decision can be made for balancing the heavy workload of an overloaded server. To overcome this difficulty, we can deploy multiple LIAs, each propagating through a portion of the server group. Figure 8.8b shows an example of using two LIAs which circulate in its own directions along each half of the overall server group, respectively. Then LIAs can cooperate to maintain the global load information by exchanging their partial results at the over-lapping servers and assembling them into the global snapshot. Using multiple LIAs increases the frequency of updating the global load information stored at each server site and is expected to lead to better performance. In fact, LIAs can be injected into the network dynamically. Since the MA can monitor and sense the changes in the environment, when an LIA observes that more LIAs have been dispatched, it will adjust its itinerary accordingly.

8.4.1.3   Job-Dispatching Agent

The JDA is an MA that implements the location policy. When the server is over-loaded, a clone of the JDA will be created by the SMA. When the JDA is activated, using the global loading snapshot stored at the local server, it will carry the jobs in the job reallocation list to the appropriate remote servers for execution.

The chaotic nature of a distributed wide-area environment may cause excessive loading fluctuation of the destination servers, which can consequently lead to server thrashing. As an attempt to prevent this from happening, conventional message-passing-based load-balancing methods implement an additional accept policy which involves a series of negotiations between the job origin site and the candidate destination sites, including a decision-making process on the acceptance/refusing of the job reallocation request. These operations require exchanging messages between the servers and will increase the network traffic, add complexity to the design of load-balancing algorithms, and yet not guarantee a desirable performance.

Using the JDA greatly simplifies the problem. Initially, the JDA will use the global load information stored at the local server site to decide which node to go for load distribution. The JDA presents itself at the receiver site and negotiates with the receiver server locally. There is no need for the sender site to wait for acknowledgment from the receiver as the JDA gets the acknowledgment on behalf of it. In the case where the JDA carries the job to a destination server and finds that the server became overloaded, the JDA can make a decision on the fly to find another suitable server by using the current system state information collected while it travels through the destination servers.

8.4.2   Preliminary Evaluation

We have carried out a preliminary study to evaluate the performance of the proposed distributed dynamic load-balancing algorithm. Preliminary experiments were set to measure the effectiveness of the MA-enabled load-balancing scheme in comparison with a server group system with no load balancing. Performance measures such as the average queue length at each server and the average throughput are used for the evaluation. Random-number generators are used to generate the job interarrival time and the job service time; both follow an exponential distribution. In the preliminary simulation, for simplicity, the workload at a server is defined as the length of the job queue, which represents the number of jobs in the queue. The threshold for dispatching a JDA was predefined between each simulation. We assume that each server can process one request at a time. Only one LIA is dispatched for maintaining the global load information.

Figures 8.10 and 8.11 illustrate the effect of load balancing on the average queue length, which reflects the variance of loads among servers, and on the average throughout a five-server group. In the figures, the first pair of bars compares the performance of the server group without and with load balancing when the job interarrival time for the servers is 2 sec. Similarly, the second pair

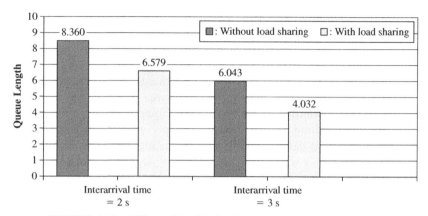

FIGURE 8.10 Effect of load balancing on average queue length.

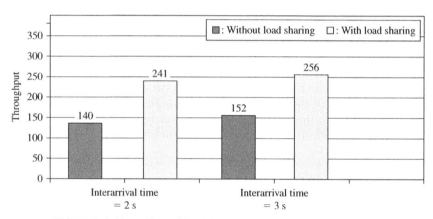

FIGURE 8.11 Effect of load balancing on average throughput.

of bars compares the performance of the server group without and with load balancing when the job interarrival time for the servers is 3.

Figure 8.12 shows the time saving due to load balancing. By time saving we mean on average the reduction in job waiting time. We observe that the average amount of time saved increases with decreasing interarrival rate. This can be explained as follows. If all the servers are very busy, it is less likely for a JDA to find a helper server. This value is also dependent on the number of visits by a JDA agent. As JDA takes time to travel around, more visits will reduce the performance gain brought by job reallocation.

As we can observe from the graphs, the MA-enabled load-balancing algorithm is effective as it improves performance compared to the no-load-balancing case, in terms of decreasing the average job waiting time and increasing the system throughput.

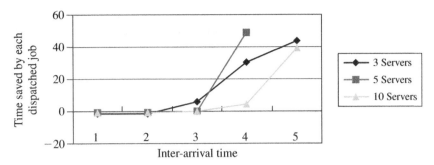

FIGURE 8.12 Average saving in waiting time.

8.5 MOBILE AGENT–ENABLED SCHEME FOR CHECKPOINTING AND ROLLBACK

Rollback error recovery is a general, powerful approach to eliminating transient errors in a system. The technique requires that a system records its state (called checkpointing) periodically during normal operation and, upon failure, restore a previous consistent state (called rollback) and restart the execution from the restored state [23, 24]. Since processes in a distributed system do not share memory, a *global state* of the system is thus composed of a set of *local states* of the processes in the system and the system state must be checkpointed distributively over all the processes. A *local checkpoint* is a saved copy of an earlier local state of one process. A *global checkpoint* of the system is a set of local checkpoints, one for each process. We say that a global checkpoint is *consistent* if the set of the local checkpoints forms a consistent global state. With an inconsistent global checkpoint, under some scenarios, cascading rollback propagation may force the system to restart from the initial state, causing the occurrence of the undesirable *domino effect* [25].

Existing distributed checkpointing and rollback algorithms are almost all based on using message passing for coordinating local process activities. This section proposes a novel approach, called MACR (mobile agent–enabled checkpointing and rollback), which uses MAs to enhance the design of distributed checkpointing and rollback algorithms [8]. In MACR, MAs act as messengers and/or monitors that travel over the network from site to site and facilitate the coordination of the distributed computation processes to carry out their checkpointing and rollback activities.

8.5.1 The Mobile Agent–Enabled Scheme

The checkpointing algorithm works as follows. Periodically, each process takes its local checkpoints independently according to its own needs. For example, a process may take a new checkpoint after t local clock ticks elapsed or after sending out k messages. The main advantage of the independent checkpointing methods is that less communication overhead is incurred during normal operation when taking checkpoints, but it may result in a domino effect upon recovery. To prevent cascade rollback, an MA, called the *coordinator agent*, is used to enforce that the number of checkpoints to rollback in case of recovery will not exceed the predefined threshold. The coordinator MA travels from one site to another, carrying updated information of previously visited sites. When it arrives at the process p, it will first read all the rollback information saved at p and then uses these to update the dependency table so that the corresponding entries will contain the necessary information about dependency between checkpoints of p and q.

Based on the checkpoint dependency information, the coordinator agent will calculate the number of rollbacks that need to be performed by the current process if a fault is detected at that moment. If that number is greater than the predefined threshold value, MaxRecovery, a coordinated check-pointing procedure will be performed to remove the possibility of a domino effect. The calculation proceeds like this. Starting from the most recent checkpoint of the current process under visit, Cp,y, until reaching the checkpoint immediately following the last consistent checkpoint Cp,x, where $x < y$, the dependency relationship is tracked to see whether Cp,y can belong to more recent consistent global checkpoint. If not, it moves to the immediately preceding checkpoint. This continues until either a more recent consistent checkpoint has been found or the last consistent checkpoint has been reached. The number of rollbacks that need to be performed by the process equals the distance between the current checkpoint and the checkpoint where the calculation is terminated.

For each checkpoint, a semiConsistentCPNO vector is used to store the partial information of the new consistent global checkpoint under construction. Each entry of this vector is a pair of values representing the ID of a site and the number of its current checkpoints under testing. Once semiConsistentCPNO proves to be a consistent global checkpoint, it will be used as the recovery line of the system. To maintain this vector, the site ID and the checkpoint number under consideration are first inserted into the semiConsistentCPNO. Then, if the checkpoint number in the corresponding entry (site ID, checkpoint number) in the dependency table is greater than that in semiConsistentCPNO, the checkpoint number is updated in the vector. With this newly updated checkpoint number, the process continues until all information has been updated and no more entries can be found in the dependency table which had a checkpoint number greater than the entries in the semiConsistentCPNO. Then the new recovery line has been found. If the value had not been found from the table, it means that consistency cannot be achieved yet and, therefore, this semiConsistentCPNO is not consistent and should be aborted.

Whenever, upon visiting a site, the coordinator agent detects that the number of rollbacks exceeds the threshold value, a forced coordinated check-pointing procedure will be initiated. The coordinator agent first generates a group of ConsistentCP agents, one for each site, and dispatches them to their corresponding sites. Algorithm 2 describes the operation carried out by a

ConsistentCP agent. When the ConsistentCP agent arrives at the remote site, it will monitor the local process to see whether there are any messages sent after its last checkpoint. If so, it will request the local process to take a forced checkpoint. Before the new local checkpoint is taken, the local process is not allowed to send out any new messages. After taking a forced checkpoint, the ConsistentCP agent sends back a complete_message to the coordinator agent. The coordinator agent will then send start_message to all remote sites and processes start the normal execution again.

Next we describe the recovery algorithm. When a process p recovers from a failure, the agent associated with the process, called CPMaster, initiates a rollback recovery. Agents associated with the processes that are in the global recovery line will be informed and need to participate in the rollback. The coordinator agent stores the recovery line at each site it visited. When the CPMaster initiates a rollback recovery, if the coordinator agent is in the middle of calculating the most up-to-date recovery line, it will inform CPMaster to wait. When the CPMaster agent receives the global recovery line, it will send a recovery_message to each site involved with the recovery, which contains the latest consistent checkpoint number of that site. Upon receiving recovery_message, the participating process will roll back to the consistent checkpoint. Then it will send back a recovery_complete message to the CPMaster agent which in turn waits until a recovery_complete message is received from all participating processes. Then, CPMaster will send restart_message to all the participating sites so they can restart their execution.

A common approach to garbage collection is to identify the recovery line and discard all information relating to events occurring before that line. To reclaim the space from the message receiving tables and local checkpoints, the coordinator agent may decide what data can be discarded. After obtaining the most up-to-date recovery line, any message header files and local checkpoints recorded before the current recovery line are deemed useless and can be safely garbage collected.

8.5.2 Performance Evaluation

The MA-enabled checkpointing and rollback algorithms proposed in this section have been implemented in the MAWSG simulator. An evaluation study has been carried out to demonstrate the effectiveness of the proposed algorithms. Experiments were set to measure the actual number of rollbacks upon failure, the probabilities for the actual number of rollbacks by each process to exceed the threshold Max_{rb}, the probabilities for delay in the detection of global recovery lines, and the number of forced checkpoints. In all experiments, each process takes 50 local checkpoints independently at random times. The threshold of the maximum number of rollbacks allowed for each process is set to be 2, 4, 6, 8, and 10, each used for 20 runs. Processes are randomly chosen to terminate to generate scenarios of failure and recovery.

Figure 8.13 compares the maximum number of checkpoints rolled back for a system failure with the traditional independent checkpointing method. The

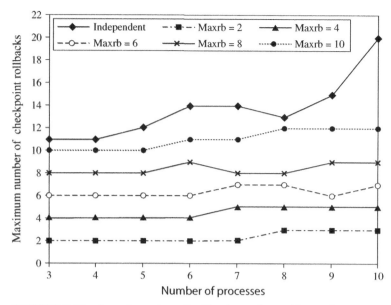

FIGURE 8.13 Actual number of rollbacks performed by each process.

results show that the hybrid approach performs better and is effective. From the figure, it can be observed that, for the independent checkpointing method, the maximum number of checkpoints rolled back upon failure increases significantly as the number of processes increases. In our hybrid algorithm, the maximum number of checkpoints needed to roll back is guarded by the user-defined parameter Max_{rb}.

Using only one coordinator agent may result in delay in the detection of recovery lines. Due to the delay in forming the recovery line, the actual number of rollbacks can be larger than Max_{rb}, but not significantly, as observed from Figure 8.13. Figure 8.14 depicts the results of the experiments measuring the probability for the actual number of checkpoints rolled back to exceed Max_{rb}. Even with 10 processes, the average probability of the actual rollback performed by each process is less than 0.09. Figure 8.15 illustrates the probability that, upon recovery, the actual recovery line used is not the most recent recovery line that can be possibly identified if no delay is incurred. The results are obtained from experiments with $Max_{rb} = 4$ with the number of processes ranging from 3 to 10. We can see that on an average, less than 10% of the recovery lines used for the restart agent exceed the specified maximum.

The last set of experiments was performed to measure the overhead of the proposed algorithms in terms of taking additional forced checkpoints. We set $Max_{rb} = 2$ and the number of processes ranges from 3 to 10. From Figures 8.16 and 8.17, it can be observed that the number of forced checkpoints in the system is not closely related to the number of processes. This is because

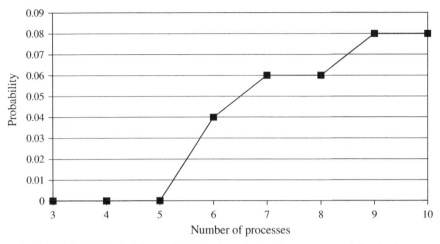

FIGURE 8.14   Probability of checkpoint rollback beyond user-defined Max$_{rb}$.

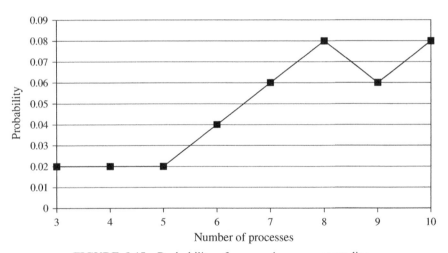

FIGURE 8.15   Probability of nonmaximum recovery line.

the forced checkpoints taken highly depend on the normal checkpointing patterns of the local processes.

8.6   CONCLUSION

In this chapter, we have proposed a novel MA-enabled framework, called MAWSG, to design distributed system functions for MA-enabled high-performance Web server groups. By high performance, MAWSG has quick response time for processing client-requested operations and provides highly available and uninterrupted services in the presence of faults. Within the

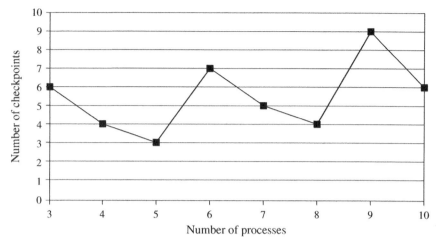

FIGURE 8.16 Average number of forced checkpoints taken within 50 checkpoints.

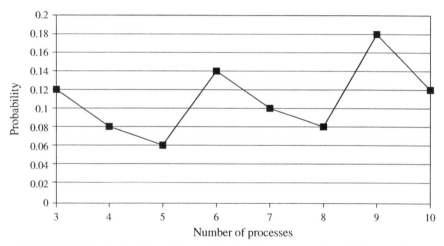

FIGURE 8.17 Probability of forced checkpoints taken within 50 checkpoints.

MAWSG framework, cooperating MAs are used to develop algorithms for various distributed control functions such as dynamic load sharing and replication management, mutual exclusion, and deadlock detection and consensus. Cooperating MAs have shown that their benefits can be very useful to meet the requirements of building flexible, adaptive, and high-performance distributed system functions.

Since the MAs are able to execute asynchronously and autonomously, this approach relieves the servers from having to wait for synchronization messages from remote servers and eliminates large amounts of message passing required in traditional algorithms. In this chapter, we demostrate the feasbility of the MAWSG framework by a fully distributed dynamic load-balancing scheme, a distributed consensus, and a checkpointing and rollback scheme. The results of the experiments showed that the proposed algorithms are effective with good performance.

REFERENCES

1. V. A. Pham and A. Karmouch, Mobile software agents: An overview, *IEEE Commun. Mag.*, July 1998, pp. 26–37.

2. D. B. Lange and M. Oshima, Seven good reasons for mobile agents, *Communi. ACM*, 42(3):88–89, 1999.

3. N. Minar, K. H. Kramer, and P. Maes, Cooperative mobile agents for dynamic network routing, in A. L. Hayzelden and J. Bigham (Eds.), *Software Agents for Future Communication Systems*, Springer, New York, 1999.

4. J. Cao, X. Wang, and S. K. Das, A framework of using cooperating mobile agents to achieve load balancing in distributed Web server groups, in *Future Generation Computer Systems*, Vol. 20/4, 2004, pp. 591–603.

5. J. Cao, Y. Sun, X. Wang, and S. K. Das, Scalable load balancing on distributed Web servers using mobile agents, *J. Parallel Distributed Comput.*, 63(10): 996–1005, 2003.

6. J. Cao, A. T. S. Chan, and J. Wu, Achieving replication consistency using cooperating mobile agents, in *Proceedings International Workshop on Wireless Networks and Mobile Computing* (held in conjunction with the International Conference on Parallel Processing—ICPP'01), Valencia, Spain, Sept. 2001, IEEE Computer Society Press, pp. 453–458.

7. J. Cao, X. Wang, and J. Wu, A mobile agent enabled fully distributed mutual exclusion algorithm, in *Proceedings 6th IEEE International Conference on Mobile Agents (MA'02)*, Spain, Oct. 2002, Lecture Notes in Computer Science, Vol. 6, Springer, 2002, pp. 138–153.

8. J. Cao, G. H. Chan, W. Jia, and T. Dillon, Checkpointing and rollback of wide-Area distributed applications using mobile agents, in *Proceedings IEEE 2001 International Parallel and Distributed Processing Symposium (IPDPS2001)*, San Francisco, CA, Apr. 2001, IEEE Computer Society Press, pp. 1–14.

9. J. Cao, J. Zhou, W. Zhu, D. Chen, and J. Lu, A mobile agent enabled approach for distributed deadlock detection, in *Proceedings 3rd International Conference on Grid and Cooperative Computing (GCC'04)*, Wuhan, China, Oct. 21–24, 2004.

10. J. Cao, X. Wang, S. Lo, and S. K. Das, A consensus algorithm for synchronous distributed systems using mobile agent, in Proc. 2002 Pacific Rim Int'l Symposium on Dependable Computing (PRDC'02), Tskuba, Japan, Dec. 2002, IEEE Computer Society Press, pp. 229–236.

11. D. B. lange and M. Oshima, *Programming and Deploying Java Mobile Agents with Aglets*, Addison Wesley, Reading, MA, 1998.

12. X. Li, J. Cao, and Y. He, A direct execution approach to simulating mobile agent algorithms, *J. Supercomput.* 29(2):171–184, 2004.

13. N. Lynch, Distributed Algorithms, Morgan Kaufmann, San Francisco, CA, 1996.

14. B. Charron-Bost and A. Schiper, Uniform consensus harder than consensus, Technical Report DSC/2000/028, École Polytechnique Fédérale de Lausanne, Switzerland, May 2000.

15. M. J. Fischer, N. Lynch, and M. S. Paterson, Impossibility of distributed consensus with one faulty process, *J. ACM*, 32(2):374–382, 1985.

16. M. Pease, R. Shostak, and L. Lamport, Reaching agreement in the presence of faults, *J. ACM*, 27(2):228–234, 1980.

17. N. Shivaratri, P. Krueger, and M. Singhal, Load sharing policies in locally distributed systems, *IEEE Computer*, 25(12):33–44, 1992.

18. T. L. Casavant and J. G. Kuhl, A taxonomy of scheduling in general-purpose distributed computer systems, *IEEE Trans. Software Eng.*, 14(2):141–153, 1988.

19. Y. Wang and R. Morris, Load sharing in distributed systems, *IEEE Trans. Computers*, C-34(3):204–217, 1985.

20. R. Mirchandaney, D. Towsley, and J. A. Stankovi, Adaptive load sharing in heterogeneous systems, in *Proceedings IEEE 9th International Conference on Distributed Computing Systems (ICDCS)*, 1989, pp. 298–305.

21. D. Eager, E. Lazowska, and J. Zahorjan, A comparison of receiver-initiated and sender-initiated dynamic load sharing, *Perform. Evalu.*, 6(1):53–68, 1986.

22. P. Krueger and M. Livny, A comparison of preemptive and non-preemptive load distributing, *Proceedings 8th International Conference on Distributed Computing Systems*, 1988, pp. 123–130.

23. J. Cao and K. C. Wang, An abstract model of distributed rollback recovery control algorithms, *ACM Operating Syst. Rev.*, 26(4):62–77, Oct. 1992.

24. E. N. Elnozahy, L. Alvisi, Y. M. Wang, and D. B. Johnson, A survey of rollback-recovery protocols in message-passing systems, Technical Report CMU-CS-99-148, School of Computer Science, Carnegie Mellon University, Pittsburgh, PA, Oct. 1999.

25. B. Randell, System structure for software fault tolerance, *IEEE Trans. Software Eng.*, SE-1(2):220–232, 1975.

9 Distributed Databases and Transaction Processing

EVAGGELIA PITOURA
University of Ioannina, Ioannina, Greece

PANOS K. CHRYSANTHIS
University of Pittsburgh, Pittsburgh, Pennsylvania, USA

GEORGE SAMARAS
University of Cyprus, Cyprus

9.1 INTRODUCTION

Mobile software agents have been used to provide an alternative way of implementing distributed applications. Mobile agents are programs that may be dispatched from a client computer and transported to a remote server computer for execution [1].

Software systems built using mobile agents allow flexibility in designing applications and extensibility, since mobile agents can be launched without requiring any preinstallation besides the existence of an efficient execution platform to host agents. They support an asynchronous mode of operation, which can be useful with intermittent connectivity, as is often the case with mobile wireless computing. Mobile agents may also reduce the communication overhead by moving the computation closer to the data. Another major advantage of mobile agents is that agents can roam the network to collect information, thus offering an attractive way to discover data in unknown networks.

The goal of this chapter is to provide an overview of the use of mobile agents in distributed database systems and applications. Mobile agents have been used to derive extended database architectures that divide functionality between database clients and servers in a more flexible way than the one achieved by traditional client−server approaches. In these approaches, mobile agents have

Mobile Agents in Networking and Distributed Computing, First Edition.
Edited by Jiannong Cao and Sajal K. Das.
© 2012 John Wiley & Sons, Inc. Published 2012 by John Wiley & Sons, Inc.

been deployed for both distributed query processing and processing of updates. The use of mobile agents allows easy parallelization of the query-processing task and reductions in communication by processing data locally at their sources. However, it makes the problem of maintaining data consistency in the presence of updates more challenging. Database transaction models and protocols have been used toward providing correctness properties to agent execution.

This chapter is structured as follows. In Section 9.2, we consider agent-based software architectures and their application in database management systems. In Section 9.3, we show how agents can be used to assist in processing queries. The focus of Section 9.4 is on transactional mobile agents. The chapter concludes in Section 9.5 with a short summary.

9.2 MOBILE AGENT ARCHITECTURES FOR DATABASE ACCESS

In this section, we first present a number of software architectures for distributed systems ranging from simple client—server configurations to architectures involving various middleware components and mobile agents. Then, we focus on the applications of these software architectures to database management systems (DBMSs). In general, mobile agents are used in these architectures to support more extensible designs and efficient database access from thin clients and through wireless communications.

9.2.1 Agent-Based Software Architectures

Various software models have been proposed to support wireless computing [2]. Most extensions of the client—server (c/s) model introduce stationary software agents between the client and the server. These agents alleviate the constraints of the wireless communications by performing various communication optimizations to reduce data transmission. They also support disconnections by deploying caching at the client and by queuing requests and responses during periods of disconnections. Further, to address resource constraints of thin mobile clients, these agents undertake part of the functionality of the resource-poor mobile clients.

A popular extension of the c/s model is a three-tier or *client—agent—server* (*c/a/s*) architecture that introduces a server-side agent (Figure 9.1a). The server-side agent either acts as a complete surrogate of the client on the fixed network or is associated with a specific service or application (such as a database or Web service). In the former case, any communication between the client and the server is through the agent, while in the latter, the agent is responsible only for the interactions related to the specific service. These middleware agents split the interaction between mobile clients and servers into two parts, one between the server and the agent and the other between the agent and the client allowing for optimizing and customizing them. However, such agents can directly optimize data transmission over the wireless link only from the server to the client and

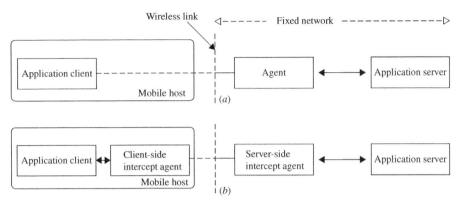

FIGURE 9.1   Extension of the client−server architecture: (a) client−agent−server and (b) client−intercept−server model

not vice versa. Furthermore, they cannot hide any potential network disconnections from the client.

To address these issues, the *client−agent−agent−server* (*c/a/a/s*) or *client−intercept−agent* (*c/i/s*) model introduces an additional agent deployed at the client device (Figure 9.1b). This way it is possible to optimize the communication from the client to the server as well. The client-side agent can also hide disconnections by caching data locally and by queuing requests that cannot be serviced immediately. This model also provides upward compatibility, since the agent pair is transparent to both the client and the server. Further, the pair of agents can adaptively split work among the client and the server based on the current availability of communication and computation resources.

Besides middleware approaches that introduce stationary agents on the path between the mobile client and the server, mobile agents have also been used to accomplish tasks required by clients, in particular in the case of mobile clients. In such architectures, mobile agents are dispatched by the (mobile) clients to perform a requested task by contacting one or more servers (Figure 9.2).By delegating the responsibility for executing a task to a mobile agent, the burden of computation is shifted from the resource-poor client to the fixed network. Further, mobile agents can migrate to follow the pattern of movement of the clients that submit them. Mobile agents can also migrate among servers until the requested task is fulfilled. Finally, such architectures support asynchronous communication which is appropriate in wireless communications.

Mobile agents are orthogonal to the extended client−server models, in the sense that mobile agents can be used not only with the c/s but also with the c/a/s and the c/i/s models. In this case, any of the components (i.e., the server, client, or agent components) of the extended model can launch mobile agents to accomplish part of their functionality or to communicate with the other components, thus enhancing the flexibility of the models. For instance, the pair of

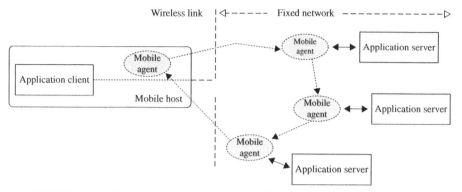

FIGURE 9.2   Client−server model with mobile agents acting as messengers.

agents of the c/i/s model may communicate with each other through mobile agents instead of simple messages or remote procedure calls (RPCs). We shall call the agents used for communication between the system component *messenger agents*. Finally, note that it is possible to use mobile agents to implement the agents of the c/a/s and the c/i/s models. For example, the server-side agent may be implemented as a mobile agent so that it can roam within the fixed network to remain close to its corresponding clients.

9.2.2   Mobile Agents for Web Database Access

Various extended client−server models using messengers have been used to provide efficient Web database access [3, 4]. The c/s with messenger model for database access employs mobile agents between the client interface and the database server machine.

The messenger, called *DBMS-agent*, provides support for database connectivity, processing, and communication. In particular, a DBMS-applet at the client creates and launches one or more mobile DBMS-agents that move to the database server. At the database server, a DBMS-agent initiates a local database driver, connects to the database, and performs any queries specified by the client. Upon completion of the queries, the DBMS-agent dispatches itself back to the client.

By using DBMS-agents to encapsulate the interaction between the client and the database server, the client applet remains lightweight and portable. Specifically, there is no need to download and initialize any drivers at the client; instead the DBMS-agent can at run time select and load the appropriate drivers from those available at the server side. Furthermore, the effect on performance is significant. Experiments conducted using IBM's Aglets platform show performance improvements by a factor of 2 in the case of wireless and dial-up environments.

To avoid the overhead of creating a mobile agent per database request, the adaptation of the c/a/s with messenger model for Web database access creates a

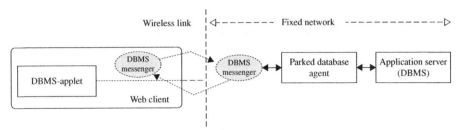

FIGURE 9.3   The c/a/s with messengers for Web database access.

database-specific agent at the server side (Figure 9.3). In particular, at the first database request, the client creates a mobile agent that is sent and parked at the database server. The parked database agent materializes the agent of the c/a/s model. Database connectivity is now the responsibility of this *parked database agent*. A database connection is established and maintained for the whole duration of the client interaction, thus eliminating the limitation of the c/s with messenger approach that requires creating one connection per request. To submit queries, a messenger agent, called *DBMS-messenger*, is created by the client. The responsibility of the DBMS messenger is to transfer the queries and their results between the client and the parked database agent or agents. The introduction of the parked database agent results in additional performance improvements even in fixed networks.

Communication through messenger agents allows flexibility as the agents can roam the network to collect information before they reach their destination. However using messengers introduces considerable overheads over a simple c/a/s approach that uses messages instead of mobile agents. To avoid such overheads, the messenger agent can be replaced by two types of messages. The first type is delivered from the DBMS-applet to the parked database agent and contains the query. The second type is delivered from the parked database agent to the DBMS-applet and contains the results of the last query.

Employing the c/i/s model introduces a database-specific agent at the mobile client as well. While the server-side agent might serve multiple clients, the client-side agent is unique to the client. The functionality of the client-side agent may include various optimizations such as client-side view materialization, caching, and queuing of queries during disconnections. The client-side agent can also be implemented as a mobile agent created at the time of the first database request. Again, communication can be performed through either messages or mobile agents, that is, messengers.

DBMS-agents can be launched to query multiple databases in parallel. The agents are launched to different hosts and cooperate with each other to perform their tasks. The simplest approach is for the DBMS-applet to create multiple agents, one per database query, and be itself responsible for combining the results. Alternatively, it can create a coordinator query agent that is assigned the responsibility of creating the agents and receiving and manipulating any intermediate results. In this case, only the final result is reported to the applet.

This model can also be extended to work with parked agents, one per database server. The coordinator query agent launches agents that roam among the parked agents.

A comparison of implementing Web-based access using two different mobile agent platforms, namely, IBM's Aglets and Mitsubishi's Concordia, is presented in [4]. The two agent platforms were used to implement all client–server variations using either messenger agents or simple messages. In addition, Concordia *service bridges* were used as an alternative nondynamic way of materializing the parked database agent. Service bridges present an efficient approach to provide services to incoming agents at the server side but lack the flexibility of "parking" the agent dynamically at run time and allowing this agent to negotiate which services to provide. The comparison of the implementations of the models is based on simple microbenchmarks and more elaborate application kernels along the lines of [5]. Microbenchmarks are short codes designed to isolate and measure basic performance properties. In the case of Web database access, microbenchmarks include the overhead of creating and launching messenger agents, the overhead of just creating and posting messages, and the overhead of agent roaming. Application kernels correspond to short synthetic codes designed to measure the basic performance properties of the application frameworks of interest, in this case, of the various models for Web database access. In summary, the Aglets workbench outperforms Concordia for the first query, while Concordia outperforms Aglets in subsequent queries. In general, Concordia provides better scalability and robustness, while Aglets offers improved flexibility.

Finally, [6] compares an implementation of the c/a/s with messenger approach using the Aglet platform with other Java-based approaches with respect to (a) performance, expressed in terms of response time under different loads, and (b) programmability, expressed in terms of the number of system calls at the client and server side.

9.2.3 Peer-to-Peer Computing

Recently, *peer-to-peer (p2p) computing* has attracted a lot of attention, mainly as a means of sharing files among dynamically formed communities of users [7]. In a p2p system, there is no clear distinction between servers and clients; instead each node acts both as a server offering data and as a client requesting data. Nodes can enter and leave a p2p system dynamically.

There are similarities in the issues addressed in agent-based systems and in p2p computing [8, 9]. From one point of view, p2p networks can be used to provide an infrastructure for deploying multiagent systems. From a modeling and design perspective, the concept of an agent and its associated properties (such as autonomy, reactivity, proactiveness, and social ability) can be understood as an extension of a peer.

The various agent-enhanced client–server models for database access are directly extensible to the peer-to-peer model. PeerDB [10] is such a network of database-enabled nodes. Each PeerDB node includes a database server and a database agent called *DBAgent*. The DBAgent encapsulates the functionalities of both the client- and the server-side agent of the c/i/s model. Each peer DBAgent has a master agent that manages the queries of the user. It clones and dispatches messenger or worker agents to neighboring nodes to collect information. Each worker agent is dispatched to the appropriate node. Answers to each query are returned directly to the master agent. Upon receiving answers, the master agent combines them and returns any results to the user. The master database agent also monitors statistics and manages reconfiguration policies. Other components of the architecture include a cache manager and a user interface component.

9.3   QUERYING WITH AGENTS

In this section, we focus on deploying mobile agents for answering queries. First, we discuss how the various stages of query processing can be assigned to mobile agents. Then, we focus on the topic of view materialization.

9.3.1   Query Processing

Query processing refers to the steps taken toward executing a database query which is most commonly expressed in some relational language, such as Structured Query Language (SQL). In centralized database systems, query processing is generally structured in three logical steps: (1) parsing, (2) rewriting and optimization, and (3) execution. In the first step, the query is parsed and translated into an internal representation. Then, query rewriting transforms the query by carrying out optimizations that are good independently of the physical state of the system, for example, the size of the relations or the presence of indexes. The rewritten query is then optimized using information from the system catalog including various statistics about the size of data and the system state. The output of the query optimizer is a query plan that specifies precisely how the query is to be executed. Finally, the query is executed by the query execution engine.

In distributed query processing, database relations, called *global*, are fragmented and distributed at various nodes. After the query involving global relations has been parsed, it is rewritten so that each relation in the query is replaced by the various fragments of the relation; this step is called *localization*. Localization results in a number of subqueries each involving relations at a single site. It is assisted by a system catalog that provides information regarding the way the relation is fragmented and the sites where each fragment is located. Subsequently, the query is rewritten and optimized globally. Each subquery involving a single site is submitted to the site and executed there. The results of the subqueries are then combined to form the final query answer.

Mobile agents may be employed to implement the various steps of query processing. Rather than providing additional functionality, mobile agents offer an alternative implementation platform for query processing that may be more efficient, easier to implement, and more extendable. In general, a single query agent can be created to coordinate the overall query execution and one *worker mobile agent* may undertake the execution of each subquery. By having one worker agent per subquery, various parts of the queries may be executed in parallel. Further the worker agent responsible for each subquery may perform locally any possible optimizations of the results attained so that the amount of data returned to the coordinator agent are reduced. Variations of this simple scheme are also possible; for example, instead of a single coordinator agent that collects the results of all worker agents, many intermediate coordinator agents may be spawned to combine the results of the subqueries incrementally. Further, in the absence of a global catalog, a *mobile catalog agent* may be used to roam the network and collect information regarding the various database fragments and their location to assist localization.

Next, we present a number of approaches that use various groups of cooperating agents for query processing.

9.3.1.1   Web Access to Multidatabases

In the mulidatabase framework of [3], DBMS-aglets are launched to query multiple Web-accessible databases. The DBMS-applet at the Web client creates a coordinator DBMS-aglet that is dispatched to the fixed network most likely to a Web server. This aglet dynamically creates and dispatches to several target sites a variable number of DBMS-aglets to work in parallel. The coordinator DBMS-aglet is also responsible for receiving and manipulating the intermediate results provided by the various DBMS-aglets. Only the final result is reported to the DBMS-applet. This approach serves the objective of keeping the client lightweight, since the coordinator DBMS-aglet does not necessarily reside on the client. Instead, processing can be done remotely with only the final result transmitted to the client.

When processing of multiple sets of queries is expected, the coordinator DBMS-aglet can be extended to create and submit parked DBMS-aglets instead of DBMS-aglets. Thus, at the submission of the first set of queries, a network of parked mobile agents is created. The coordinator DBMS-aglet uses this infrastructure of parked agents already set up to process any subsequent set of queries. If a subsequent query needs to access a new site, the coordinator aglet may choose, based on the current system conditions, either to create a new aglet or to instruct a nearby one to move to the new site and execute the query. In the latter case, the query is sent to the existing aglet along with the move instruction. Instead of waiting to receive all results, the coordinator agent may employ multithreading to receive and manipulate results in parallel.

9.3.1.2   PeerDB

PeerDB [10] provides support for querying a dynamically formed network of database-enabled nodes where each node serves as both a database client and server. There is no catalog information available; thus information regarding the location of the various relation fragments must be collected before the query is rewritten. To this end, query processing is performed in two phases. In the first phase, a relation-matching strategy is applied to locate relations of interest. These relations are returned to the query node, allowing the user to select the appropriate relations. In the second phase, the queries are submitted for execution to nodes containing the selected relations. Note that the two phases can be interleaved providing the user with intermediate results. Both phases are assisted by agents.

For the first phase, PeerDB adopts a flooding strategy: a query for relations is forwarded to the neighbors of the query node, which in turn forward the query to their own neighbors, until a specified maximum number of steps is reached. In particular, when a user issues a query (SQL-like selection query), a master (i.e., coordinator) agent is created to oversee the evaluation of the query. The agent parses the query, clones relation-matching agents, dispatches them to all neighbors of the node, and waits for their replies. When a relation-matching agent arrives at a node, it searches the local catalog of the node and returns promising relations to the master agent. If the maximum number of steps is not reached, the relation-matching agent clones more relation-matching agents and dispatches them to the neighbors of the current node; otherwise, the agent is dropped. Upon receiving any answers, the master agent returns them to the users for selection.

In the second phase, the master agent clones one worker agent for each relation selected during the first phase. Each worker agent reformulates the query so that it matches the schema at the target node. If the target relations are found locally, the worker agent submits a reformulated SQL query to the local DBMS. If the target relations are on a remote node, then the worker agent is dispatched to this node, reformulates the SQL query, and submits it to the DBMS of this node. Once the answers are retrieved, they are returned to the master node directly. If the retrieved data need to be processed further before being returned, then the worker agent performs this task as well (with the code that it carries along) and returns only the summarized data. The worker agent may then be dropped.

9.3.1.3 ACQUIRE

ACQUIRE (Agent-based Complex QUerying and Information Retrieval Engine) [11] is a framework for querying heterogeneous and distributed data sources. ACQUIRE implements the following three phases. In the first phase, a user query is decomposed (i.e, localized) appropriately into a set of subqueries using site and domain models of the distributed data stores. In the second phase, an optimized plan is created for retrieving answers to these subqueries over the Internet and a set of intelligent mobile agents is spawned to delegate these tasks. During the last phase, the answers returned by the mobile agents are merged and returned to the user.

The system was tested on simulated NASA Earth Science data sources under three test conditions. In the first condition, multiple mobile agents were spawned (one for each data source) using an ACQUIRE decomposition and planning system. The second condition also used the standard ACQUIRE retrieval mechanism, except that a single hopping mobile agent was used to retrieve data by visiting each data source in turn. Finally, a reference case was conducted in which the planning, optimization, and remote data-processing features were disabled; thus all data required for the query was transferred to the host application without any remote data processing. However, data were still retrieved by ACQUIRE mobile agents. The best performance was attained by the multiple mobile agents case, since data were processed and retrieved in parallel, whereas the single agent was required to retrieve data in a serial manner. The reference case required the most time and had a much larger total download size than either of the two other methods.

Instead of using multiple single-hop agents that move to a site, retrieve some data, and return to the coordinator, ACQUIRE also suggests to allow agents to send data to other data sources (via the mobile agent agencies located there) rather than sending all retrieved data to a single coordinator. Furthermore, adaptive query optimization is proposed. Typically, query optimization is static, that is, an optimized execution order of the subqueries (and thus the itinerary for the corresponding agents) is determined at query planning and remains unchanged during query execution. The query optimization method in ACQUIRE has been enhanced to dynamically optimize the retrieval strategy as it is carried out. This requires equipping each spawned agent with the full query execution plan instead of just the part delegated to it as well as with the necessary code to execute the retrieval plan at any data site in the network. The spawned agents communicate, collaborate, and negotiate with each other to dynamically decide where to migrate, send data, and perform necessary computations to complete query execution. These decisions depend on current conditions such as network speed, data size, and the computational capabilities of the data servers and agents involved in the retrieval. For example, an agent that finishes retrieval earlier than other agents producing a small-size dataset may decide at run time to migrate to the site of another agent to complete there a computation that requires data produced by both agents.

9.3.1.4 PaCMAn

Employing mobile agents for query processing can be seen as a specific instance of a general framework of executing complicated tasks in parallel using mobile agents. Such a framework is provided by the PaCMAn (Parallel Computing with Java Mobile Agents) middleware [12]. PaCMAn launches multiple Java mobile agents that communicate and cooperate to solve problems in parallel. Each mobile agent can travel anywhere in the network to perform its tasks. A number of brokers/load forecasters keep track of the available resources and provide load forecasts to clients. Clients select the servers that they will utilize based on their specific resource requirements and the load forecast. The PaCMAn mobile agents are modular; the mobile shell is separated from the specific task code of the target application. This is achieved through TaskHandlers, which are Java objects implementing a particular task of the target application. TaskHandlers are dynamically assigned to PaCMAn's mobile agents. In the case of query processing, TaskHandlers perform the task of data connection and retrieval.

As in ACQUIRE, the implementation of query processing using PaCMAn was tested for a mode in which a single agent moves from database server to database server gathering and joining all results and for a mode in which there is one agent per subquery, that is, one agent per database server. In addition, a third mode was implemented in which, instead of having a single coordinator agent joining all results, the results are joined in a tree reduction order. Using multiple agents achieves a considerable speedup over the case of a single agent which corresponds to a serial execution. The additional improvement of using multiple coordinators comes from the extra parallelism attained during the result join phase.

9.3.1.5 Location-Dependent Queries
A distributed architecture based on mobile agents has also been proposed for the continuous processing of location-dependent queries [13]. In the proposed architecture, a set of fixed computers, called proxies, manage the location information of moving objects within a certain geographic area, called proxy area. For example, in a cellular network, a proxy area is obtained as the union of the coverage areas of one or base stations. At each proxy, a data management system (DMS) module is in charge of storing and providing location data about moving objects within that proxy area. Mobile agents are used to track interesting moving objects, optimize the wireless part of the communication, and adapt to the movement of relevant objects. Each continuous location query is parsed into a set of standard queries about the location of objects referenced in the query.

A QueryMonitor agent is created for each location query at the query submission site. The QueryMonitor agent sends a MonitorTracker mobile agent to the proxy in charge of its location, which is the proxy that provides it with wireless coverage. The MonitorTracker performs three main tasks: (1) follows the monitor wherever it goes (moving from proxy to proxy with the goal of staying close to the user), (2) stores any data requested by the user in case of disconnection of the monitor, and (3) refreshes the data that the QueryMonitor must present to the user minimizing wireless communications with the monitor. For the third task, the MonitorTracker creates a Tracker mobile agent to track the location of each reference object and process the standard queries related to such a reference object.

Trackers perform three main tasks continuously: (1) stay on the proxy that manages the location of its reference object, traveling from proxy to proxy when needed in order to request its location locally, (2) detect and process new locations of its reference object by querying the corresponding DMS at its proxy, and (3) detect and process, with the help of Updates agents, the location of target objects that satisfy the constraints where such a reference object appears.

The Tracker agent creates one Updater agent on each proxy whose area intersects the extended area of the reference object; these proxies are obtained by querying a proxy catalog which stores information about the proxies [such as their Internet Protocol (IP) address and proxy area]. Updater agents are initialized with the corresponding standard query. The goal of Updaters is to detect the location of target objects by executing its standard query against the DMS on its proxy at the required refreshment frequency.

Every agent in the network correlates the results received from its underlying agents and communicates its own results to its creator agent. Note that the only wireless data transfer occurs at query initialization when the QueryMonitor sends the MonitorTracker to the proxy of the monitor and at refreshment when the MonitorTracker sends a new answer to the QueryMonitor.

9.3.2   View Materialization

In database systems, views are virtual relations defined using a query over a set of stored database relations, called base relations. Views are used for security reasons to allow specific parts of base relations to be hidden from users. Aside from security considerations, views also allow the creation of personalized collections of data that better match the interests of users. When the results of the query that defines the view are stored, the view is called *materialized*. A materialized view needs to be updated to reflect any changes at the base relations. This process is called *view maintenance*.

Views Supported by Mobile Agents (ViSMA) [14] provides the functionalities of defining, materializing, and maintaining views over multiple data sources by taking advantage of mobile agent technology. The role of mobile agents in ViSMA is twofold: First, views are carried within mobile agents called *view agents* (VAs) that may relocate themselves to reduce the distance from users that frequently request them. Second, mobile agents are used for migrating to a remote data source and locally execute update propagation and query materialization operations, relieving remote clients and local data sources from performing this task while saving network resources.

ViSMA is based on a multitier architecture including components at the client, middleware, and data source site. ViSMA handles view definitions consisting of select, project, and join (SPJ) queries as well as set operations over SPJ queries. Views are categorized as simple or complex. A *simple view* can be broken down to a sequence of SPJ operations that represent a linearized query evaluation plan. *Complex views* consist of at least two simple views combined with a set operator. Views can also be defined to extract data from a single data source (single views) or from multiple data sources (multiviews). The user can define a view to be either sharable or private. *Sharable* views are visible to all ViSMA clients, while *private* views are available only to the clients that have created them. Additionally, personalized views can be defined. Personalized views are materialized subviews of existing views.

Client-side components implement the interface between the user and ViSMA. At the middleware layer, the agents provide catalog information regarding participating data sources and created views. With respect to views, metadata include the view definition as well as the current location of the mobile agent carrying the view. In particular, the ViSMA middleware server consists of four types of agents: view dictionary agent (VDA), VA, data holder (DH), and personal agent (PA). The VDA is the central coordinator agent. Each time a user defines a view, the VDA creates a VA. A VA is responsible for creating, materializing, and maintaining the view. A VA allows its materialized view to be queried by external entities if the view is defined as sharable. Since the VA is a mobile agent, its data are carried with it as it moves. Therefore, view migration is achieved. When a VA is created, it also initializes and dispatches a number of DHs to support the materialization and maintenance of its view. The DHs created by a VA are configured in a hierarchy with the VA as the root. Each DH in the hierarchy is responsible for handling a single view fragment. The VA is responsible for combining the fragments and producing the final view. The personal agent (PA) is a mobile agent that is used for creating personalized, nonsharable-subviews. It derives its subview from the view of a VA and maintains its subview by issuing queries to the VA. The PA may communicate directly with its client and may move as its client moves.

Components at the data source site are agents that generally reside at a data source location (e.g., at a remote database server location). These components may be created at the ViSMA middleware server but they migrate and execute the bulk of their operations at one or more data source locations. The components at the data source site are two mobile agents, namely, the DBInfo agent and the ViewEvaluator agent, and two stationary agents, namely, the Assistant agent and the Monitor agent. The Assistant agent is a stationary agent that maintains a pool of connections to the data source to serve visiting agents. These agents provide transparent connectivity between the data sources and the agents that require access to these data sources. Changes to data source connectivity settings need only be made known to the corresponding Assistant agent. The DBInfo agent migrates to the remote data source, collects its metadata (schema and data types), and sends them to the VDA. The Monitor agent is a stationary agent created and dispatched to a data source by a DH to enable view maintenance. The Monitor agent parks at the remote site and periodically resubmits queries to the data source and sends changes to the DH. A view evaluator agent (VEA) is a mobile agent that is sent by a DH to roam from one data source to another to collect the data required for the materialization of the view fragment for which the DH is responsible. A VEA incrementally materializes the view fragment as soon as data are collected.

To cope with disconnected operation, view holders may use view versioning to allow application sessions to access more current data without invalidating work previously done [15]. A data validation process detects inconsistencies with newer versions of data upon reconnection. Essentially, these agents compute the period of time or consistency window, measured in versions, for which the results of a mobile client application are consistent. Rules are supplied that govern the creation and sharing of results and show how inconsistencies can be detected to offer a higher availability of data while organizing and gracefully degrading the amount of consistency achieved between the mobile clients and the data sources.

Because the view holder combines and computes the necessary derived data, it is also able to offer different levels of view consistency between the data available and the derived data given to the clients. Work in [16] considers ways to customize view currency and discuss how they affect view consistency. Two types of view maintenance algorithmic approaches are examined: (1) recomputational maintenance that constructs an entirely new version of a materialized view and (2) incremental view maintenance that allows updates to be slowly incorporated within an existing version.

9.4 TRANSACTION MANAGEMENT

During execution, agents access their local data, data carried by other agents, and data at the remote sites that they visit. Maintaining data consistency despite failures and under concurrent execution of many programs accessing them is the focus of transaction management. In this section, we consider transaction management in the case of mobile agents. After a short introduction, we present issues related to a fault-tolerant execution of agents. Then, we focus on transaction models for agents and protocols for maintaining the transaction properties of agents even in the case of failures.

9.4.1 Agents and Transactions

Transactions are used to model the execution of a program that accesses and possibly modifies various data items. To ensure data integrity, transactions are required to have the *ACID properties*, namely Atomicity, Consistency, Isolation, and Durability [17, 18]. *Atomicity* guarantees that either all or none of the operations of a transaction are executed, despite any failures. Thus, a failure does not leave data in a state where a transaction is partially executed. *Consistency* ensures that the execution of each transaction preserves the consistency of data. Even though multiple transactions may be executed concurrently, *isolation* guarantees that each of the concurrently executing transactions is unaware of the others; each transaction gets the same view as if it were the only one executing in the system. Finally, *durability* ensures that, after a transaction completes successfully, any modifications made by the transaction are permanent and persist even if there are failures.

The ACID properties are maintained through a set of system protocols. *Concurrency control protocols* schedule the operations of each transaction so as to achieve isolation usually by employing locks or timestamps. *Commit protocols* are used to coordinate the execution of a transaction so that atomicity and durability are ensured even in the case of failures. A commonly used protocol for committing the results of transaction accessing distributed resources is the *two-phase commit (2PC)* protocol. In the first phase of 2PC, all participants of a transaction reply to a request to commit, either with an abort or with a prepare-to-commit vote. In the second phase, the transaction is committed only if all participants have replied with a prepare-to-commit vote. Otherwise, the transaction is aborted. After a transaction is committed, its results become permanent. The recovery manager guarantees atomicity and durability in the case of failures, by ensuring that the data items are left in a consistent state and modifications made by committed transactions are not lost.

In general, during its execution, a mobile agent accesses and modifies a number of data items including its own and other agents' data (internal agent states) as well as data at the various sites that it visits (external states). Attributing to the execution of an agent all or some of the ACID transaction properties offers a systematic way to ensure integrity even in the case of failures.

9.4.2 Fault Tolerance

In general, a mobile agent executes on a number of machines. In each machine, a place provides the logical execution environment for the agent. During the execution of an agent, the agent itself, the places that execute the agent, the machines hosting the places, or the communication links may fail for a variety of reasons. When a place fails, all agents running on it fail to continue their execution. A machine failure causes all places and agents running on it to fail as well. Finally, a communication link failure causes the loss of messages or agents currently on transmission on this link. A desirable property of a mobile agent system is fault tolerance, meaning that the system can provide its services even in the presence of failures.

To avoid loss of information, in the case of an agent failure, an agent should regularly checkpoint its state to stable storage. After a failure, the agent may be retrieved from stable storage and computation may be restarted from the point that the checkpoint was taken. For mobile agents, checkpoints may be taken, for instance, when an agent arrives at a place and is maintained there until the agent successfully completes its operation at this place, as, for example, in Concordia [19]. Taking frequent checkpoints results in both faster recovery and losing smaller parts of computation than taking less frequent ones; however, it increases the cost of normal operation. For example, in the extreme case of taking checkpoints only at the beginning of an agent computation at a place, the associated overhead is small; however, the agent should be restarted from the beginning of its execution at the place.

To avoid blocking, the method in [20] exploits another form of redundancy in agents systems: the fact that a mobile agent can achieve a task in more than one way. Alternative ways to accomplish a task are captured through nondeterministic constructs in the agent language. During the execution of the agent, the actual computational path taken by the agent is maintained in its possible computational tree. Upon detection of a failure, the recovery manager rolls back computation and restarts the agent from a previous point in its computational tree down to a different but equivalent computational path without waiting for the actual failure to be repaired.

Restarting an agent may lead to multiple executions of the same operations. This may be acceptable for idempotent operations, for instance, searching for the best travel itinerary, but may lead to incorrect results in other cases, for instance, when booking a hotel room. The *exactly-once* or *safety* property requires that the code of an agent is executed only once.

Based on the number of failures that a system can mask, a *t-nonblocking* system ensures that progress is made, that is, agent execution is not blocked, despite the occurrence of up to *t* failures. A common way to avoid blocking is by introducing redundancy through replication. Again, one potential problem with replication is that it may lead to multiple executions of the same task and thus to a violation of the exactly-once property. There are many forms of redundancy. For instance, machines can be added to make it possible for the system to tolerate the loss of some of its components. Similarly, additional information such as extra bits can be included in communication protocols to allow recovery from grabbled bits. There are two types of redundancy specific to mobile agents: having more than one agent executing the same task and replicating places.

In terms of replication at the agent level, a task may be assigned to a group of agents instead of a single agent. In terms of replicating places, we may distinguish three types of places, isoplaces, heteroplaces, and heteroplaces with witnesses [21, 22]. *Isoplaces* correspond to traditional server replication, where all places reflect the same state. The places run a replication protocol that ensures consistency among the place replicas. *Heteroplaces* are not exact replicas of each other, instead, they correspond to a set of places that all provide similar services. *Heteroplaces with witnesses* are a generalization of heteroplaces where only a subset of the places provide the service. The places that do not provide the service are the witnesses. In general, a witness is a place that can execute the mobile agent but does not provide the particular service requested by the agent. Hence, the agent request fails, but the agent can continue its execution despite an infrastructure failure of a place.

In the case of mobile agents, it is also important to provide exactly-once delivery of an agent despite failures. Often fault-tolerant agent systems utilize transactional message queues for communication. Transactional message queues provide for persistent messages. They also ensure exactly-once delivery; that is, once a queue manager has accepted a message, the message will be delivered exactly once despite any node or communication failures. This is ensured by performing the put-and-get queue operations within ACID transactions.

The discussion above considered fail-stop failures, that is, failures that cause a component to stop processing. Even more challenging are arbitrary failures where no assumptions can be made about the behavior of faulty components. For example, in this case, executing on a faulty place may alter the results of an agent in a nondeterministic way. Further, a faulty place may interfere with the execution of an agent even if the agent is never executed at this place, for instance, by generating bogus agents that interfere with it. In this case, replication with voting can be used to mask the effects of executing an agent on a faulty place or machine [23].

9.4.3 Transactional Models

In many applications, especially in the case of long-running transactions, it is not possible or desirable to maintain all four ACID properties [24]. To address such cases, a number of advanced transaction models have been proposed that relax one or more of the ACID properties. Most often a transaction is viewed as a collection of related subtransactions allowing the flexibility of specifying the atomicity and isolation properties at the subtransaction level. In *multilevel transactions*, modifications made by a subtransaction become visible to all other subtransactions when it completes its operations. In *nested* transactions, modifications made by a subtransaction become visible only to its parent (sub) transaction when the subtransaction completes. In *open nested* transactions, these results can also be made visible to other nonconflicting subtransactions. A subtransaction may abort without forcing its parent to abort. Instead, the parent transaction may ignore the subtransaction or invoke a compensating transaction. *Compensating* transactions are transactions designed to semantically reverse the effects of another transaction. They are used whenever atomic rollback is not possible because results have already been committed.

Open nested transaction models are more suitable for modeling the execution of mobile agents when compared to flat ACID transactions, since agent computations correspond to long-running activities. In this case, the top-level transaction of the nested transaction model corresponds to the overall mobile agent execution. The overall agent execution is then divided into subtransactions. Most commonly, each subtransaction corresponds to the execution of the mobile agent at a single place. We call the execution of an agent at a single place a *step*. Thus, the agent execution proceeds in a number of steps, each one executed at a single machine. Note that some of the challenges regarding agent transactions are similar to those also faced when supporting transactions in mobile wireless computing [25].

In [22], transactional approaches are classified into commit-after-stage and commit-at-destination protocols based on when the modifications made by an agent become permanent and visible to other agents. The execution of an agent is modeled as a sequence of stages where each *stage* corresponds to the execution of an agent at a single place. With *commit after stage*, modifications of an agent are made permanent and visible immediately after each stage execution, while with *commit at destination*, modifications are committed after the agent finishes its entire computation. A commit-after-stage execution corresponds to an open-nested-transaction model where each subtransaction corresponds to a stage. A commit-at-destination execution corresponds to viewing each agent execution as a flat ACID transaction. Commit-at-stage executions are further classified based on whether the decision to commit is taken by a single or multiple places.

The transaction model for agents in [26] provides the semantics of open nested transactions. It is based on a single wandering agent moving from site to site executing all tasks. Tasks either offer an external prepare-to-commit state or not. Each local subtransaction (i.e., step) is executed as an ACID transaction. The approach in [27] builds on the distributed transaction processing of the Object Management Group (OMG) to guarantee exactly-once semantics for the migration of a mobile agent, that is, of a step. The migration of an agent is contained within a transaction.

A protocol for preserving the exactly-once execution property of mobile agents in the case of failures is introduced in [28]. This is achieved by using transactional message queues to implement steps as subtransactions. There is a message queue at each place. Each place uses a *get* operation to remove an agent from its input queue. Then, it executes the agent locally. Finally, it uses *put* to place the agent directly to the input queue of the place to be visited next. The *get, execute,* and *put* operations are executed within a transaction and hence build an atomic unit. The protocol reduces the blocking probability through heterospaces with witnesses. In particular, for each step, there is a set of nodes, called a *stage,* that can perform the step. The nodes of a stage are called *observers* and either provide the same or alternative services, in which case they are called *workers,* or provide just an environment for running agents. A voting algorithm along with a 2PC protocol is run among the observers of each stage, so that a single worker is committed per step. This protocol is extended in [29] with a mechanism to partial rollback of an agent. The mechanism is based on using savepoints (i.e., checkpoints) along with compensation operations and considers both the internal state of the agent (i.e., its private data) and the external state (i.e., data local at a place). The protocol is also extended in [30] to allow for an agent to execute more than one (sub)transaction at each step. Using a three-phase commit (3PC) protocol is suggested to avoid the blocking problem of 2PC when the coordinator fails.

The transactional agents of [31, 32] are mobile agents whose task is to implement transactional access to objects. A transactional mobile agent moves through a number of machines and, locally at each machine, manipulates a number of objects. In the case of two agents accessing the same object in a conflicting manner, the two agents negotiate with each other on whether to hold or release the object. When an agent leaves a machine, it creates a surrogate agent and assigns it all its locks on objects in this machine. The surrogate agent holds the locks until the agent finishes its execution. In the case that the agent detects that the next place or machine to visit is faulty, it finds a new place or machine to move. If the agent is faulty, its surrogate can create a new instance of it or bypass it. The fact that the surrogate of an agent is faulty is detected by some preceding surrogate and the faulty agent is again replaced or ignored. Transactional agents implemented in Aglets are used to support transactions in a relational multidatabase where objects correspond to relations.

The work in [33] shows how transaction support can be provided in a mobile agent system. The focus is on applications in which resources change over time. In this case, using multiple, mobile, and autonomous agents enables monitoring resources, even those not known in advance, and reacting to changes in a very flexible manner. Transactional support builds on the OMG Object Transaction Service. A transaction involves a group of mobile agents where each mobile agent is responsible for a subtask (subtransaction) that involves only one resource. There are also compensating subtransactions per subtransaction (resource) as well as contingency transactions that are used to specify alternative paths. Agents observe the resources in a preparation phase and execute the actual transaction in a later phase. The goal is to minimize the probability of aborts and rollback/compensation. This is similar to optimistic protocols or preexecution protocols proposed for real-time systems. In general, the semantics achieved are those of open nested transactions. In the case that all resources are cooperative and offer an external prepare-to-commit state, then the semantics of closed nested transactions can also be offered. Exactly-once semantics are achieved for more than one step by performing all operations in one global transaction.

In the general case, an agent does not just correspond to a sequence of subtransactions (steps) each one executed at a single site. Instead, an agent may model complex computational activities, with several dependencies among the various parts (subtransactions) of an agent execution. The agent transaction models of [34–36] aim at expressing such complex dependencies within the agent as well as among agents.

The work in [34] introduces an object-oriented programming model for representing the plan of an agent. A plan is modeled using workflows to represent the activities that comprise the plan and the control dependencies that specify the order of execution of these activities. Special mediator nodes are used to represent control flow activities, including an ANDSplit mediator that spawns activities to be executed in parallel and an XORJoin mediator that awaits a successful completion of one of its preceding activities; otherwise it aborts. An activity may be defined as compensatable in which case it can commit independently of other activities. A plan is carried out by assigning it to a transactional mobile agent. The agent has access to its entire plan locally. Control flow and transaction semantics are encapsulated by mediators. Transaction processing involves a forward and a backward process. The forward phase is similar to the preparation phase of [33]. During the forward process, mobile agents migrate to activity locations (places) to execute activities and to mediators locally or to propagate an internal abort until the final activity is reached. During the backward process, the agents return to these locations to locally commit, abort, or compensate activities. The protocol is carried out by the original agent and its clones. A transaction completes once its plan has a path from the top to the final activity along which all activities complete successfully.

Similarly the work in [35, 36] allows the specification of control flow dependencies both within a single agent, called *intra-agent dependencies,* and among agents, called *interagent dependencies*. Intra-agent dependencies are expressed through structural dependencies that define the flow of control between the tasks of an agent. Interagent dependencies are based on breakpoints that define points in the execution of an agent where the modifications made by an agent can be made visible to other agents. Each agent encapsulates a transaction manager that schedules its task according to the specified dependencies and uses a timestamp-based concurrency control method to ensure local serializability of all executions involving access to its own private data. For accessing database resources, there is also a local database agent that is responsible for all tasks involving access to the database that in addition uses a ticket-based method [37] to ensure local serializability.

9.5 SUMMARY

In this chapter, we have discussed issues related to mobile agents and databases. Several agent-based architectures have been developed to provide access to multiple, distributed, and Web databases. The main advantages of these architectures are as follows:

a. They can be deployed at run time, thus requiring minimum setup and preinstallation as well as being easily adjustable to support new components.
b. They can dynamically reallocate load among the various system components.
c. They provide an asynchronous mode of operation and can minimize the network by moving the computation closer to the data sources.

Mobile agents offer an ideal platform for developing cooperative computations including database ones. We have also shown how query processing can be performed by groups of cooperating agents and discussed their applicability.

Finally, we have considered the challenging problem of data consistency when mobile agents not only query data but update them as well. We have discussed models and methods for the execution of agents with correctness guarantees through transaction properties.

Besides the above advantages, an important concern with mobile agents is the potential of compromising security by allowing foreign code to execute at a machine. We have not discussed this issue in this chapter, since this is not unique to database applications. Security in agent-based applications is a general problem and a major challenge of the agent-based applications and needs to be addressed at the level of the agent-based platform.

In conclusion, mobile agents may offer no advantage in database applications which are rather static and where simpler communication protocols with small transmission overheads would suffice. However, mobile agents offer a great promise for developing distributed, highly dynamic, and collaborative database applications. Currently, there are no large-scale agent-based applications mainly because the use of mobile agents assumes the existence of an efficient and secure agent platform to host mobile agents. We expect that large mobile agent applications will be deployed soon in environments where security is not an issue, for example, in processing data in emergency and disaster management.

REFERENCES

1. D. M. Chess, C. G. Harrison, and A. Kershenbaum, Mobile agents: Are they a good idea? in J. Vitek and C. Tschudin (Eds.), *Mobile Object Systems—Towards the Programmable Internet, Second International Workshop, MOS'96*, Linz, Austria, July 8–9, 1996, selected presentations and invited papers, Lecture Notes in Computer Science, Vol. 1222, Springer, Berlin, Heidelberg, 1997, pp. 25–45.

2. C. Spyrou, G. Samaras, E. Pitoura, and P. Evripidou, Mobile agents for wireless computing: The convergence of wireless computational models with mobile-agent technologies, *MONET*, 9(5):517–528, 2004.

3. S. Papastavrou, G. Samaras, and E. Pitoura, Mobile agents for world wide web distributed database access, *IEEE Trans. Knowledge. Data Eng.*, 12(5):802–820, 2000.

4. G. Samaras, M. D. Dikaiakos, C. Spyrou, and A. Liverdos, Mobile agent platforms for web databases: A qualitative and quantitative assessment, paper presented at the 1st International Symposium on Agent Systems and Applications/3rd International Symposium on Mobile Agents (ASA/MA '99), Palm Springs, CA, Oct. 3–6, 1999.

5. M. D. Dikaiakos and G. Samaras, Performance evaluation of mobile agents: Issues and approaches, in R. R. Dumke, C. Rautenstrauch, A. Schmietendorf, and A. Scholz (Eds.), *Performance Engineering*, Lecture Notes in Computer Science, Vol. 2047, Springer, Berlin, Heidelberg, 2001, pp. 148–166.

6. S. Papastavrou, P. K. Chrysanthis, G. Samaras, and E. Pitoura, An evaluation of the java-based approaches to web database access, *Int. J. Cooperative Inf. Syst.*, 10 (4):401–422, 2001.

7. J. Risson and T. Moors, Survey of research towards robust peer-to-peer networks: Search methods, *Computer Networks*, 50(17): 3485–3521, 2006.

8. M. Koubarakis. Multi-agent systems and peer-to-peer computing: Methods, systems, and challenges, in M. Klusch, S. Ossowski, A. Omicini, H. Laamanen (Eds.), *Cooperative Information Agents VII, 7th International Workshop*, CIA 2003, Helsinki, Finland, Aug. 27–29, 2003, Lecture Notes in Computer Science, Vol. 2782, Springer, Berlin, Heidelberg, 2003, pp. 46–61.

9. G. Moro, A. M. Ouksel, and C. Sartori, Agents and peer-to-peer computing: A promising combination of paradigms, in G. Moro and M. Koubarakis (Eds.), *Agents and Peer-to-Peer Computing, First International Workshop*, revised and invited paper, AP2PC 2002, Bologna, Italy, July, 2002, Lecture Notes in Computer Science, Vol. 2530, Springer, Berlin, Heidelberg, 2002, pp. 1–14.

10. W. S. Ng, B. C. Ooi, K.-L. Tan, and A. Zhou, Peerdb: A p2p-based system for distributed data sharing, in U. Dayal, K. Ramamritham and T. M. Vijayaraman

(Eds.), *Proceedings of the 19th International Conference on Data Engineering (ICDE)*, Bangalore, India, IEEE Computer Society, Washington, DC, USA, Mar. 5–8, 2003, pp. 633–644.

11. S. Kumar Das, K. Shuster, C. Wu, and I. Levit, Mobile agents for distributed and heterogeneous information retrieval, *Inform. Retrieval*, 8(3):383–416, 2005.

12. P. Evripidou, G. Samaras, C. Panayiotou, and E. Pitoura, The pacman metacomputer: Parallel computing with java mobile agents, *Future Generation Comp. Syst.*, 18(2):265–280, 2001.

13. S. Ilarri, E. Mena, and A. Illarramendi, Location-dependent queries in mobile contexts: Distributed processing using mobile agents, *IEEE Trans. Mobile Comput.*, 5(8):265–280, 2006.

14. K. Karenos, G. Samaras, P. K. Chrysanthis, and E. Pitoura, Mobile agent-based services for view materialization, *Mobile Comput. Commun. Rev.*, 8(3):32–43, 2004.

15. S. Weissman Lauzac and P. K. Chrysanthis, View propagation and inconsistency detection for cooperative mobile agents, in R. Meersman and Z. Tari (Eds.), *On the Move to Meaningful Internet Systems, 2002 – DOA/CoopIS/ODBASE 2002 Confederated International Conferences, Irvine, California*, Lecture Notes in Computer Science, Vol. 2519, Springer, Berlin, Heidelberg, Oct. 30–Nov. 1, 2002, pp. 107–124.

16. S. Weissman Lauzac and P. K. Chrysanthis, Personalizing information gathering for mobile database clients, in G. B. Lamont, H. Haddad, G. Papadopoulos and B. Panda (Eds.), *Proceedings of the 2002 ACM Symposium on Applied Computing (SAC)*, Madrid, Spain. ACM, New York, NY, Mar. 10–14, 2002, pp. 49–56.

17. G. Weikum and G. Vossen, *Transactional Information Systems: Theory, Algorithms, and the Practice of Concurrency Control and Recovery*, Morgan Kaufmann, San Francisco, CA, 2001.

18. K. Ramamritham and P. K. Chrysanthis, *Advances in Concurrency Control and Transaction Processing*, IEEE Computer Society Press, Washington, DC, USA, 1998.

19. N. Paciorek T. Walsh, and D. Wong, Security and reliability in concordia, in *Proceedings of the Thirty-First Hawaii International Conference on System Sciences*, Kohala Coast, HI, Jan. 6–9, 1998, IEEE Computer Society, Washington, DC, 1998, pp. 44–53.

20. A. Mohindra, A. Purakayastha, and P. Thati, Exploiting non-determinism for reliability of mobile agent systems, in *Proceedings of the 2000 International Conference on Dependable Systems and Networks (DSN)*, June 25–28, 2000, IEEE Computer Society, Washington, DC, USA, New York, NY, 2000, pp. 144–156.

21. S. Pleisch and A. Schiper, Fault-tolerant mobile agent execution, *IEEE Trans. Computers*, 52(2):209–222, 2003.

22. S. Pleisch and A. Schiper, Approaches to fault-tolerant and transactional mobile agent execution—an algorithmic view. *ACM Comput. Surv.*, 36(3):219–262, 2004.

23. F. B. Schneider, Towards fault-tolerant and secure agentry, in M. Mavronicolas and P. Tsigas (Eds.), *Distributed Algorithms, 11th International Workshop*, WDAG '97, Saarbrucken, Germany, Sept. 24–26, 1997, Lecture Notes in Computer Science, Vol. 1320, Springer, Berlin, Heidelberg, 1997, pp. 1–14.

24. A. Elmagarmid, *Database Transaction Models for Advanced Applications*, in M. Mavronicolas and P. Tsigas, (Eds.), Morgan Kaufmann, Springer, Berlin, Heidelberg, 1992.

25. P. K. Chrysanthis, Transaction processing in a mobile computing environment, in B. Bhargava (Ed.), *Proceedings of IEEE Workshop on Advances in Parallel and Distributed Systems*, IEEE Computer Society Press, Princeton, NJ , USA, Oct. 6, 1993, pp. 77–82.

26. F. M. de Assis Silva and S. Krause, A distributed transaction model based on mobile agents, in K. Rothermel, R. Popescu-Zeletin (Eds.), *Mobile Agents, First International Workshop, MA'97*, Berlin, Germany, Apr. 7–8, 1997, Lecture Notes in Computer Science, Vol. 1219, Springer, Berlin, Heidelberg, 1997, pp. 198–209.

27. H. Vogler, T. Kunkelmann, and M.-L. Moschgath, An approach for mobile agent security and fault tolerance using distributed transactions, *paper presented at the 1997 International Conference on Parallel and Distributed Systems (ICPADS '97)*, Seoul, Korea, Dec. 11–13, 1997, IEEE Computer Society, Washington DC, 1997, pp. 268–274.

28. K. Rothermel and M. Strasser, A fault-tolerant protocol for providing the exactly-once property of mobile agents, *paper presented at the Symposium on Reliable Distributed Systems.*, West Lafayette, Indiana, USA, IEEE Computer Society, Washington, DC, USA, Oct. 20–22, 1998, pp. 100–108.

29. M. Strasser and K. Rothermel, System mechanisms for partial rollback of mobile agent execution, in M. G. Gouda (Ed.), *Proceedings of the 20th International Conference on Distributed Computing Systems*, Taipei, Taiwan. IEEE Computer Society, Washington, DC, April 10–13, 2000, pp. 20–28.

30. F. M. de Assis Silva and R. Popescu-Zeletin, An approach for providing mobile agent fault tolerance, in *Mobile Agents, Second International Workshop, MA'98*, Stuttgart, Germany, Sept. 1998, Lecture Notes in Computer Science, Vol. 1477, Springer, 1998, pp. 14–25.

31. T. Kaneda, M. Shiraishi, T. Enokido, and M. Takizawa, Mobile agent model for transaction processing on distributed objects, in L. Barolli (Ed.), paper presented at the 18th International Conference on Advanced Information Networking and Applications (AINA 2004), Fukuoka, Japan, Mar. 2004, 29–31, IEEE Computer Society, Washington, DC, 2004, pp. 506–511.

32. T. Kaneda, Y. Tanaka, T. Enokido, and M. Takizawa, Transactional agent model for fault-tolerant object systems, in H. Haddad, L. M. Liebrock, A. Omicini and R. L. Wainwright (Eds.), *Proceedings of the 2005 ACM Symposium on Applied Computing (SAC)*, Santa Fe, NM. ACM, New York, NY, Mar. 13–17, 2005, pp. 1133–1138.

33. H. Vogler and A. P. Buchmann, Using multiple mobile agents for distributed transactions, in *Proceedings of the 3rd IFCIS International Conference on Cooperative Information Systems (CooPIS)*, Aug. 20–22, 1998, New York, IEEE Computer Society, Washington, DC, USA, 1998, pp. 114–121.

34. R. Sher, Y. Aridor, and O. Etzion, Mobile transactional agents, in M. G. Gouda (Ed.), *Proceedings of the 21st International Conference on Distributed Computing Systems (ICDCS 2001)*, Phoenix, Arizona, USA. IEEE Computer Society, Washington, DC, USA, April 16–19, 2001, pp. 73–80.

35. E. Pitoura, Transaction-based coordination of software agents, in G. Quirchmay, E. Schweighofer and T. J. M. Bench-Capon (Eds.), *Database and Expert Systems Applications, 9th International Conference*, DEXA '98, Vienna, Austria, Aug.

24–28, 1998, Lecture Notes in Computer Science, Vol. 1460, Springer, Berlin, Heidelberg, 1998, pp. 460–469.

36. E. Pitoura and B. K. Bhargava, A framework for providing consistent and recoverable agent-based access to heterogeneous in mobile databases, *SIGMOD Record*, 24(3):44–49, 1995.

37. D. Georgakopoulos, M. Rusinkiewicz, and A. P. Sheth, Using tickets to enforce the serializability of multidatabase transactions, *IEEE Trans. Knowledge Data Eng.*, 6(1):166–180, 1994.

10 Mobile Agents in Mobile and Wireless Computing

PING YU and JIAN LU

State Key Laboratory for Novel Software Technology at Nanjing University, Nanjing, Jiangsu, P.R. China

JIANNONG CAO

Department of Computing, Hong Kong Polytechnic University, Hung Hom, Kowloon, Hong Kong

10.1 INTRODUCTION

Mobile and wireless computing raises new challenging problems due to the diverse types of devices used, user mobility, and dynamic nature in network conditions and execution context [1]. Mobile devices [e.g., laptop, personal digital assistant (PDA), mobile phone] do not have a permanent connection to the network and are often disconnected for long periods of time. When a device is reconnected to a network, the performance of the network connection can vary dramatically from the previous connection; the connection often has low bandwidth and high latency and is prone to sudden failures.

To provide support for building mobile computing applications, research in the field of middleware systems has proliferated in recent years. A middleware is a software system that connects two otherwise separate distributed entities, serving as the glue between application and service components in a distributed system. It handles network communication and distributed processing issues, providing application developers with a higher layer of abstraction [2]. A number of research projects have been exploring how to build mobile applications with the support of a middleware [3]. Several efforts focus on how to deliver information/services to mobile users regardless of

Mobile Agents in Networking and Distributed Computing, First Edition.
Edited by Jiannong Cao and Sajal K. Das.
© 2012 John Wiley & Sons, Inc. Published 2012 by John Wiley & Sons, Inc.

where they are and how they are connected in a suitable format [4–10] as well as on how to support weakly connected or disconnected operation on the user's device and maintain consistency after providing mechanisms for mobile host/agents to interact and coordinate their activities [11, 12]. Recently, more research has been carried out on developing middleware techniques for supporting context-aware and adaptive mobile applications [13–17].

As a paradigm for code mobility in large-scale distributed settings like the Internet, the mobile agent (MA) introduces several advantages, including reducing network load by accessing resources locally, executing task asynchronously and autonomously, and being hardware and transport layer independent [18]. Many applications can benefit from using a MA. Mobile agents can migrate through the network of sites to search, filter, and process information they need to accomplish their tasks. They can also cooperate with each other by sharing and exchanging messages and partial results and collectively making decisions. The MA paradigm is especially suited in mobile computing environments where disconnection from the network often occurs. Recently, many proposals have suggested using MA technology for mobile computing middleware [8, 19, 20]. Mobile agent–based middleware shows promise for providing an advanced infrastructure that integrates support protocols, mechanisms, and tools to permit communication and coordination of mobile entities [21]. Implementing MA based middleware to support mobile computing requires extending the mobile agent platform architecture, which is generally organized in layered services.

In this chapter, we cover MA-based mobile computing middleware and an MA-enabled platform on wireless hand-held devices. We will first provide a survey on the state-of-the-art in mobile computing middleware. Second, we will concentrate on MA technology in mobile and wireless computing. We will also discuss our own work. We will investigate the MA for pervasive/ubiquitous computing as well.

10.2   MOBILE COMPUTING MIDDLEWARE

Mobile computing is distinguished from traditional distributed computing by hand-held devices, wireless network connection, and dynamic execution context [3, 22]. Hand-held devices (except laptop) usually have limited battery capability and memory size, low central processing unit (CPU) speed, small screen size, and other constraints. These devices can not afford to run traditional distributed computing middleware such as Common Object Request Broker Architecture (CORBA) and Java 2 Enterprise Edition (J2EE) on them because these middleware always involve a lot of services or modules even though they are not required by most on-top applications. In middleware running on resource-scarce devices, the middleware should be lightweight with a configurable architecture. Not all functionalities of the middleware will be installed in the device. The user can choose a subset of these functionalities based on device profile or user preferences. Wireless networks also vary from type to type, such as wireless local-area networks (WLANs), General Packet Radio Service (GPRS), and Universal Mobile Telecommunication System (UMTS). These different networks coexist today to provide the foundational network infrastructure for mobile users. The bandwidth of wireless networks has been greatly improved. However, wireless network bandwidth outdoors is still much lower than WLAN bandwidth indoors. In addition, with wireless networks, users experience network disconnection when moving between different areas. To minimize packet or session loss, asynchronous communication should be enabled for mobile computing applications. Specifically, mobile computing middleware should support disconnected operations when the wireless network is unstable or network handoff occurs. Moreover, when a user moves from one area to another, not only will his or her location change but also the whole execution environment, especially the network properties (e.g., bandwidth or connectivity), will change accordingly. This requires middleware and applications to behave differently to cope with dynamic changes in the environment. Context awareness is desired for building a flexible and efficient mobile computing middleware. As Anind Dey defined, system is context aware if it can adapt to its location of use, the collection of nearby people and objects, as well as changes of these objects over time [23]. The context of a mobile unit is usually determined by its current location. Besides location information, user context also includes information about device characteristics, a user's profile or preferences, and any physical or social environmental status (e.g., temperature, noisy level, people nearby, user activity) which impacts application execution.

A lot of middleware has been implemented for mobile computing, addressing different aspects such as lightweight, asynchronous communication and context awareness.

10.2.1 Lightweight

Compared with traditional distributed computing middleware, the middleware for mobile computing should be designed as lightweight as possible. In general, *separation of concerns* is the basic principle when designing such a lightweight middleware. How to make the middleware dynamically configurable is a challenge to software engineers. Non configurable middleware always hides implementation details from users. This feature is recognized as transparency. Otherwise, the middleware should expose some of its implantation to the users and provide a mechanism for modifying middleware settings in a principled way, that is, openness. The reflection technology satisfies these requirements nicely. Initially, reflection was proposed in programming languages for the program introspecting itself during run time [38]. Then reflection is exploited in operating systems and middleware. A reflective system is basically separated into two levels: metalevel and base level. The base level is in charge of modeling the real-world problem domains, whereas the metalevel is responsible for modeling the base-level elements. The metalevel can intercept base-level status and modify the base-level structure or behavior by natural means. Consequently, the middleware base level can be initialized with a minimal subset of function-alities. Through metainterfaces between the metalevel and base level, the mid-dleware can dynamically load or unload services as current requirements indicate. Examples of middleware based on the principle of reflection include but are not limited to OpenORB [24], OpenCorba [25], dynamicTAO [26], MobiPADs [16], and CARISMA [27].

10.2.2   Asynchronous Communication

Communication in a mobile or wireless environment is characterized by fre-quent disconnection and limited bandwidth [22]. Synchronous communication, such as message passing and remote procedure call/remote method invocation (RPC/RMI), depends on a tight coupled relationship between client-side devices and server-side hosts. In contrast, asynchronous communication does not require the sender and receiver of a request to be connected at the same time. So it is more competitive compared to the synchronous paradigm especially in an unstable and fragile network. As we know, tuple space is a widely accepted facility for asynchronous interaction. It decouples participants by a globally shared and associatively addressed memory space, which acts as a repository of tuples that can be seen as a vector of typed values [3]. Tuples are anonymous; thus their selection is based on a pattern-matching mechanism. So this paradigm fits well in mobile settings where logical and physical mobility are both involved. LIME [11], T Space [12], and JavaSpace [28] all fall in this category. Besides the tuple space communication paradigm, some other technologies coexist as competitors. Message-oriented middleware [e.g., IBM's MQSeries and Sun's Java Message Service (JMS)] is a popular middleware which also decouples senders and receivers by publishing and subscribing topics.

10.2.3   Context Awareness

To let applications adapt to the user's environment, middleware should make applications capable of sensing the context which is used to characterize the execution environment. In general, there are three levels of activity between a mobile computing device and its user: personalization, passive context aware-ness, and active context awareness [29]. Personalization provides facilities for the users to customize his or her own settings or profiles of how the application should behave in a given situation. Passive context awareness makes the con-text change explicit to the users who will take charge of the application adaptation. Active context awareness is similar to automatic reconfiguration or self-adaptation which will change application behaviors according to the sensed context information without user intervention. Many research groups focused on location-aware middleware which depends on the underlying network operating systems to obtain location information and generate a suitable for-mat to be used by the system. Different locating technologies, such as the global positioning system (GPS) outdoors and infrared and radio frequency sensors indoors, will provide absolute or relative location information. Other kinds of context information can also be accessed by external public services like weather forest service or internal context providers which will acquire context value directly from ubiquitous sensors located in a specific environment or from system status monitors. To provide context awareness, context-aware middleware should include a context processing module which is responsible for detecting, collecting, interpreting, filtering, reasoning, predicting, and disseminating context [23, 30, 31].

10.3 MOBILE AGENT IN MOBILE COMPUTING

The MA has been a promising technology in the distributed and mobile computing society since the late 1990s. In general, MAs are self-contained and identifiable computer programs that can move autonomously within the network and act on behalf of the user or other entities. Mobile agent platforms provide migration service that serializes both MA code and execution state into streams suitable for network transfer and storage persistency. Consequently, a MA can execute at a host for a while, halt execution, dispatch itself to another host, and resume execution there [32]. Today, the MA has been exploited in network management, mobile e-commerce, as well as mobile computing middleware.

Since we have discussed the MA for networking in previous chapters, in this chapter, we will focus on how to make use of an MA in mobile computing middleware and application layers.

10.3.1 Mobile Agent–Based Mobile Computing Middleware

Mobile agent–based middleware provides an advanced infrastructure that integrates support protocols, mechanisms, and tools to permit communication and coordination of mobile entities [21]. As discussed in Section 10.2, mobile computing middleware should be at least lightweight and context aware and support asynchronous communication. Generally, the MA performs well in wireless network where bandwidth is always limited and disconnection happens frequently. As a widely known scenario in MA-based mobile computing, a mobile user can dispatch a MA to perform computation tasks autonomously over the network. Once the agent is dispatched, the user can disconnect from the network. Later the user reconnects to the network to collect the result returned by the MA. Consequently, disconnected operation can be supported by an MA in a more principled and natural way. Moreover, as an autonomous entity roaming in the network, an MA also has the capability of adapting itself when its execution environment changes. That is, mobile agents can be naturally programmed as context-aware entities. In terms of a MA-based system, the system can be configured and reconfigured by dynamically deploying specific agents. However, due to the lack of serialization support and other constraints in resource-limited mobile devices, MA systems encounter a big obstacle when being deployed in these devices. So we proposed a lightweight

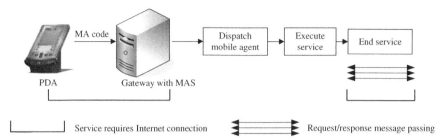

FIGURE 10.1   PDAgent agent–proxy–server approach.

and highly portable middleware (*PDAgent*) for deploying and developing MA-enabled applications in wireless hand-held devices.

10.3.1.1   Agent–Proxy–Server Approach

In PDAgent, we do not install any MAS (mobile agent server) on hand-held devices but introduce a midtier, a gateway (i.e., the proxy), to bridge the mobile user with the MA server deployed in the wired network (Figure 10.1). We name this approach agent–proxy–server to distinguish from other MA-based methods. The client-side device interacts with the gateway using the protocol defined in PDAgent middleware. The gateway parses and interprets the user's requests and interacts with MA servers. The server-side hosts provide services for incoming MA to perform their tasks. From the user's point of view, the gateway is treated as a proxy of a MA server. Users can control this remote proxy to deploy and manage their agents directly from their hand-held devices. Moreover, the middleware on the client side supports multiple channels for the users to get the MA codes. This means the MAs that can be deployed are provided or designated by the user and not bounded by the gateway. On the gateway side, various MA servers can be linked. So the deployable MAs in this approach are not constrained with a specific MA system. It is more flexible than other monolithic MA-based systems.

Furthermore, the middleware enables application developers to specify the way to customize the initialization of the agent instances using so-called parameters. In addition to those parameters from the end users, special parameters, called context parameters, can also be specified by the MA developers. While deploying the MAs on PDAgent, the middleware will fill in those special parameters according to the current user's profile and the status of the current environment. Thus, context-aware deployment is achieved.

10.3.1.2   Design of PDAgent

The design of PDAgent middleware is based on the following model as shown in Figure 10.2. A set of MA servers are dispersed over the network to support the running of the MAs and provide services. Mobile users, holding the wireless hand-held devices, can roam in the network. The client-side middleware enables the user to manage MA code, issue commands for dispatching and controlling

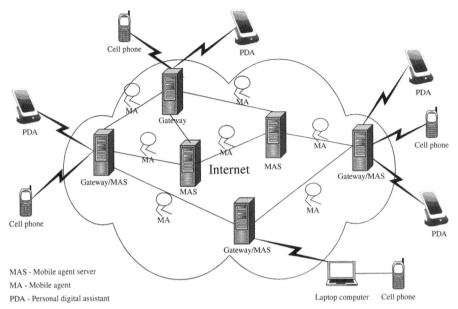

FIGURE 10.2   System model of PDAgent.

agent instances, and deal with results. To bring the MA technology to those users without burdening the carried devices with a whole ponderous system, MA servers for roaming users to run their mobile agents are still placed within the wired network instead of being ported into the hand helds. Gateways are deployed at the edge of the wired network, bridging the mobile users and the MA servers. Users are able to select one of the gateways currently reachable to serve as a proxy for them to dispatch or manage their MAs. The gateway can be set by the user who prefers a particular trustful gateway or be dynamically discovered by the underlying system. The gateway-side middleware deploys MAs to the MA servers as the user designated, and then manages the MAs' life cycle according to user's commands. In addition, it provides additional mechanisms, such as running a database to store final results that have to be kept for the user.

Thus, the service-accessing procedure becomes convenient, reliable, and affordable for those mobile users. They can import MA codes and store them in their hand-held devices locally. On dispatching an agent, the user should fill in all the customizing information for the instance and send it with the corresponding codes to the gateway. Then the gateway will create an instance on behalf of this user. Future management on this instance can also be performed by the roaming user remotely issuing management commands to the gateway. The agent instance will travel from the gateway to its destination host for the desired service and bring the result, if any, back. The user can get results by "pull" or "push" mechanisms. In terms of pull, the user can get connected again to the gateway and fetch the result. In contrast, when a user reconnects to a network, the device will send an "online" signal to the gateway and then the cached result in the gateway can send the result to the client. Basically, the agent can be deployed when the user is offline. After the codes and parameters are delivered, the user can disconnect again until the next time he or she wants to communicate with the gateway to manage the agent instances. The requirement for network connection is minimized.

In general, PDAgent middleware is featured as follows:

- *Lightweight* The PDAgent client side is able to run smoothly on resource-constrained devices because the client side will not include a heavyweight mobile agent server as most existing MA middleware requires. The agent is actually initiated and activated at the gateway. The client side only needs to send a request to the gateway to trigger the action. Though no agents run in the client side, the client also has full control of the agent by management functions provided by the PDAgent client-side services, such as querying agent status, retracting the agent, and destroying the agent.

- *Connectivity* The PDAgent provides mobile users with reliable and efficient network access minimizing network connection. Besides the disconnected operation supported in MA, the gateway provides a flexible, extendable, and scaleable bridge between the client and the MA environment. Network traffic is encoded/decoded by XML Engine and compressed/decompressed by Network. Thus, the network traffic for the running of applications is greatly reduced.

- *Flexibility* The PDAgent allows users to subscribe to the MA-enabled services in a loose manner. The client side is not aware of implementation details of these services. The service's self-description is encapsulated in XML file for the client to dynamically subscribe and download MA-enabled service. Moreover, users can import MA code from various origins into their hand helds. The customizing information for one instance, which we call parameters, can configure the outgoing instance with not only the user requirements but also the context factors. We also address information security and network optimization in this platform. These features make the PDAgent platform a novel solution to extend the use of MAs to the mobile computing environment.

- *Heterogeneity* Because of the gateway architecture, a user may deploy MAs on heterogeneous MA systems. User only needs to care about the business logic of MAs. There is a "connector" in the gateway which decouples the design of the platform on its client side from the MA server side. By using multiple connectors to link different MA servers in a gateway, the user of this gateway can transparently deploy MAs of different types.

These properties make it feasible to deploy PDAgent middleware in mobile devices for mobile computing. Mobile users can access network services in different locations from all wireless devices that install PDAgent middleware. Moreover, security and network optimization have been taken into account when implementing PDAgent middleware:

- *Security* Most security issues in the MA environment, such as shielding the service provider from malicious MAs and protecting MAs from evil servers, are still open research issues in this distributed computing model. To avoid potential threats in this area, we advocate that the user download MA code from trusted places and MAs be executed on trusted servers in a secure environment, such as within an intranet or over a virtual private network (VPN). We try to address security problems from the point of view of PDAgent middleware, as an enabling platform for the mobile user to use MA technology. A secure communication channel is established between the client and the gateway once an online interaction is needed. Currently, all information exchange is based on the HyperText Transfer Protocol (HTTP) and HTTP Secure (HTTPS) is adopted to ensure the information transferred is neither eavesdropped nor tampered with. Besides, on sending any information package out, the sender will attach the digital signature of the outgoing message. The receiver can verify this signature; the authentication and authorization at the application level can follow after the identity of the sender is extracted. The gateway will only serve those authorized users in a predefined list. Thus information security and the mutual authentication/authorization are achieved in our PDAgent platform.

- *Network Optimization* Reducing network connection time and probability of network errors is an important issue in designing mobile computing middleware. To ensure that communication between wireless devices and the gateway will not affect overall operational performance, we employ the following two techniques to optimize networking services. First, all the traffic between the hand-held device and the gateway will be compressed before transmission. Some compressing algorithms are used to reduce the amount of data needed to transmit over the air. Second, a probing procedure can be carried out manually or automatically, according to the preference set by the user. The round-trip time (RTT) from the current reachable gateways to the client will be detected by sending a probing request. On receiving the responses, the RTT can be calculated and the gateway with the smallest RTT value is appointed to be the active one to communicate with the client. Assuming that the network bandwidth is the same, this can minimize data transfer times, with the minimum transfer time being dependent on the shortest RTT from the wireless device to the gateway.

10.3.2 Mobile Agent for Mobile Applications

A lot of mobile applications can benefit from using mobile agents. Mobile e-commerce is the most popular one. For example, when you are using wireless devices to browse an electronic book store, it must displease you if you have to visit a lot of Web pages until you finally get what you want. In the case of an MA, you only need to dispatch an MA which encapsulates your requests of the book name, preferred price, and publisher name, to name a few. The agent will help you search in the back-end servers (e.g., book store's database) or even other book stores which provide similar services. When the agent is dispatched, you do not need to stay online as you would with traditional m-ecommerce applications. After the agent gets satisfactory results, it will return to you when you are online again. This not only saves time but also saves network usage fees. Other applications, such as yellow page searcher, Web page searching, and email assistant, will also benefit from using MA, and most have been implemented on top of PDAgent middleware.

Another appealing application is cooperative agents for scheduling meetings. Suppose you want to hold a meeting. It is more natural and efficient for each person to dispatch an agent to a common space, discussing the meeting time and place. Generally, mechanisms for cooperative MAs are developed with the concept based on a multiagent system, which enables an agent to communicate and coordinate with other agents of different types or dispatched by different users automatically and intelligently, in either a reactive or a cognitive way. *Tuple Space* [12] seamlessly integrated into PDAgent middleware supports cooperative agent-based mobile applications such as scheduling a meeting.

Mobile agents are also suitable for more advanced mobile applications such as workflow management. A workflow can be easily organized by deploying and dispatching a multitask, multihop agent. This agent will migrate to different sites performing tasks one by one as defined. Parallel execution is also allowed by simultaneously dispatching an agent and its clones to different sites. The agent is also aware of different situations and capable of automatically adjusting the migration route and tasks during run time. When all tasks are finished, it will provide you with a complete report.

There are many other MA-based applications in the mobile computing society. For example, Hermes supports MAs for bioinformatics and industrial control where MAs are used to support data collection, service discovery and self-healing and so on [33]. With the advent of pervasive computing, MAs are also finding their way in pervasive computing. We will discuss this in the next section.

10.4 MOBILE AGENT FOR PERVASIVE COMPUTING

Mobile computing is followed by pervasive or ubiquitous computing where computation can be performed anywhere, anytime. Since pervasive computing has evolved from mobile computing, properties in mobile computing such as heterogeneous hand-held devices, various and unstable networks, and a changeable environment are all included in pervasive computing. In particular, pervasive computing emphasizes seamless migration in smart spaces and self-adaptation without user intervention. In general, users are almost unaware of the computing technologies they are enjoying, as the technologies have been weaved into the fabric of everyday life [34].

However, currently, development of the application of pervasive computing falls behind the technology advances in hardware and communication infra-structures. The pervasive computing community is expecting applications that can make better use of the underlying networking facilities and computing devices without requiring complicated interactions with the users. In addition to the requirements already mentioned in mobile computing, the following are of particular importance for pervasive computing.

10.4.1 Seamless Mobility

One of the most challenging issues in pervasive computing is application migra-tion. Traditional application-to-computer association will disappear in pervasive computing environments [35]. The user will benefit from application mobility in at least two aspects: (1) User can transfer application to any available device before the device is running out of power or when more powerful devices are available to make good use of various hardware resources and (2) When usermoves to another place, the application running on his original unmovable device can automatically migrate to any computing device available at his destination place to continue the computation. Such an application is well known as a "follow-me" application.

10.4.2 Self-Adaptation

Being context aware, applications and the underlying supporting system should be able to adapt to the users needs. Adaptation includes changing application content, user interface, and other application or system-level behavior for switching between different devices and environments.

The MA has been demonstrated to be suitable for mobile computing. Consequently, it is natural for us to consider exploiting the MA in pervasive computing, especially for movable applications. In fact, we have developed MA-enabled applications, such as "smart" notepad and media player, for pervasive computing. A corresponding middleware, MDAgent, which is based on the previous PDAgent platform, is implemented to provide application with mobility and self-adaptability. Generally speaking, MAs provide a more principled and elegant approach to move the application from site to site compared with other proposals designed for application mobility.

In MDAgent middleware, an application is wrapped by an MA that has full control (e.g., suspending, migrating, resuming) of the application. Aided by the underlying middleware services, MA-enabled applications can be made

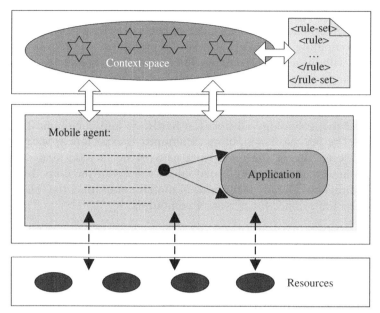

FIGURE 10.3 Abstract model of agent enable application.

sensitive to the user's execution context, including location, activity, preference, and the profile of the device being used. With regard to self-adaptation, each MA is attached with a rule set which consists of some adaptation rules. Figure 10.3 shows the abstraction of such an MA-enabled application.

The general life cycle of an MA-enabled application is illustrated in Figure 10.4. When the middleware detects that the user is leaving the current site, the agent will automatically be suspended and then cached to secondary storage (deactivated). When the user logs into another device, the middleware will activate the agent, which will exit the current site and migrate to the device where the user is working. When it arrives at the new site, the agent first checks in (or registers) at the middleware. After successful registration (e.g., authentication, authorization), it will restore from the application snapshot and resume execution from exactly where it was suspended. The MDAgent middleware consists of an agent container for agent execution and management, a context space for collecting and processing internal and external context information, a rule engine for making decisions when the context changes, and a discovery and directory component for looking up hosts and services when the agent migrates to another site.

Pervasive computing is distinguished by invisibility, which means computation will happen silently without distracting users. We believe the MA is suitable for this kind of invisible computing as we have practiced in MA-enabled "follow-me" applications. In the future, people will hold a mobile device where a lot of MAs with different tasks are installed. These agents are

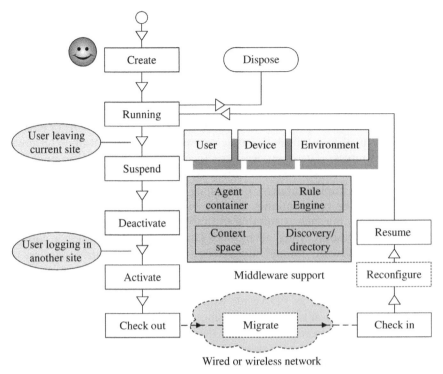

FIGURE 10.4 Life cycle of "follow-me" application in MDAgent.

customized by users or automatically configured by underlying context-aware services, keeping the user's intention in mind. Whenever they are dispatched out to execute computation tasks, they will work according to their user's needs. An MA will perform boring and complicated searching, typing, or other unnecessary tasks. Mobile agent–based smart spaces have also been proposed.

10.5 RELATED WORK

Currently researchers in distributed and mobile computing areas have presented a number of architectures and systems that provide lightweight, asynchronous, and context-aware computation, communication, and coordination for mobile devices and wireless networks. In this section, we discuss some related work in detail and compare the work with ours:

- **SOMA** SOMA is MA-based middleware which provides three different kinds of mobility—user mobility, terminal mobility, and mobile access to resources—on top of a secure and open MA-distributed programming framework [21]. Concretely, a mobility-enabled naming service based on a directory and discovery protocol [37] is provided as a foundational service for tracing entities that move in the global Internet environment. The user virtual environment (UVE) lets users connect to the Internet at different locations while maintaining their personal configurations as indicated in user profiles. The mobile virtual terminal (MVT) service enables continuous execution while moving by virtue of underlying MA platforms. For mobile access to resources, virtual resource management (VRM) is implemented to establish dynamic connections between mobile terminals and resources. These services are incorporated into an original MA platform to satisfy requirements from mobile computing. Moreover, Java-based implementation overcomes the platform heterogeneity and applies to the open Internet. SOMA middleware demonstrated that the MA-based approach is suitable to mobile computing where users, terminals, and resources are all mobile. However, SOMA is not a context-aware middleware. Though it provides a UVE for customizing services the as user's profile indicates, it cannot fetch real-time context information from the environment and adapt the services dynamically. Moreover, It is not as lightweight as PDAgent, which can be run in a personal digital assistant (PDA) or mobile phone, since it relies on a standard JVM which is not available in resource-limited devices.

- **LIME** LIME is the abbreviation of Linda in a Mobile Environment [11]. To exploit Linda in the mobile computing environment, the traditional Linda tuple space is broken down into many tuple spaces (interface tuple space), each permanently associated to a mobile unit. The mobile unit refers to the mobile agent or mobile host. The mobile agent provides logical mobility, while the mobile host indicates physical mobility. Two novel operations—engagement and disengagement—are introduced for attaching and detaching distributed interface tuple spaces. Coordination between different mobile units is asynchronous due to the intrinsic properties of Linda. Moreover, LIME extends Linda operations with tuple location parameters that allow operating on different projections of the transiently shared tuple space. Besides LIME, other tuple space–based middleware coexist. T Space and JavaSpaces are two famous ones supported by IBM and Sun, respectively. L2imbo is another tuple-based middleware, which is featured as quality-of-service (QoS) aware by attaching QoS information to tuple fields [36]. In general, in view of asynchronous communication and coordination for mobile computing, the tuple space is a competitive and preferred choice. In PDAgent, though we adopt the MA to provide disconnected operation for wireless network communication, when we consider cooperative agent-based applications such as meeting scheduling, the tuple space is also incorporated into PDAgent as a basic service for agent coordination. Otherwise, LIME and L2imbo extend the original Linda model and introduce new characteristics for mobile computing.

LIME supports dynamic reconfiguration of MAs and mobile hosts within the same programming interface (i.e., interface tuple space), whereas L2imbo provides QoS-aware services by associating agents to monitor network conditions and power consumption.

- **CARISMA** As we mentioned above, reflection is a principled mechanism to provide middleware lightweight-ness and context awareness. CARISMA is such a reflective middleware exploiting the principle of reflection to enhance the construction of adaptive and context-aware applications in a mobile computing environment. Being aware of user preferences and application profiles which are defined in a platform-independent extensible Markup Language (XML) format, middleware can deliver services in a more suitable way. Reactive adaptation and proactive adaptation are both enabled by virtue of introspection and intercession, two basic operations in a reflective system. Users can define their preference to applications through metainterfaces exposed by the middleware. Meanwhile, middleware can adapt applications dynamically through built-in reflective application programming interfaces (APIs). In short, CARISMA is capable of deployment time configurability and run time adaptability. In fact, in PDAgent and MDAgent, we also exploit reflection techniques to achieve elegant adaptation. The MA is attached to metadata specifying properties of the agent itself and the intention of agent holders. These metadata are used by the underlying middleware when context or user requirements change. The middleware will adapt the agent according to the user's preferences. Moreover, the agent can also adapt itself by reflecting on its metadata during run time.

- **JADE** JADE [19] is a middleware for development and run-time execution of peer-to-peer agent applications which can work and incorporate both in the wired and wireless environment. JADE is a Foundation for Intelligent Physical Agents (FIPA) specification-compliant implementation. It enables an agent to operate with other FIPA-compliant agents. JADE is independent of the Java version. Thus it can be easily deployed into mobile devices without worrying about Java virtual machine (JVM) difference. Moreover, JADE can be configured to adapt to the characteristics of the deployment environment. JADE architecture, in fact, is completely modular and, by activating certain modules instead of others, it is possible to meet different requirements in terms of connectivity, memory, and processing power. Agents in JADE are very suitable for those coordination- and negotiation-oriented applications due to the underlying agent-based peer-to-peer architecture. Agent mobility is also supported in JADE. Moreover, agents are capable of controlling themselves without human intervention. In another words, agents are self-adaptive according to our definition above. JADE is an ongoing project in the open-source society today, with new features incorporated continuously. JADE provides a lightweight agent container and a flexible peer-to-peer mechanism for upper layer applications which are adopted by a few mobile computing and pervasive computing proposals.

10.6 CONCLUSION

In this chapter, we discussed the state-of-the-art of mobile and wireless computing middleware. Generally, lightweight, asynchronous communication and context-aware services are recognized as most important issues for enabling reliable and flexible computation and communication from mobile devices. In accordance, several different kinds of middleware satisfy these new requirements. The reflection technique becomes promising in constructing a lightweight and context-aware middleware by run time introspection and intercession. A tuple space is a preferred facility for enabling loosely coupled communication and coordination in the mobile computing environment. An MA provides a principled paradigm for computation mobility and communication asynchronicity in both the local network and World Wide Web settings. Based on existing knowledge, we proposed a lightweight and highly portable platform named PDAgent for programmers to develop and for mobile users to deploy MA-enabled applications from wireless devices. It has several advantages over existing solutions in providing support for the development of highly portable, efficient, flexible, and context-aware mobile applications.

The PDAgent platform, developed on the agent−proxy−server approach, provides efficient and reliable network services to mobile users with hand-held devices without requiring a large amount of resources. PDAgent also has low network latency and low network connectivity costs. This is achieved by providing a lightweight platform for running MAs that can execute services autonomously on the Internet on behalf of the user. Mobile clients can go offline after they submit the execution plan to the MA, thereby reducing network connection times. Security and optimizations are also considered to get close to practical application.

New challenges continually emerge as the pervasive computing era comes near. Among these challenges, seamless migration and minimized user distraction are most distinguished. Smartness is emphasized by the pervasive computing society as well. However, as a follower of mobile computing, pervasive computing also shares a lot of properties with mobile computing, such as lightweight and heterogeneous devices, variable and unstable networks, and a continuously changing environment. Naturally, in view of building a middleware for pervasive computing, the experiences from mobile computing are most helpful. From this opinion, we try to exploit MAs in this new area. We believe that the autonomy and reactivity of MAs is most suitable to the new paradigm. Mobile agents, similar to virtual surrogates of human beings, can perform tasks intelligently by bearing user intention in mind and sensing environmental change over time. As a result, the MDAgent middleware, which is based on the original PDAgent platform, becomes our touchstone to pervasive computing. This enables applications to follow the users from one place/device to another. Application states are conserved during migration due to the mobility service provided by MAs. Context awareness is also achieved by exposing environmental conditions to agents, which can subscribe the context they are interested in. The middleware is in charge of collecting and processing real-time context and notifying agents to adapt themselves when necessary.

However, many issues require further investigation. For example, fault tolerance is such a common issue which should never be ignored. In particular, MA-based middleware should guarantee reliable agent migration when the destination is unreachable suddenly or when the server is damaged during agent execution. Security is another critical issue which usually includes protecting the server from malicious agent attach and protecting agents from untrustful services. For agents in a pervasive computing environment, the issue of how to let agents from different sources recognize and cooperate with each other also challenges researches. The goal of organizing a smart space by mobile agents is identified as our next step.

REFERENCES

1. M. Satyanarayanan, Fundamental challenges in mobile computing, in J. E. Burns (Eds.), *Proceedings Symposium on Principles of Distributed Computing*, Philadelphia, PA, USA, May 23–26, 1996, pp. 1–7.

2. P. A. Bernstein, Middleware: A model for distributed system services, *Commun. ACM*, 39(2):86–98, Feb. 1996.

3. C. Mascolo, L. Capra, and W. Emmerich, Mobile computing middleware, in E. Gregori et al. (Eds.), *Networking 2002 Tutorials*, Lecture Notes in Computer Science, 2497, Springer, Berlin, Heidelberg, 2002, pp. 20–58.

4. A. T. S. Chan, S.-N. Chuang, and J. Cao, Dynamic service composition for wireless Web access, in T. S. Abdelrahman (Ed.), *Proceedings of the 31st International Conference on Parallel Processing (ICPP-2002)*, Vancouver, BC, Aug. 2002, pp. 429–436.

5. D. Chakraborty et al., Middleware for mobile information access, in A. M. Tjoa (Ed.), *Proceedings of the 13th International Workshop on Database and Expert Systems Applications (DEXA'02)*, Aix-en-Provence, France, Sept. 2002, pp. 729–733.

6. J. Jing, A. Helal, and A. Elmagarmid, Client-server computing in mobile environments, *ACM Comput. Surv.*, 31(2):117–157, 1999.

7. A. D. Joseph, J. A. Tauber, and M. F. Kasshoek, Mobile computing with the Rover Toolkit, *IEEE Trans. Comput.*, 46(3):337–352, Mar. 1997.

8. Q. H. Mahmoud, MobiAgent: A mobile agent-based approach to wireless information systems, in G. Wagner, K. Karlapalem, Y. Lesperance, and E. Yu (Eds.), *Proceedings of the 3rd International Bi-Conference Workshop on Agent-Oriented Information Systems*, Montreal, Canada, May, 2001, pp. 87–90.

9. S. Saha, M. Jamtgaard, and J. Villasenor, Bringing the wireless Internet to mobile devices, *IEEE Computer*, 34(6):54–58, June 2001.

10. M. Satyanarayanan, Accessing information on demand at any location Mobile information access, *IEEE Personal Commun.*, 3(1):26–33, 1996.

11. A. L. Murohy, G. P. Picco, and G.-C. Roman, LIME: A middleware for physical and logical mobility, in *Proceedings 21st International Conference on Distributed Computing Systems (ICDCS-21)*, Mesa, AZ , USA, Apr 2001, pp. 524–533.

12. P. Wyckoff, S. W. Mclaughry, T. J. Lehman, and D. A. Ford, T spaces, *IBM Syst. J.*, 37(3):454–471, 1998.

13. G. Chen and D. Kotz, A survey of context-aware mobile computing research, Technical Report TR2000-381, Dept. of Computer Science, Dartmouth College, Hanover, NH, Nov. 2000. Available at: http://www.cs.dartmouth.edu/reports/abstracts/TR2000-381/.

14. L. Capra, Mobile computing middleware for context-aware applications, in *Proceedings of the 24th International Conference of Software Engineering (ICSE 2002)*, Orlando, FL, May 2002, pp. 723–724.

15. S-N. Chuang, A. T. S. Chan, J. Cao, and R. Cheung, Dynamic service reconfiguration for wireless Web access, in *Proceedings of the 12th International World Wide Web Conference*, Budapest, Hungary, May 2003, ACM, New York, NY, USA 2003, pp. 58–67.

16. A. T. S. Chan and S. N. Chuang, MobiPADS: A reflective middleware for context-aware computing, *IEEE Trans. Software Eng.* 29(12):1072–1085, Dec. 2003.

17. S. S. Yau and F. Karim, A context-sensitive middleware-based approach to dynamically integrating mobile devices into computational infrastructures, *J. Parallel Distrib. Comput.*, 64(2):301–317, 2004.

18. D. B. Lange and M. Oshima, Seven good reasons for mobile agents, *Commun. ACM*, 42(3):88–89, 1999.

19. F. Bellifemine, G. Caire, A. Poggi, and G. Rimassa, JADE: A white paper, Telecom Italia EXP in search of innovation, Journal special issue on JADE, 3(3):6–19, Sept. 2003.

20. J. Cao, D. C. k. Tse, and A. T. S. Chan, PDAgent: A platform for developing and deploying mobile agent enabled applications for wireless devices, in R. Eigenman (Ed.), *Proceedings of the International Conference on Parallel Processing 2004*, Montreal, Canda, Aug. 2004, pp. 510–517.

21. P. Bellavista, A. Corradi, and Cesare, Mobile agent middleware for mobile computing, *IEEE Comput.*, 34(3):73–81, Mar. 2001.

22. A. Gaddah and T. Kunz, A survey of middleware paradigms for mobile computing, Carleton University Systems and Computing Engineering Technology Report SCE-03-16, July 2003. Available at: http://www.sce.carleton.ca/wmc/middleware/middleware.pdf.

23. A. K. Dey and G. D. Abowd, Towards a better understanding of context and context-awareness, in H.-W. Gellersen (Ed.), *Proceedings of HUC'99*, Lecture Notes in Computer Science, Vol. 1707, Springer, Berlin, Heidelberg, 1999, pp. 304–307.

24. G. S. Blair et al., The design and implementation of open ORB 2, *IEEE Distributed Syst. Online*, 2(6):1–40, 2001.

25. T. Ledoux, OpenCorba: A reflective open broker, in P. Cointe (Ed.), *Proceedings of Reflection 99*, Lecture Notes in Computer Science, Vol. 1616, Springer, London, UK, 1999, pp. 197–214.

26. F. Kon et al., Monitoring, security, and dynamic configuration with the dynamic-TAO reflective ORB, in J. Sventek and G. Coulson (Eds.), *Proceedings of Middleware 2000*, Lecture Notes in Computer Science, Vol. 1795, Springer, Berlin, Heidelberg, Apr. 2000, pp. 121–143.

27. L. Capra, W. Emmerich, and C. Mascolo, CARISMA: Context-aware reflective middleware system for mobile applications, *IEEE Trans. Software Eng.*, 29(10): 929–945, Oct. 2003.

28. E. Freeman, S. Hupfer, and K. Arnold. *JavaSpaces™ Principles, Patterns, and Practice*. Addison-Wesley as part of the Jini Technology Series from Sun Microsystems, Inc. San Antonio Road, Palo Alto, California 94303 USA, 1999.

29. L. Barkhuus and A. Dey, Is context-aware computing taking control from the user, in A.K. Dey (Ed.), *Proceedings of UbiComp 2003*, Lecture Notes in Computer Science, 2864, Springer, Berlin, Heidelberg, 2003, pp. 150–156.

30. B., Schilit, N. Adams, and R., Want, Context-aware computing applications, in *Proceedings of the Workshop on Mobile Computing Systems and Applications*, Santa Cruz, CA, 1994, pp. 85–90.

31. T. Gu, H. K. Pung, and D. Q. Zhang, Toward an OSGi-based infrastructure for context-aware applications, *IEEE Pervasive Comput.*, 3(4):66–74, Oct.–Dec. 2004.

32. D. Kotz and R. S. Gray, Mobile agents and the future of the Internet, *SIGOPS Oper. Syst. Rev.*, 33(3):7–13, 1999.

33. F. Corradini and E. Merelli, Hermes: Agent-based middleware for mobile computing, M. Bernardo and A. Begliolo (Eds.), *SFM-Moby 2005*, Lecture Notes in Computer Science, Vol. 3465, Springer, Berlin, Heidelberg, 2005, pp. 234–270.

34. M. Weiser, The computer for the twenty-first century, *Sci. Am.*, 265:94–101, 1991.

35. M. Roman, H. Ho, and R. H. Campbell, Application mobility in active spaces, in *Proceedings of the 1st International Conference on Mobile and Ubiquitous Multimedia*, Oulu, Finland, 2002.

36. N. Davies, A. Friday, S.P. Wade, and G.S. Blair. L2imbo: A distributed system platform for mobile computing, in I. Chlamtac, D. Lee, and M.Schwarts (Eds.), *ACM Mobile Networks and Applications, Special Issues on Protocols and Software Paradigms of Mobile Network*, Kluwer Academic Publishers, Hingham, MA, USA, 1998, pp. 143–156.

37. S. Helal, Standards for service discovery and delivery, *IEEE Pervasive Comput.*, 1(3): 95–100, 2002.

38. B. Smith, Reflection and semantics in a procedural programming language, Ph.D. thesis, Massachusetts Institute of Technology, Cambridge, MA, Jan. 1982.

PART IV
Design and Evaluation

11 Naplet: Microkernel and Pluggable Design of Mobile Agent Systems

CHENG-ZHONG XU

Department of Electrical and Computer Engineering, Wayne State University, Detroit, Michigan

11.1 INTRODUCTION

An agent is a sort of special object that has autonomy. It behaves like a human agent, working for clients in pursuit of its own agenda. A mobile agent has as its defining trait the ability to travel from machine to machine proactively on open and distributed systems, carrying its code, data, and running state. The proactive mobility of autonomous agents, particularly their flow of control, leads to a novel distributed processing model on the Internet.

Mobile agents grew out of early code mobility technologies such as process migration and mobile objects in distributed systems. Process migration deals with the transfer of code as well as data and running state between machines for the purpose of dynamic load distribution, fault resilience, eased system administration, and data access locality [1, 2]. In distributed object systems such as Common Object Request Broker Architecture (CORBA) and Java remote method invocation (RMI), object mobility is realized by passing objects as arguments in remote object invocation. Object migration makes it possible to move objects among address spaces, implementing a finer grained mobility with respect to process-level migration.

Early process and object migration frameworks are mostly targeted at a "closed" system in which programmers have complete knowledge of the whole system and full control over the disposition of all system components. They assume the process/object authority is under the control of a single administrative domain and hence deal with the transfer of authority in a limited way. In contrast, a mobile agent is tailored to "open" and distributed environments. Its

Mobile Agents in Networking and Distributed Computing, First Edition.
Edited by Jiannong Cao and Sajal K. Das.
© 2012 John Wiley & Sons, Inc. Published 2012 by John Wiley & Sons, Inc.

autonomy lends itself to be migratable, with a delegation of resource access privileges, between loosely coupled systems. Mobile agents can also survive intermittent or unreliable connections between the systems. What is more, mobile agents feature a proactive mobility in the sense that the migration can be initiated by the agent itself according to its predefined itinerary.

The proactive mobility of mobile agents was first demonstrated by Telescript [3]. Since then, mobile agent technology has received much attention in both academia and industries. Java language and its dynamic class loading model, coupled with several other important features such as serialization, remote method invocation, and reflection, greatly simplify the construction of mobile agent systems. Examples of the systems include Aglet [4], Ajanta [5], Concordia [6], D'Agent [7], JADE [8], JACK [9], and FarGo [10].

Naplet system [11, 12] was initially designed as an educational package for students to develop understanding of advanced concepts in distributed systems and to gain experience with network-centric mobile computing. Although distributed systems have long been in the core of computer science curriculum, there are few educational software platforms available that are small but full of key concepts and principles. A mobile agent system consists of a group of agent servers that accomplish agent migration between the servers and perform agent execution in a confined environment. Each agent server not only needs to protect itself from attacks by malicious or misbehaved agents but also requires to isolate the performance of the agents that are concurrently running in the same server and ensure reliable communication between collaborative agents while they are moving. Systems with such support serve as an ideal middleware platform for experiments with related concepts such as naming, process migration, resource management, reliable communication, and security in distributed systems.

An educational system must be organized in a way that mechanisms are separated from policies so that various algorithms can be implemented without changes in system infrastructure. For example, multiple agents running in the same server can be scheduled in First Come First Served (FCFS), round-robin, or other more complex policies. The system framework should be able to accommodate different policies at run time. More importantly, the system should be in a microkernel and pluggable architectural design so that its functionalities can be extended by reconfiguring present modules or plugging in new functionality modules. For example, agent execution in a remote server often requires to access certain local server resources (e.g., printer, data file, system status). The resources should be provided as run time plug-ins so that new or modified resources can be offered without shutting down the server. It is known that different applications have requirements for different types of agent communication: synchronous versus asynchronous, reliable versus unreliable. The communication module of an agent server should be able to be reconfigured at installation time or run time to meet the needs of different applications.

We reviewed a number of representative systems in public domains and none of them met our needs. The Naplet system was then developed. Since its alpha release in 1998, it has been used by hundreds of students at Wayne State

University and a number of other schools around the world as a platform for programming laboratories and term projects of advanced distributed system courses. Over the years, it has also evolved into a research testbed for the study of mobility approaches for scalable Internet services. The latest release of the software package is available at www.cic.eng.wayne.edu/software/naplet.html. Part of this chapter draws on [12].

Like other mobile agent systems, Naplet provides constructs for agent declaration, confined agent execution, and mechanisms for agent monitoring, control, and communication. It has the following distinct features: structured navigation facility, reliable interagent communication mechanism, open resource management policies, and agent-oriented access control. The system is in a microkernel and pluggable architectural design. It separates a minimal functional core for navigation and remote execution from extended functionalities such as interagent communication and application-specific local resource provisioning. The extensions can be plugged in at run time or configured at installation time. This design makes it easy to accommodate future development by plugging in new functionality modules and enhances system extensibility and availability.

The rest of this chapter is organized as follows. Section 11.2 presents the design goal and Naplet architecture. Section 11.3 gives system support for structured navigation. Section 11.4 presents a hybrid of centralized and decentralized mechanisms for agent tracking. The tracking mechanism enables location-independent internaplet communication. Two mechanisms for synchronous and asynchronous communication are discussed. Section 11.5 is dedicated to Naplet security architecture and related resource management strategies. Section 11.6 demonstrates the system programmability in network management. Related work is reviewed in Section 11.7. Section 11.8 concludes the chapter with remarks on the potential and limitation of Naplet and mobile agent technology in general.

11.2 DESIGN GOALS AND NAPLET ARCHITECTURE

11.2.1 Design Goals

Code mobility introduces a new dimension to traditional distributed systems and opens vast opportunities for new network-centric distributed applications. In a networked world, one is obligated to specify not only how to execute designated tasks of an agent but also where to execute them. A primary goal of the Naplet system is to support an agent-oriented programming paradigm. It is centered around two first-class objects: Naplet and NapletServer. The object Naplet is an abstraction of agents, defining a set of basic attributes pertaining to agents, hooks for application-specific functions to be performed on visited servers, and itineraries to be followed by the agent. We refer to each instance of the object Naplet as a *naplet*. NapletServer defines a docking place that provides naplets with a protected run-time environment within a Java virtual machine (JVM). We refer to each object of NapletServer as a Naplet server (or server).

The Naplet system was designed with two goals in mind—*microkernel* and *pluggable*:

- **Microkernel** To support the execution of naplets in a confined environment, Naplet server must provide an array of mechanisms for agent migration, agent communication, resource management, security control, and so on. The server is designed in a microkernel architectural pattern. It is represented by a highly modular collection of powerful abstractions, each containing handlers to deal with different but specific aspects of the mobility support. The microkernel design requires a separation of core functionalities in the form of a microkernel from extended functionalities and application-dependent services.

- **Pluggable** Most of the modules in the Naplet server are made as external services to the microkernel. Their default implementations can be easily replaced or enhanced with new implementations. Moreover, application services accessible to naplets should also be installed, configured, reconfigured, and removed dynamically without shutting down the server.

In principle, a microkernel architectural pattern defines five kinds of participating components in a software system: microkernel, internal server, external server, clients, and adapter [13]. The microkernel implements the minimal core functionalities; the internal server extends the core functionalities; and the external server exposes the functionalities to clients via an interface defined in the adapter.

When the microkernel pattern is applied to the Naplet system, each naplet is viewed as a client which interacts with local and remote Naplet servers in pursuit of its agenda on behalf of its owner (or program that dispatches the naplet). Naplet server comprises seven key components, as shown in Figure 11.1: NapletSecurityManager, ResourceManager, NapletMonitor, Navigator, Messenger, Locator, and NapletManager. Among them, the first three modules form a microkernel in support of remote execution in a confined environment and the other three modules are internal servers, providing extended functionalities for multihop migration and internaplet communication. NapletManager is an external server that provides an interface between naplets and other functionality modules. The interface is defined in an adapter, NapletContext.

In the following, we first present the class Naplet in detail. We then delineate the microkernel and pluggable design of NapletServer.

11.2.2 Naplet Class

Naplet is a template class that defines a generic mobile agent. Its primary attributes include a systemwide unique immutable identifier, an immutable codebase uniform resource locator (URL), and a protected serializable container of an application-specific agent running states in the following:

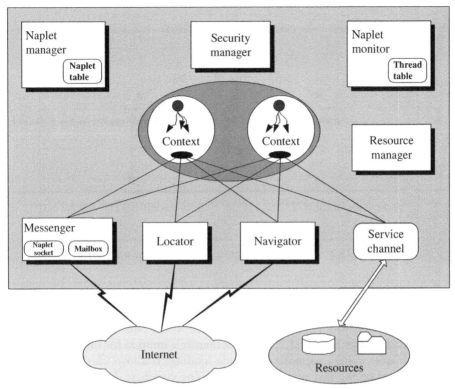

FIGURE 11.1   The Naplet server architecture.

```
public abstract class Naplet implements Serializable, Cloneable {
    private NapletID nid;
    private URL codebase;
    private Credential cred;
    private NapletState state;
    private transient NapletContext context;
    private Itinerary itin;
    private AddressBook aBook;
    private NavigationLog log;
    public abstract onStart();
    public void onInterrupt() {};
    public void onStop() {};
    public void onDestroy() {};
}
```

The naplet identifier (ID) contains the information about what the naplet is and when and where it is created. In support of the naplet clone, the naplet ID also includes version information to distinguish the cloned naplets from each other. Since a naplet can be recursively cloned, we use a sequence of integers to encode the pedigree information and reserve 0 for the originator in a generation.

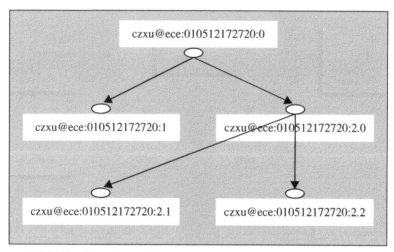

FIGURE 11.2   Hierarchical naming of naplet ID of a HelloNaplet.

For example, a naplet ID czxu@ece.wayne.edu:010512172720:2.1/HelloNaplet represents the hello naplet that was cloned from the original one created by a user czxu at 17:27:20 May 12, 2001 in the host ece.wayne.edu. The pedigree information is shown in Figure 11.2; the naplet name is omitted for brevity.

The Naplet system supports lazy-code loading. It allows classes loaded on demand and at the last moment possible. The codebase URL points to the location of the classes required by the naplet. The naplet classes and their associated resources, such as texts and images in the same package, can be zipped into a Java archive (JAR) file so that all the classes and resources the naplet needs are transported at a time.

Note that both the naplet ID and codebase URL are immutable attributes. They are set at the creation time and cannot be altered in the naplet life cycle. To ensure their integrity, they can be certified and signed by the naplet owner's digital signature. The naplet credential is used by naplet servers to determine naplet-specific security and access control policies.

As a generic class, Naplet is to be extended by agent applications. Application-specific agent states are contained in a NapletState object. Any object within the container can be in one of the three protected modes: *private*, *public*, and *protected*. They refer to the states accessible to the naplet only, any Naplet servers in the itinerary, and some specific servers, respectively. For example, a shopping agent that visits hosts to collect price information about a product would keep the gathered data in a private access state. The gathered information can also be stored in a protected state so that a Naplet server can update a returning naplet with new information.

The naplet executes in a confined environment, defined by its NapletContext object. The context object provides references to dispatch proxy, messenger, and stationary application services on the server. The context object is a transient attribute and is to be set by a resource manager on the arrival of the naplet. It cannot be serialized for migration.

In addition to the attributes, Naplet class also provides a number of hooks for application-specific functions to be performed in different stages of the agent life cycle: onStart, onStop, onDestroy, and onInterrupt. onStart is an abstract method which must be instantiated by extended agent applications. It serves as a single entry point when a naplet arrives at a host. onStop and onDestroy are event handlers when respective events occur during the execution of the agent. The agent behavior can also be remotely controlled by its owner via the onInterrupt method. Details of these will be discussed in Section 11.2.3.

Mobile agents have a defining characteristic of proactive mobility. Each naplet is associated with an Itinerary object for the way of traveling among the servers. It is noted that many mobile applications can be implemented in different ways by the same agent associated with different travel plans. We separate the business logic of an agent from its itinerary in Naplet class. Each itinerary is constructed based on five primitive constructs: singleton, sequential, parallel, alternative, and loop. Complex constructs can be composed recursively. In addition to the way of traveling, itinerary constructs also allow users to specify a postaction after each visit. The postaction mechanism facilitates interagent communication and synchronization. Details about the itinerary mechanism will be discussed in Section 11.3.

Many mobile applications involve a group of agents and the agents need to communicate with each other. In addition, an agent in travel may need to communicate with its home server from time to time. In support of interagent communication, we associate with each naplet an AddressBook object. Each address book contains a group of naplet IDs and their original locations. The locations may not be current, but they provide a starting point for tracking. The address book of a naplet can be altered as the naplet grows. It can also be inherited from naplet cloning. We restrict communications between naplets whose IDs are known to each other.

The last attribute of Naplet class is NavigationLog for naplet management. It records the arrival and departure time information of the naplet at each server. The navigation log provides the naplet owner with detailed travel information for postanalysis.

11.2.3 NapletServer Architecture

NapletServer is a class that implements a docking place of naplets within a JVM. It is responsible for executing naplets in a confined environment and making local host resources available to them in a controlled manner. It also provides mechanisms to facilitate resource management, naplet migration, and naplet communication.

Naplet servers are run autonomously and cooperatively to form a *naplet sphere*, where naplets are launched, migrated, and executed during their life cycles in pursuit of their agenda on behalf of their owners. Each naplet has a home server in the sphere where it is launched through NapletManager. The manager provides local users or application programs with an interface to launching naplets, monitoring their execution states, and controlling their behaviors. In addition, the manager maintains the information about its locally launched naplets in a naplet table. Footprints of all past and current visiting naplets are also recorded for management purposes. Each JVM can contain more than one Naplet server. This feature greatly simplifies the task of code debugging.

At the heart of a Naplet server are three modules: NapletSecurityManager, ResourceManager, and NapletMonitor. They together form a microkernel in support of remote execution. A Naplet server can be configured or reconfigured with various hardware, software, and data resources available at its host. Hardware resources such as central processing unit (CPU) cycles, memory space, and network input–output (I/O) constitute a confined basic execution environment. The software and data resources are largely application dependent and often configured as services. For example, naplets for distributed network management need to access local network-monitoring services; naplets for distributed high-performance computing need access to various math libraries.

NapletSecurityManager defines a suite of agent-oriented access control policies for the protection of the Naplet server from illegitimate access to local resources. Its focus is on authentication and authorization. It leaves monitoring of the naplet execution and control of resource and service consumption to NapletMonitor and ResourceManager.

NapletMonitor schedules the execution of multiple naplets from the same or different owners in a Naplet server. It defines a thread-based scheduling mechanism which enables the implementation of scheduling policies, such as FCFS and weighted fair queueing, for performance optimization.

ResourceManager provides a local service access control and allocation mechanism to request naplets, with the assistance of ServiceProxy. The proxy contains references to local services open to visiting naplets. Because of their diversity, services are defined in a reflection pattern so that any type of services can be accessed via the proxy. Moreover, the reflective design and the indirection of access to services via the proxy make it possible to "plug in" (or install) new services or reconfigure the existing ones at run time, without shutting down the Naplet server. More details about the run time pluggable service microkernel will be presented in Section 11.5. Section 11.6 gives an example of such services in network management.

The microkernel implements the core functionalities for naplet remote execution. It is enhanced by three other components: Navigator, Messenger, and Locator. Navigator is responsible for the realization of naplet migration between Naplet servers. Naplet migration can be initiated by the naplet itself according to its own itinerary or requested by the server where the naplet resides. On receiving a request for migration, the local navigator consults its associated security manager for a LAUNCH permission. Then, it contacts its counterpart in the destination server for a LANDING permission. Success of a launch will release all of the resources occupied by the naplet.

On receiving a naplet migration request from a remote server, the navigator consults its security manager and resource manager to determine whether a LANDING permission should be granted. When the naplet arrives, the navigator reports the arrival event to the managers and then passes the control over the naplet to the local naplet monitor. We note that single-hop migration is not a big issue in Java-based systems. The challenge is the implementation of multihop migration according to the naplet's own itinerary. In Section 11.3, we will define a set of structured itinerary patterns and present an implementation for various itineraries.

Each Naplet server contains a Messenger for internaplet asynchronous persistent communication. There are two types of messages: system and user. System messages are used for naplet control (e.g., callback, terminate, suspend, and resume); user messages are for communicating data between naplets. On receiving a system message, the messenger casts an interrupt onto the running naplet thread. How the control message should be reacted by the naplet is application dependent and left for programmers to specify. The interrupt handler is given in method onInterrupt when a naplet is created. On receiving a user message, the messenger puts it into a mailbox associated with the receiving naplet. The naplet decides when to retrieve the message from its mailbox.

The messenger relies on a Locator for naplet tracking and location services and supports location-independent communication. Naplet ID-based message addresses are resolved through a centralized or distributed naplet directory service. Due to the mobility nature of naplets and network communication delay, the location information provided by the directory service may not be current. The messenger provides a PostOffice mechanism to handle messages passing between mobile naplets.

In addition to support for asynchronous communication, each Naplet server provides a NapletSocket mechanism for complementary synchronous transient communication between naplets. NapletSocket bears much resemblance to Java Socket in application programming interfaces (APIs), except it is naplet oriented. Conventional Transmission Control Protocol (TCP) has no support for mobility. To guarantee message delivery, an established socket connection must migrate with naplets continuously and transparently. Section 11.4.2 gives the details about NapletSocket and Messenger.

Recall that NapletManager provides an interface for naplet launch and remote control. It also provides an interface for visiting naplets to interact with the microkernel and extended migration and communication modules. This is implemented by NapletContext. It serves as an adapter which contains references to instances of Navigator, Messenger, and ServiceProxy. The context object is created on receiving a naplet by the Naplet monitor. Because of the indirection of access, migration and communication functionality can be enhanced by plugging in new modules at run time.

11.3   STRUCTURED ITINERARY MECHANISM

Mobility is the essence of naplets. A naplet needs to specify functional operations for different stages of its life cycle in each server as well as an itinerary for
its way of traveling among the servers. The functional operations are mainly
defined in the methods of onStart() and onInterrupt() in an extended Naplet
class. The itinerary is defined as an extension of Itinerary class. Separation of
the itinerary from the naplet's functional operations allows a mobile application to be implemented in different ways following different itineraries. This
section presents the design and implementation of a set of primitive constructs
in support for easy representation of itineraries.

11.3.1   Primitive Itinerary Constructs

The itinerary of a naplet is mainly concerned with visiting order among the
servers. Each visit is defined as the naplet operations from the arrival event
through the departure event. The visiting order encoded in the itinerary object
is often enforced by departure operations at servers. Correspondingly, we
denote a visit as a pair $\langle S; T \rangle$, where S represents the operations for server-
specific business logic and T represents the operations for itinerary-dependent
control logic. For example, consider a mobile agent–based information collection application. One or more agents can be used to collect information from
a group of servers in sequence or in parallel. At each server, the agents perform
information-gathering operations (S) (e.g., work load measurement, system
configuration diagnosis) as defined by the application. The operations are
followed by agent movement-dependent operations (T) for possible interagent
communication and exception handling. Different itineraries would lead to
different communication patterns between the naplets. Different itineraries
would also have different requirements for handling itinerary-related exceptions. For example, in a parallel search over a number of servers, naplets needs
to communicate with each other about their latest search results before they
move forward. Success of the search in a naplet may need to terminate the
execution of the others.

   We note that servers listed in a journey route may not be necessarily visited
in all cases. Many mobile applications involve conditional visits. For example,
in a mobile agent–based sequential search application, the agent will search
along its route until the end of its route or the search is completed. That is,
all visits except the first one should be conditional visits. We denote a conditional visit as $\langle C \rightarrow S; T \rangle$, where C represents the guard condition for the
visit $\langle S; T \rangle$.

   Based on the concepts of visit and conditional visit, we define an itinerary in
recursively constructed visiting orders. Its base is a singleton, comprising a
single visit or conditional visit. Assume P and Q are two itineraries. We define
four primitive construct operators seq, alt, and par over P and Q for con-
structions of sequential, alternative, and parallel patterns. Specifically:

- seq(P, Q) refers to an itinerary that the visits of *P* are followed by the visits of Q by one naplet;
- par(P,Q) refers to an itinerary that the visits of *P* and *Q* are carried out in parallel by a naplet and its clone;
- alt(P,Q) refers to an itinerary that the visits of either *P* or *Q* are carried out by one naplet;
- loop(C → P) refers to an itinerary that the visits of *P* are repeated until the guard condition *C* becomes false.

Formally, itinerary *P* is defined in Backus–Naur Form (BNF) syntax as

$$\langle V \rangle ::= \langle S \rangle | \langle S; T \rangle | \langle C \rightarrow S; T \rangle$$
$$\langle P \rangle ::= \text{singleton}(V) | \text{seq}(P, Q) | \text{par}(P, Q) | \text{alt}(P, Q) | \text{loop}(C \rightarrow P)$$

11.3.2   Itinerary Programming Interfaces

Each itinerary construct defines an itinerary pattern. Its abstraction, guard condition, and postaction are expressed respectively in the following public programming interfaces in the Naplet system.

```
public interface ItineraryPattern extends Serializable, Cloneable {
    public void go (Naplet nap) throws UnableDispatchException;
}
public interface Checkable extends Serializable, Cloneable {
    public boolean check (Naplet nap);
}
public interface Operable extends Serializable, Cloneable {
    public void operate (Naplet nap);
}
```

The Naplet system contains five built-in ItineraryPattern implementations: Singleton, Seq, Alt, Par, and Loop. Their class diagrams are shown in Figure 11.3. In the following, we give two itinerary examples constructed from visits and conditional visits to demonstrate itinerary programming. Consider a mobile agent–based information collection application. One or more agents can be used to collect information from a group of servers s_1, s_2, ..., s_n in sequence or in parallel. At each server, the agents perform information-gathering operation (e.g., work load measurement, system configuration diagnosis) as defined by the application. They are followed by itinerary-dependent operations for possible interagent communication and exception handling. Different itineraries would lead to different communication patterns. In a parallel search, naplets need to communicate with each other about their latest search results. Success of the search by a naplet may need to terminate the execution of the others.

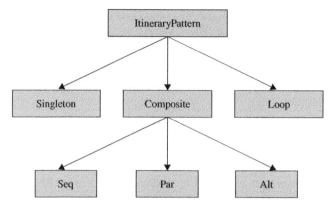

FIGURE 11.3   Built-in itinerary constructs in the Naplet system.

Example 11.1   The class MyItinerary1 defines a sequential itinerary using the SeqPattern construct for MyNaplet, indicating a sequential information collection. We define the MyNaplet class as an extension of the base class Naplet (line 1). The method onStart (line 3) is one of the hooks of the Naplet class for application-specific functions to be performed on agent arrival at a server. It contains a location-aware business logic collectInfo (line 4). After completion of this function, the agent travels according to its itinerary (line 6). The itinerary is defined in a private class MyItinerary1 (line 10). It is a simple sequential visiting order over an array of servers (line 13). At the end of its itinerary, the agent reports its collected results back to its home by a post-action as defined in the class ResultReport (line 9). Since the itinerary class MyItinerary1 is declared as a private inner class of the naplet, the postaction can be defined on the naplet states. The itinerary is set via a setItinerary method of the Naplet class (line 13).

```
1)  public MyNaplet extends Naplet {
2)      . . . . . .
3)      public void onStart() {
4)          collectInfo();      // Location-aware business logic
5)          try {
6)              getItinerary().travel( this );
7)          } catch (UnableDispatchException node) {};
8)      }
9)      private class ResultReport implements Operable { . . . }
10)         private class MyItinerary1 extends Itinerary {
11)             public MyItinerary1(String[] servers) {
12)                 Operable act = new ResultReport();
13)                 setItinerary(new SeqPattern(servers, act));
14)             }
15)     }
16) }
```

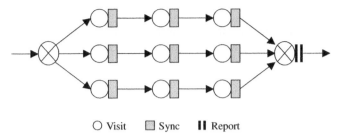

○ Visit   ■ Sync   ‖ Report

FIGURE 11.4   Example of parallel-search itinerary using three cloned naplets.

Example 11.2   The class MyItinerary2 defines a parallel-search itinerary, as shown in Figure 11.4, by the use of k cloned naplets, each for an equal number of servers (for simplicity, we assume n can be divided by k). Let $m = n/k$. Totally $m \times k$ visits are defined (lines 12–22). Each visit is of class Singleton, comprising a checkable object ResultVerify as its guardian precondition (line 12). Whenever a naplet finds the target, it will skip the rest of its servers and meanwhile inform the others. The synchronization is realized by a collective operation defined in the operable object DataComm (line 13). A cloned naplet i, $0 \le i \le k$, will visit m servers in sequence, as defined in a SeqPattern object journeys[i] (line 21). The naplets report their results to their home at the end of their journeys via postaction ResultReport. All the journeys together form parallel itinerary using the ParPattern construct (line 23). The itinerary object is set via a setItinerary method of the base class Naplet. Naplet cloning is due to the ParPattern construct.

```
10)   private class MyItinerary2 extends Itinerary {
11)      public MyItinerary2(String[] servers, int k) {
12)         Checkable guard = new ResultVerify();
13)         Operable sync = new DataComm();
14)         Operable report = new ResultReport();
15)         int n = servers.length; int m = n/k;
16)         Singleton[][] visits = new Singleton[k][m];
17)         SeqPattern[] journeys = new SeqPattern[m];
18)         for (int i=0; i < k; i++) {
19)            for (int j=0; j < m; j++)
20)               visits[i][j] = new Singleton(guard, servers[i*k+j], sync);
21)            journeys[i] = new SeqPattern(visits[i], report);
22)         }
23)         setItinerary(new ParPattern(journeys));
24)      }
25)   }
```

11.3.3   Implementations of Itinerary Constructs

In [14], we showed that the set of primitive constructs in Figure 11.3 is regular complete in the sense that any itinerary in a regular trace can be constructed based on these primitive constructs. However, they are insufficient to express itineraries like "Visiting site s_1 for x times, followed by visiting of s_2 for the same number of times." In the following, we show the implementation details of Singleton and SeqPattern as programming examples of user-defined itinerary constructs.

In Naplet system, each customized itinerary associated with a naplet is extended from a serializable Itinerary class. The itinerary contains a reference to the current itinerary pattern and keeps in stack the naplet trace for recursive traverse.

```
public class Itinerary implements Serializable, Cloneable {
    private ItineraryPattern cur;     // Current itinerary pattern the
                                          naplet is on.
    private Stack patterns;              // Stacked itinerary patterns the
                                          naplet was on.
    . . .  . . .
    protected ItineraryPattern popPattern() { return patterns.pop(); }
    protected void pushPattern(ItineraryPattern itin) { return patterns.
    push(itin); }
    public final void setCurPattern( ItineraryPattern itin ) { cur = itin; }
    public final void travel( Naplet nap ) { current.go( nap ); }
}
```

Singleton class defines the visit of a single server coupled with a precondition and postaction. Each itinerary has an associated iterator, indicating the next itinerary pattern to be visited. ItineraryIterator is defined as serializable to replace Java built-in iterators because the Java iterators are nonserializable and their index information will be lost after migration. The go() method shows the details of a naplet visit of a single server, including guard condition checking, postaction execution, and migration. Recall that the execution of a naplet is confined to a context created by Naplet Monitor, as shown in Figure 11.1. The context contains a reference to local navigator service. The migration is accomplished by the navigator.

Because an itinerary is recursively constructed, backtrack() method returns the itinerary pattern following the current singleton. For example, consider an itinerary seq(s1, s2, s3). Internally, it is transformed into a representation of seq (singleton(s1), singleton(s2), singleton(s3)). The visit of singleton(s1) will end up with a backtrack to the initial visit of the seq(. . .) pattern with its iterator pointing to singleton(s2). Details about the visit of a composite pattern like seq will be discussed later.

```
public class Singleton implements ItineraryPattern {
    private URN server;                    // Site to be visited
    private Checkable guard;                // Precondition of the visit
    private Operable action;               // Postaction of the visit
    private ItineraryIterator iter; // A serializable iterator defined over
                                          itinerary
              . . .  . . .
    public void go( Naplet nap ) throw UnableDispatchException {
      if ( iter.hasnext() ) {
          URN next = ((Singleton)iter.next()).server;
          if ( guard == null || guard.check(nap) )
            nap.getNapletContext().getNavigator().toDispatch( next, nap );
```

```
            else { backtrack(nap); }
      } else {
          if (action != null)
           action.operate( nap );      // Perform postaction
           iter.reset();  // Reset the iterator for possible itinerary loop
           backtrack(nap);
      }
  }
  private void backtrack(Naplet nap) {
      ItineraryPattern itin = nap.getItinerary().popPattern();
      if (itin !=null) {
          nap.getItinerary().setCurPattern( itin );
          itin.go( nap );
      }
  }
}
```

SeqPattern class is defined as an extension of CompositePattern class. A CompositePattern object contains a number of itineraries together with a pre-condition and a postaction. The itineraries are stored in an ArrayList data structure indexed by an itinerary iterator. CompositePattern class is defined as an abstract class because it leaves the visiting order of the servers in ArrayList unspecified. SeqPattern class defines a sequential visit order in its go() method. It recursively traverses each itinerary recorded in the agent itinerary until a Singleton is reached. Due to space constraints, other composite itinerary patterns such as ParPattern and Loop are omitted.

```
public abstract class CompositePattern implements ItineraryPattern {
   private ArrayList path;        Composite itinerary is stored in an array list
   protected Checkable guard;   Terminate condition for the loop
   protected Operable action;    Postaction after the loop
   protected ItineraryIterator iter;      A serializable iterator over the
                                          array list
   public abstract void go(Naplet nap) throws UnableDipatchException;
}

public class SeqPattern extends CompositePattern {
     . . .  . . .
    public void go( Naplet nap ) throw UnableDispatchException {
       if (iter.hasNext() ) {
            ItineraryPattern next = iter.next();
            if (guard == null ‖ guard.check(nap)) {
            nap.getItinerary().pushPattern( this );
            nap.getItinerary().setCurPattern( next );
            next.go( nap );  // traverse next pattern recursively
       } else { backtrack( nap ); }
    } else {
       if (action ≠ null) action.operate( nap );
       iter.reset();
       backtrack( nap );
    }
  }
}
```

11.4   NAPLET TRACKING AND INTERNAPLET COMMUNICATION

Mobility is a defining characteristic of mobile agents. Mobility support poses a basic requirement for tracking agents and finding their current locations dynamically. The agent location information is needed not only for home servers to contact their outstanding agents for agent management purposes but also for interagent communication. In this section, we first present a Naplet location service. Then, we show two mechanisms for synchronous and asynchronous communication between naplets.

11.4.1   Naplet Location Service

In general, there are three approaches for the tracking and location-finding problem: *broadcast, location directory*, and *forward pointer*.[1] A broadcast approach sends a location query message to all servers. It is simple in concept and easy to implement if the system supports broadcasting in network and transport layers. In large-scale networks with unreliable communication links, reliable broadcasting is nontrivial by any means. Moreover, any agent location change during the process of broadcasting makes the approach impractical.

A location directory approach is to designate one or more directory servers to keep track of agent locations. The directory service is advertised so that any location query message is directed to the directory. Because naplets may keep moving and the communication delay between naplet servers and the directory is nonnegligible, the location information provided by the directory service is not necessarily current.

Agent tracking based on forward pointers relies upon agent footprints left on visited servers to chain the servers together in a visiting order. In this approach, a location query message will first be sent to the agent home server. The message will then be passed down the agent visiting path. Since the approach requires no updates of agent locations, it incurs no extra overhead in migration. Its downside is a long delay for tracking.

The Naplet location service is based on a hybrid of directory and forward pointer mechanisms. The location service interface is defined by Locator in the following:

```
public interface Locator {
    public URN lookup(NapletID nid) throws NapletLocateException;
    public URN lookup(NapletID nid, long timeout) throws
    NapletLocateException;
}
```

Recall that NapletServer can be running in one of the two modes: with and without naplet directory service. In a system without a directory service,

---

[1]In [15], the author discussed five messaging models, which mixed agent tracking with messaging services together.

naplets are tracked by using naplet footprint information recorded in each naplet manager in the servers. NapletFootPrint, as defined in the following, contains the source and destination information about each naplet visit.

```
public class NapletFootPrint {
        private URN source;              // Where the naplet comes from
        private Date arrivalTime;
        private URN dest;                // Where the naplet leaves for
}       private Date departTime;
```

On receiving a location-finding request from Messenger or other high-level location-independent services, the local Locator first checks with its local naplet manager to find out whether the target naplet is in. If not, the Locator then retrieves the home server address of the target naplet, as encoded in its NapletID object, and sends a query message to the home server. The query message is forwarded along the path, starting from the home server, according to the agent footprints left in visited servers until the message catches the target naplet. Note that a query message may arrive at a server before its target because the query message and the naplet may be transferred in different physical routes and the naplet may be blocked in the network. If the query message arrives after the naplet's landing and before its departure, it is responded with the current server location; otherwise, the message needs to be buffered for a certain time period. Readers are referred to Section 11.4.2.1 for details about in-order message/agent delivery on non-FIFO (first in–first out) communication networks.

Since a query message for an agent needs to traverse its whole path, the lookup service will be timed out as the agent path stretches out. This is an inherent problem with the forward-pointer approach. As a remedy, the Naplet system requires updating the home server with the new location whenever a query is responded to. Another solution is to use a forward pointer together with a location directory.

In a system with an installation of NapletDirectory, each locator can position naplets by looking up the directory. Although the location information from the directory may not be current due to the communication delay between a naplet server and the directory, it can be used as a starting point for tracking via the complementary forward-pointer approach. Note that we distinguish between two types of naplets: long lived and short lived in terms of their expected lifetime at each server. For stability, the naplet tracking and location service is limited to long-lived naplets.

```
public interface NapletDirectory extends Remote {
    public void register (NapletID nid, URN server, Date time, int event)
                        throw DirectoryAccessException
    public URN lookup( NapletID nid ) throw RemoteException;
}
```

On launching or receiving a naplet, it is the navigator that registers the ARRIVAL and DEPARTURE events with the directory. The departure event is reported after a naplet is successfully dispatched. However, there is no guarantee of the time when the naplet arrives at the destination. The arrival event is reported after the naplet lands. We postpone the execution of the naplet until the arrival registration is acknowledged. This guarantees that the directory keeps the current location information about the naplets. If the latest registration about a naplet in the directory is a departure from a server, the naplet must be in transmission out of the server. If its latest registration is an arrival at a server, the naplet can be either running in or leaving the server (departure registration may not be needed). NapletDirectory is currently implemented as a component of the naplet server, although its installation is optional. In fact, it can be realized as a stand-alone Lightweight Directory Access Protocol (LDAP) service. One LDAP server can be installed for each naplet sphere (i.e., a group of naplet servers), independent of Naplet servers. To access the LDAP service, each Naplet server must authenticate itself to the service.

We note that a location directory is not necessarily implemented in a centralized manner. The Naplet directory services can be provided collaboratively by the Naplet manager of each server. Since each naplet has its own home server and the home information is encoded in naplet IDs, the naplet location information can be maintained in their home managers. Correspondingly, any naplet tracking and location requests are directed to respective home managers.

The Naplet location service is demanded by a messenger for internaplet communication or by a Naplet manager for naplet management. The location service also caches recently inquired locations so as to reduce the response time of subsequent naplet location requests. The buffered naplet location information can be updated on migration either by home Naplet managers in systems with distributed Naplet directory services or by remote-residing Naplet servers in systems with forward pointers.

11.4.2   Internaplet Communication

The Naplet location service enables location-independent communication between naplets. That is, a Naplet can take messages from a specific naplet, any naplet from a naplet server, or any naplet in the sphere if the message sender is not specified. In the following, we present a PostOffice mechanism for asynchronous communication and a NapletSocket mechanism for synchronous communication.

11.4.2.1   PostOffice Messaging Service
Messenger in a Naplet server supports asynchronous message passing between naplets. The messages are naplet ID oriented and location independent. The asynchrony is realized based on a mailbox mechanism. On receiving a naplet, the messenger creates a mailbox for its subsequent correspondences with other naplets or its home naplet manager. Recall that we distinguish messages into two classes: system message for naplet control and user message for data communication. System messages are delivered to their target naplets immediately via interrupts, while user messages are stored in respective mailboxes for

the target naplets to retrieve. For flexibility in communication, each messenger also keeps the mailbox open to its naplet so that the naplet can access the mailbox directly, bypassing the send/receive interface.

```
public interface Messenger {
  public void send(NapletID dest, Message msg) throws NapletCommException;
  public void receive(URN server, Message msg) throws NapletCommException;
  public void receive(Message msg) throws NapletCommException;
  public Mailbox getMailbox(NapletID nid);
}
```

The mailbox-based scheme provides a simple and reliable way for asynchronous communication between naplets. Under the hood is a PostOffice delivery protocol that implements message forwarding to deal with agent mobility. Each messenger maintains a mailbox cabinet to contain all mailboxes of the residing naplets. In addition, it has a special system mailbox, called *s-box*, to temporarily store undelivered messages and deal with a *rendezvous* issue between forwarded messages and moving agents.

Assume naplet A residing on server Sa is to communicate with naplet B. Naplet A makes a request to Sa's messenger. The messenger checks with its associated locator to find out naplet B's most recent server or its home server. Due to the mobility nature of naplets and communication delay, this server information is not necessarily current. Without loss of generality, we assume naplet B used to be in server Sb. Messenger in server Sa sends the message to its counterpart in server Sb. On receiving this message:

1. If naplet B is still running in the server, Sb's messenger replies to Sa with a confirmation and meanwhile inserts the message into naplet B's mailbox. The confirmation message is kept in Sa's messenger only for further possible inquiry from naplet A.
2. If naplet B is no longer in server Sb, Sb's messenger checks with its naplet manager against its naplet trace and forwards the message to the server to which the naplet moved. The forwarding continues until the message catches up to naplet B, say in server Sc. Sc's messenger replies to Sa with a confirmation and inserts the message onto B's mailbox.
3. If naplet B has not arrived in server Sb yet (it is possible because the naplet might be temporarily blocked in the network), Sb's messenger checks with its naplet manager against its naplet trace and finds no record of naplet B. The messenger will insert the message into the s-box, waiting for the arrival of naplet B. On receiving naplet B, Sb's messenger creates a mailbox and transfers B's messages in the s-box to B's mailbox.

11.4.2.2 NapletSocket for Synchronous Communication
Asynchronous persistent communication is not sufficient for applications that require agents to cooperate closely. Synchronous transient communication would keep the agents working more closely and efficiently.

The PostOffice-based asynchronous communication mechanism aside, Messenger provides a Naplet socket service for interagent synchronous communication. The Naplet socket service is built on a pair of classes: NapletSocket and NapletServerSocket. They are implemented as wrappers of Java Socket and ServerSocket, respectively, providing similar APIs to the Java socket service:

```
public interface NapletSocket {
    public NapletSocket(NapletID dest, boolean isPersistent);
    public void close();
    public void suspend();
    public void resume();
    public InputStream getInputSteam();
    public OutputStream getOutputStream();
}
public interface NapletServerSocket {
    public NapletServerSocket(NapletID dest, boolean isPersistent);
    public NapletServerSocket(boolean isPersistent);
    public NapletSocket accept();
    public void close();
    public void suspend();
    public void resume();
}
```

Unlike the Java socket service, which is network address oriented, the Naplet socket service is oriented toward location-independent NapletID. Assume naplet B runs a server socket and naplet A wants to establish a synchronous communication channel with B. Naplet A creates a NapletSocket connecting to naplet B. The local messenger locates naplet B via its associated internal tracking service provided by locator and establishes an actual socket connecting to the destination.

An established socket can be closed by either side. In addition, the Naplet socket service supports connection migration. We distinguish communication channels between *persistent* and *transient*. Persistent channels need to be maintained during migration, while transient channels are not. To support connection migration NapletSocket provides two new methods: suspend and resume. They can be called either by agents for explicit control over connection migration or by a messenger for transparent migration. Assume there is an established socket connected between naplet A and naplet B. If naplet A is to migrate and naplet B is stationary, the socket is simply suspended before A's migration and resumed after it arrives at its destination. If naplet B is about to leave, its messenger needs to suspend all of its outstanding sockets and inform them of its destination for reconnection.

Socket hand-off in connection migration must hold until the servers are assured no messages are in transmission. A challenge is how a naplet monitors the status of its naplet sockets. By default the close() method returns immediately, and the system tries to deliver any remaining data. By setting a socket option SO_LINGER, the system is able to set up a zero-linger time. That is, any unsent packets are thrown away when the socket is closed.

NapletSocket module relies on an out-band control channel for the management of connections and operations that need access right to socket resources. Details of the mechanism for synchronous persistent Naplet socket services are available in [16].

11.5 SECURITY AND RESOURCE MANAGEMENT

A primary concern in the design and implementation of mobile agent systems is security. Most existing computer security systems are based on an identity assumption. It asserts that whenever a program attempts some action we can easily identify a user to whom that action can be attributed. We can also determine whether the action should be permitted by consulting the details of the action and the rights that have been granted to the user running the program. Since mobile agent systems are hard to keep this important assumption valid, their deployment involves more security issues than for traditional stationary code systems.

11.5.1 Naplet Security Architecture

Over the course of agent execution on a server, server resources are vulnerable to illegitimate access by residing agents. On the other hand, the agents are exposed to the server, their carried confidential information can be breached, and their business logic can even be altered on purpose. The design and implementation of a mobile agent system need to protect agents and servers from any hostile actions from each other. These two security requirements are equally important in an open environment because mobile agents can be authored by anyone and executed on any site that has docking services. However, server protection is more compelling in an environment where the sites are generally cooperative and trustworthy, although mobile code from different sites may have different levels of trustiness.

The Naplet system assumes naplets are run on trustworthy servers. Security measures focus on protection of servers from any possible naplet attacks. The Naplet Security Architecture (NSA) is based on the standard Java security manager to prevent untrusted mobile code from performing unwanted actions. Unlike Java's early sandbox security model, which hard coded security policies together with its enforcement mechanism in a SecurityManager class, the Java 2 security architecture separates the mechanism from policies so that users can configure their own security policies without having to write special programs.

A security policy is an access control matrix that says what system resources can be accessed, in what manner, and under what circumstances. Specifically, it maps a set of characteristic features of naplets to a set of access permissions granted to the naplets. It can be configured by a Naplet server administrator. It is our belief that any Naplet server should be prepared to run an overwhelming number of alien agents from different places. It is cumbersome, if not impossible, to manage the security needs of each individual agent. NSA supports the concept of agent group. A group of agents represents a collection of agents that share certain common properties. For example, cloned agents belong to a group which should be granted similar access permission; agents from the same owner, organization, or geographical region may form a group that shares the same access control policies. Moreover, agents can also be grouped in terms of their functionalities/responsibilities or particular resources that the agents need to access. Such a group is often referred to an agent role. The administrator, anonymous agents, and normal agents are examples. NSA defines security policies for agents as well as groups and roles. Following is a policy example that grants agents from an Internet domain ece.wayne.edu to look up a yellow page service.

```
grant Principal NapletPrincipal "ece.wayne.edu/*" {
    permission NapletServicePermission("yellow-page", "lookup");
}
```

Early Java security architecture was targeted at code source. That is, authorization is based on where the code in execution comes from, regardless of the subject of code execution. Subject-based access control is not supported until JDK 1.2. Naplet is one of the primary subjects we defined in NSA. Other subjects include Administrator and NapletOwner. Their authentication is based on a username/password login module defined in Java Authentication and Authorization Service (JAAS). The principal for naplet authentication is Naplet ID, which encodes the information about its owner, birth place, and birth time. This information remains unchanged under naplet migration. The information is signed by a private credential of the home Naplet manager on behalf of its owner.

Agent-oriented access control is realized via an array of additional permissions: NapletServicePermission, NapletRuntimePermission, and NapletSocketPermission. They grant access control privileges to system resources as well as application-level services in a flexible and secure manner. Details of the access control model can be seen in [17].

11.5.2 Resource Management

NSA supports policy-driven and permission-based access control to prevent visiting agents from illegitimate access to local services of a server. It leaves monitoring of the Naplet execution and control of resource consumption to Naplet monitor and resource manager components of the Naplet server.

11.5.2.1 NapletMonitor
Critical resources to be monitored are CPU cycles, memory sizes, and network bandwidth. A mobile agent system without appropriate resource management is

vulnerable to denial-of-service (DoS) attacks. The Naplet system supports migration and remote execution of multithreaded naplets. Also, multiple naplets may be run concurrently in a server. It is the Naplet monitor that monitors and schedules the execution of multithreaded agents on a Naplet server.

On receiving a naplet, the monitor creates a NapletThread and a thread group for the execution of the naplet. The NapletThread object assigns a run time context, NapletContext, to each naplet thread and sets traps for its execution exceptions. Each context contains references to the internal services of a server, including navigator, locator, messenger, and service proxy. All the threads created by the naplet are confined to the thread group. The group is set to a limited range of scheduling priorities so as to ensure that all alien threads are running under the control of the monitor. The monitor maintains the running state of each thread group in a NapletThreadTable data structure.

Notice that newly arrived agents are not necessarily activated immediately. Their threads are first buffered in a ready-to-run queue. Their execution order is determined by an internal Scheduler service. A default implementation is a FIFO best effort model, in which naplets are started to run in their arrival order. Naplet servers in this model are vulnerable to DoS attacks. A challenge is to schedule the execution of multithreaded naplets in a fair-share manner.

Thread scheduling is one of the troublesome problems in Java because it is closely dependent on the scheduling mechanisms in the underlying operating system (OS). A JVM uses fixed-priority scheduling algorithms to decide the thread execution order. If more than one thread exists with the same priority, the JVM would switch between the threads in a round-robin fashion only if the OS uses time-slicing scheduling. The JVM on Microsoft Windows supports fair-share scheduling between threads with the same priority. But a Java thread on Sun Solaris would continue to run until it terminates, gets blocked, or is preempted by a thread of higher priority. In Linux, the Java thread priority information is totally ignored. Because of this platform-dependent effect, thread priority is unreliable to support fair-share scheduling. In fact, Java language specification recommends that thread priority information be used as guides to efficiency only.

Thread priority aside, another performance factor of multithreaded agents is the number of threads. Since JVM cannot distinguish threads from different naplets, a malicious agent can block the execution of other agents by spawning a large number of threads. To ensure fair-share scheduling between naplets, Scheduler needs to implement a scheduling policy to isolate the performance of naplets, regardless of their priorities and number of threads.

We implemented a CPU fair-sharing policy by using the Java thread suspend/resume mechanism with priority control to switch the execution of threads in each scheduling epoch. Although we were able to achieve the goal of fair sharing of CPU utilization on the time scale of seconds in all of the platforms that we tested, including Windows, Solaris, and Linux, it came at a high scheduling overhead. The Java suspend/resume primitives have been deprecated because they are deadlock prone. It remains an open issue to provide performance isolation between concurrent agents in Java-based middleware platforms.

11.5.2.2 Access to Local Services

Naplets can do few things without access to local services installed on servers and external to the Naplet system. The services include those provided by local operating systems, database management, and other user-level applications. They may be implemented in legacy code and most likely run in a privileged mode. Although such local services can open to visiting naplets by setting appropriate permissions in NapletSecurityManager, visiting naplets should not be allowed to access these services directly. To prevent any threats from misbehaved naplets, resource access operations must be monitored all the time. This is realized by the use of ServiceProxy and ServiceChannel objects inside the resource manager, as shown in Figure 11.5.

ServiceChannel class defines a bidirectional communication channel between local services and accessing naplets. Each channel is created by the resource manager and attached to a local service by assigning a pair of input/output endpoints, ServiceInputStream and ServiceOutputStream, to the local service. The other pair of endpoints, NapletOutputStream and NapletInput Stream, are left open. The open endpoints will be assigned to a visiting naplet after its service access permission is granted. Once the naplet receives the endpoints, it can start to communicate with the local service under the auspices of the proxy.

Note that the service channel is essentially a synchronous pipe. But it is different from a Java built-in pipe facility. Java pipe is symmetric in the sense

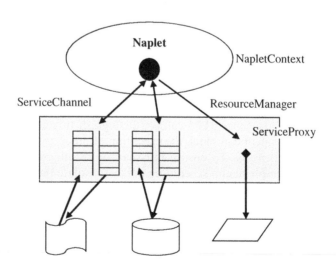

FIGURE 11.5   Access control over privileged and nonprivileged service.

that both ends rely on each other and the pipe can be destroyed by any party. In contrast, a service channel is asymmetric in that the channel can be allocated by the service proxy to any authorized naplet as long as the service provider is alive. The asymmetry of service channels enables dynamic installation and reconfiguration of application services.

Local privileged services are accessed via service channels. A naplet server may also be configured with nonprivileged services. Examples are small utility services and math libraries. Nonprivileged services are published with access handlers (e.g., math library function calls). It is also the responsibility of ServiceProxy to allocate the handlers to requesting naplets.

11.6 PROGRAMMING FOR NETWORK MANAGEMENT IN NAPLET

Agent-based mobile computing has been experimented with in various applications, such as distributed information retrieval, high-performance distributed computing, network management, and e-commerce; see [18–23] for examples. It provides a number of advantages over conventional distributed computing paradigms [24, 25]. For example, in the client–server model, clients are limited to services that are predefined in a server. Agent-based mobile computing overcomes this limitation by allowing an agent to migrate carrying its own new service implementation. Due to its unique property of proactive mobility, a featured agent should also be able to find necessary services and information in an open and distributed environment. Other advantages include low network bandwidth due to its on-site computation property, resumed execution after a disconnection from the network is reconnected, ability to clone itself to perform parallel computation, and migration for performance scalability or fault tolerance.

In this section, we demonstrate Naplet programmability in a network management application. We also gives an example of internal services that are pluggable at run time due to the microkernel architectural design. Network management involves monitoring and controlling the devices connected in a network by collecting and analyzing data from the devices. Conventional network management is mostly based on the Simple Network Management Protocol (SNMP). It assumes a centralized management architecture and works in a client–server paradigm. SNMP demon processes (i.e., SNMP agents) reside on network devices and act like servers. They communicate on request device data to a network management station. The device data are stored in a management information base (MIB) and accessible to local SNMP agents. The management station requests remote MIB information through a pair of fine-grained get and set operations on primitive parameters in MIBs. This centralized micromanagement approach for large networks tends to generate heavy traffic between the management station and network devices and heavy computational overhead on the management station.

The performance issues in the centralized network management architecture can be resolved in many ways. One of the attempts is a mobile agent–based

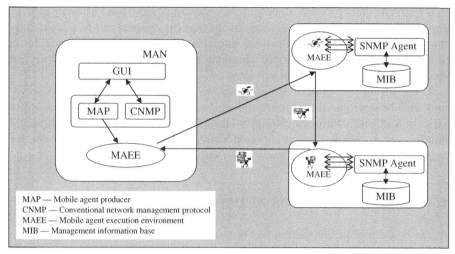

FIGURE 11.6 Architecture for mobile agent–based network management.

distributed approach. Instead of collecting MIB information from SNMP agents, in this approach, the network management station programs required device statistics or diagnostics functions into an agent and dispatches the agent to the devices for on-site management. Figure 11.6 shows a Naplet-based network management framework. The mobile agent–based framework, namely MAN, is in a hybrid model [22]. It gives the manager the flexibility of using mobile agents or SNMP according to the requirements of management activities.

The MAN management system relies on privileged services provided by the local SNMP agent in each device. In the following, we first present an implementation of the privileged services and then define a naplet class for network management.

11.6.1 Privileged Service for Naplet Access to MIB

In the MAN framework, a network management station creates naplets and dispatches them to target devices. The naplets access the MIB through local SNMP agents of the devices. For communication between Java-compliant naplets and SNMP agents, we deployed an AdvenNet SNMP package on each managed device. The AdventNet SNMP packages provide a Java API for network management.

Following is a NetManagement class extended from a PrivilegedService base class. It is instantiated by a resource manager and associated with a pair of ServiceReader and ServiceWriter channels: in and out. Through the input channel, the naplet server gets input parameters from naplets and reorganizes them into an AdventNet SNMP format (lines 6–10). It then conducts a sequence of operations, as shown in the private "retrieve" method, to communicate with the AdventNet SNMP for required MIB information. The information is returned to the naplet through the out channel (line 12). The whole process can be repeated for a number of inquiries from the same naplet or different naplets.

```
1)  import naplet.*;
2)  import com.adventnet.snmp.beans.*;
3)  public class NetManagement extends PrivilegedService {
4)      public void run() {
5)          for (;;) {
6)              String parms = in.readLine();
7)              Vector values = new Vector();
8)              StringTokenizer paramTokenizer = new StringTokenizer
                (parms,";");
9)              while ( paramTokenizer.hasMoreElements() )
10)              values.addElement( (String)paramTokenizer.nextToken() );
11)             String result = retrieve( values );
12)             out.writeLine( result );
13)         }
14)     }
15)     private String retrieve( Vector parameters ) {
16)         StringBuffer result = new StringBuffer();
17)         String result = null;
18)         SnmpTarget target = new SnmpTarget();
                                    // Create an enquiry SNMP target
19)         target.loadMibs("RFC1213-MIB");
                                    // Load MIB
20)         target.setTargetHost(InetAddress.getLocalHost());
                                    //Set the SNMP target host
21)         target.setCommunity("public");
22)         Enumeration enum = parameters.elements();
23)         while(enum.hasMoreElements()) {
24)             target.setObjectID((String)enum.nextElement()+".0");
25)             result = target.snmpGet();  // Issue an SNMP get request
                                                    on managed node
26)         }
27)         return result.toString();
28)     }
29)}
```

11.6.2 Naplet for Network Management

The privileged service defined reflectively in NetManagement can be plugged in at run time. It is accessed by requesting naplets through its registered name serviceImpl.NetManagement. The run time service pluggability greatly enhances the system stability because new services can be installed or reconfigured without terminating the naplets in execution and shutting down the server. Also, the service reflective design increases the system programmability because there is no need for users to dramatically change their naplet code to access new services.

Following is a naplet example for network management. The NMNaplet class is extended from the Naplet base class with name, list of servers to be visited, and MIB parameters. It is also instantiated with a NapletListener object to receive information retrieved from the servers. All the information will be stored in a reserved ProtectedNapletState space. At last, the newly created NMNaplet object is associated with a custom-designed parallel itinerary pattern shown in lines 37–45.

On arrival at a server, the naplet starts to execute its entry method: onStart (). It gets a handler to predefined NetManagement privileged service (lines 16 and 17). It then sends parameters to the server through a NapletWriter channel and waits for results from a NapletReader channel. Notice that NapletWriter and ServiceReader are two ends of a data pipe from naplets to servers. Another pipe links a ServiceWriter to a NapletReader.

When the naplet finishes work on a server, it travels to the next stop (line 27). At the end of its itinerary, the naplet executes an operate() method (lines 30–34) to report the results back to its home. Since NMItinerary defines a broadcast pattern (lines 40–43), the naplet will spawn a child naplet for every server. The spawned naplets will report their results individually.

```
1)   import naplet.*;
2)   import naplet.itinerary.*;
3)   public class NMNaplet extends Naplet {
4)     private String parameters; // MIB parameters to be accessed
5)     public NMNaplet(String name, String[] servers, String param,
          NapletListener ch) throws InvalidNaplet
          Exception, InvalidItineraryException {
6)         super(name, ch);
7)         parameters=param;
8)         setNapletState(new ProtectedNapletState()); // Set space to keep
                                                       device info.
9)         getNapletState().set("DeviceStatus", new HashTable
          (servers.length));
10)        setItinerary (new NMItinerary (servers));
           // Associate an itinerary with NMNaplet
11)  }
12)  // Entry point of a naplet at each server
13)  public void onStart() throws InterruptedException {
14)      String serverName=getNapletContext().getServerURN().
          getHostName();
15)      Vector resultVector=new Vector();
16)      HashMap map=getNapletContext().getServiceChannelList();
17)      ServiceChannel channel=map.get("serviceImpl.NetManagement");
18)      NapletWriter out=channel.getNapletWriter();
19)      out.writeLine( parameters ); // Pass parameters to servers
20)      NapletReader in=channel.getNapletReader();
21)      String result=null;
22)      while ( (result=in.readLine()) !=EOF ) {
23)        resultVector.addElement( result );
24)      }
25)      Hashtable deviceStatus= (Hashtable) getNapletState().get
          ("DeviceStatus");
26)      deviceStatus.put( serverName, resultVector );
```

```
27)        getItinerary().travel( this );
28)  }
29)  private class ResultReport implements Operable {
30)        public void operate( Naplet nap ) {
31)          if ( nap.getListener()!=null ) {
32)              Hashtable messages = (Hashtable) nap.getNapletState().get
                  ("message");
33)              nap.getListener().report( deviceStatus );
34            }
35)        }
36)  }
37)  private class NMItinerary extends Itinerary {
38)     public NMtinerary( String[] servers) throws
              InvalidItineraryException {
39)        Operable act = new ResultReport();
40)        ItineraryPattern[] ip = new ItineraryPattern[servers.length];
41)        for (int i=0; i < servers.length; i++)
42)            ip[i] = new SingletonItinerary(servers[i], act);
43)        setRoute( new ParItinerary(ip) );
44)     }
45)  }
46)  }
```

11.7 RELATED WORK

The software agent technology has found its way in two main areas: artificial intelligence (AI) and distributed systems. AI research is mainly on the exploitation of agent autonomy in human–agent interaction and multiagent applications [26]. JADE [8] and JACK [9] are two recent examples of autonomous agent development systems. Although agents in such systems can be moved from one host to the other on demand, their migration is limited to reactive mobility. In contrast, research in distributed systems emphasizes system support for agent proactive mobility. In this section, we provide a concise review of current mobile agent systems with a focus on their treatments of four design issues related to proactive mobility: migration, naming, communication, and security.

The defining trait of mobile agents is their proactive mobility, which allows them to migrate from server to server in their own itineraries. Itinerary representation is a major undertaking of mobile agent systems. Various itinerary programming constructs were developed in different systems. Aglet [4] implemented only a single itinerary pattern for a list of servers to be visited in sequence. Mole [27] provided *sequence, set*, and *alternative* constructs as well as a priority assignment facility in support of flexible travel plans. Ajanta [5] implemented two additional constructs: *split, split–join*, and *loop*. They demonstrated the programmability and expressiveness of the constructs mainly by examples. The Naplet system defines five core structural constructs: *singleton, sequence, parallel, alternative*, and *loop*. Since each itinerary pattern is associated with a precondition and a postaction, the parallel construct provides flexible support for set, split, and split–join itinerary patterns. In [14], we extended the core itinerary constructs into a general-purpose mobile agent itinerary language, namely, MAIL. We analyzed its expressiveness based on its operational semantics and showed MAIL to be amenable to formal methods to reason correctness and safety properties regarding mobility.

Naplet deals with proactive mobility of individual agents. In contrast, FarGo [10] provided system support for reference-based group mobility. It assumes that mobile components of a distributed application be interconnected by different types of references (pull, stamp, etc). Migration of a component may need to pull other components along or leave behind a copy in the original server so as to preserve the validity of its incoming and outgoing component references.

Agent naming determines the way of agent tracking, as required by agent communication and management. Systems such as Agent Tcl [28], Aglet [4], and Tacoma [29] deploy a location-dependent naming scheme in which an agent changes its name to reflect the new location whenever it migrates. Agents are simply tracked by the use of the domain name system (DNS). Albeit simple in agent tracking, the scheme comes at extra cost for agent renaming in migration. In contrast, Naplet uses a location-transparent naming scheme and implements a directory-based name resolution mechanism to find current agent locations.

The directory can be organized in a centralized or distributed way. A centralized organization maintains all location information in a single server. Due to its simplicity in management, this centralized directory approach has been widely used in today's mobile agent prototypes, such as Mobile Objects and Agents (MOA) [30], Grasshopper [31], and Aglet [4]. A major drawback of this approach is poor scalability. In highly mobile agent systems, the centralized directory is prone to bottleneck jams.

A distributed directory organization maintains the agents' location information in their respective home servers. Whenever an agent migrates, its home server is updated with the new location. This approach was used in the Object Management Group (OMG) Mobile Agent System Interoperability Facilities (MASIF) [32]. Because this approach binds each agent to a home directory statically, the home server address must be retrievable from the agent name. The Naplet system can be deployed in two modes: with and without centralized directory. Considering the fact that the location information in the directory is not necessarily current due to the presence of communication delay, the Naplet system provides an additional mechanism, *agent footprint*, to help keep track of agent locations.

Communication is a crucial issue in multiagent systems. Agent communication languages such as Knowledge Query and Manipulation Language (KQML) [33] and the Foundation for Intelligent Physical Agents (FIPA) Agent Communication Language (ACL) [34] were designed for interactions between autonomous agents that originate in different places. Communication in mobile agent systems is concerned more with connectivity of agents on the move. Mailbox is a basic location-independent mechanism for asynchronous communication between mobile agents, in which each agent is associated with a mailbox. Aglet and JADE are two examples relying on mailbox-based communication protocols. But their frameworks cannot deal with the rendezvous issue between forwarded messages and moving agents, as illustrated in Section 11.4.2.1. Cao et al. [35] argued that mailboxes could be decoupled from their agents and migrated independent of agents. They suggested a synchronization scheme between mailbox migration and message delivery in support of reliable communication. In contrast, the PostOffice mechanism in the Naplet system creates a temporary mailbox for each visiting naplet. The mailbox is revoked after its naplet is left. Message forwarding is implemented by the use of agent footprint log information in Naplet servers.

In addition to asynchronous communication, Aglet supports synchronous message passing between agents similar to the message-passing interface (MPI). Its implementation requires the two agents in communication to remain stationary during the session. NapletSocket is our solution that supports synchronous communication between agents on the move during a session. Each persistent communication channel between a pair of moving agents preserves their logical relationship all the time. Mishra et al. [36] proposed a communication mechanism which is similar to NapletSocket in application semantics. But they achieved persistent connectivity via a centralized clearinghouse which matches sender and receiver and passes their addresses to each other for their subsequent direct communication. This may incur long message delivery latency because it requires at least twice the one-way message delay plus processing time.

Security is one of the primary concerns in mobile agent systems. Many past studies were devoted to the protection of agents and servers from hostile actions from each others; see [12, 37] for a recent comprehensive review of mobile code and security measures. The NapletState attribute of the Naplet class is designed in a similar way to the security measure of Ajanta [5] to protect the confidentiality of agent-carried application-specific information. Like other Java-based mobile agent systems, such as Concordia [6] and Aglet [4], the security measures of Naplet is built on the foundation of the Java security architecture [38] for server protection. However, early Java-based systems defined permissions on the origin of the mobile code. The static code source information is insufficient for identification of the agent owner in access control. In contrast, the Naplet system defines each naplet as an authenticated subject representing the source of a request. Its agent-oriented access control permissions allows naplets to exercise their own role-based privileges [17].

11.8 CONCLUDING REMARKS

In this chapter, we have presented the microkernel architectural design and implementation of the Naplet mobile agent system. The system features a structured navigation facility, synchronous and asynchronous agent communication mechanisms, open resource management policies, and agent-oriented access control. We have demonstrated its programmability in an agent-based network management application.

It is not our intention to sell the Naplet system as a versatile platform containing all necessary features for various mobile agent applications. It remains an experimental middleware prototype in support of research in mobile distributed computing. Due to its microkernel and pluggable architectural design, the system can be further enhanced easily by plugging in new internal services. For example, the current messaging service implements the send and receive methods in the Messenger interface with no consideration of message loss or out-of-order delivery. It can be replaced by implementation of a reliable messaging service at system configuration time, if needed, without the need for any change in other system components. New application-specific services such as NetManagement can be plugged in at run time without shutting down a server.

Mobile agent technology has been the focus of much speculation and hype in the past decade. Research on mobile agents has recently cooled down, as the technology awaits the emergence of killer applications. Security concerns also impede the acceptance of mobile agents in real-world applications. The Java security architecture provides a reasonably solid foundation to protect the server from hostile actions of mobile code. But current JVM provides very limited support for preemptive thread scheduling. As a result, JVM-based agent servers are prone to DoS attacks from visiting agents. Without support for performance isolation, an aggressive agent can also block the execution of others. Recent research on resource management interface for the Java platform [39] has opened a path to fine-grained control of resource consumption and construction of a more secure agent execution environment.

ACKNOWLEDGMENTS

This research was supported in part by U.S. National Science Foundation Grant Nos. CCF-0611750, DMS-0624849, CNS-0702488, CRI-0708232, CNS-0914330, and CCF-1016966.

REFERENCES

1. D. Milojicic, F. Douglis, Y. Paindaveine, R. Wheeler, and S. Zhou, Process migration, *ACM Comput. Surv.*, 32(3):241–299, Sept. 2000.

2. C.-Z. Xu and F. Lau, *Load Balancing in Parallel Computers: Theory and Practice*, Springer/Kluwer Academic, 1997.

3. J. E. White, Mobile agents make a network an open platform for third-party developers, *IEEE Computer*, 27(11):89–90, Nov. 1994.

4. D. Lange and M. Oshima, *Programming and Deploying Java Mobile Agents with Aglet*, Addison-Wesley, Reading, MA, 1998.

5. A. Tripathi, N. Karnik, M. Vora, T. Ahmed, and R. Singh, Design of the Ajanta system for mobile agent programming, *J. Sys. Software*, May 2002, pp. 123–140.

6. D. Wang et al., Concordia: An infrastructure for collaborating mobile agents, in *Proceedings of the 1st International Workshop on Mobile Agents (MA'98)*, 1997, Springer, Apr. 11, 2006, pp. 86–97.

7. R. Gray, D. Kotz, G. Cybenko, and D. Rus, D'Agents: Security in a multiple-language, mobile-agent system, in G. Vigna (Ed.), *Mobile Agents and Security*, *Lecture Notes in Computer Science*, Vol. 1419, Springer, London, 1998.

8. F. Bellifemine, G. Caire, and D. Greenwood, *Developing Multi-Agent Systems with JADE*, Wiley, Hoboken, NJ, 2007.

9. N. Howden, R. Roennquist, A. Hodgson, and A. Lucas, Jack intellent agents: A summary of an agent infrastructure, in *Proceedings of the 5th International Conference on Autonomous Agents*, 2001, Agent Oriented Software Pty. Ltd.

10. O. Holder, I. Ben-Shaul, and H. Gazit, Dynamic layout of distributed applications in FarGo, in *Proceedings of the International Conference on Software Engineering*, ACM New York, NY, USA, 1999.

11. C.-Z. Xu, Naplet: A flexible mobile agent framework for network-centric applications, in *Proceedings of IEEE IPDPS Workshop on Internet Computing and E-Commerce (ICEC)*, Fort Lauderdale, Florida, Apr. 2002, pp. 219–226.

12. C.-Z.Xu, *Scalable and Secure Internet Services and Architecture*, Chapman & Hall/ CRC, June 10, 2005.

13. F. Buschmann, R. Meunier, H. Rohnert, P. Sommerlad, and M. Stal, *Pattern-Oriented Software Architecture: A System of Patterns*, Wiley, New York, 1996.

14. S. Lu and C.-Z. Xu, A formal framework for agent itinerary specification, security reasoning, and logic analysis, in *Proceedings of IEEE ICDCS* Workshop on Mobile *Distributed Computing*, Columbus, Ohio, USA, 2005, pp. 580–586.

15. D. Deugo, Mobile agent messaging models, in *Proceedings of Int. Symp. on Autonomous Decentralized Systems*, Dallas, TX , USA, 2001, pp. 278–286.

16. X. Zhong and C.-Z. Xu, A reliable connection migration mechanism for synchronous communication in mobile codes, in *Proceedings of International Conference on Parallel Processing (ICPP)*, Montreal, Quebec, Canada, 2004, pp. 431–438.

17. C.-Z. Xu and S. Fu, Privilege delegation and agent-oriented access control in Naplet, in *Proceedings of IEEE ICDCS Workshop on Mobile Distributed Computing (MDC)*, Providence, RI, USA, Apr. 2003, pp. 493–497.

18. A. Bieszczad, T. White, and B. Pagurek, Mobile agents for network management, *IEEE Commun Surv.*, 1(1), 1998, pp. 2–9.

19. B. Brewington, R. Gray, K. Moizumi, D. Kotz, G. Cybenko, and R. Rus, Mobile agents in distributed information retrieval, in M. Klusch (Ed.), *Intelligent Information Agents*, Springer, Heidelberg, 1999.

20. P. Dasgupta et al., MAgNET: Mobile agents for networked electronic trading, *IEEE Trans. Knowledge Data Eng.*, 11(4):509–525, July/Aug. 1999.

21. K. Kato et al., An approach to mobile software robots for the WWW, *IEEE Trans. Knowledge Data Eng.*, 11(4):526–548, July/Aug. 1999.

22. M. Kona and C.-Z. Xu, A framework for network management using mobile agents, *Int. J. Parallel Emergent Distributed Comput.*, 20(1):39–55, 2005.

23. C.-Z. Xu and B. Wims, Mobile agent based push methodology for global parallel computing, *Concurrency: Practice and Experience*, 14(8):705–726, July 2000.

24. C. Harrison, D. Chess, and Kershenbaum, Mobile agents: Are they a good idea? Technical report, IBM Watson Research Center, Yorktown Heights, NY, Mar. 1995.

25. D. Lange and M. Oshima, Seven good reasons for mobile agents, *Commun. ACM*, 42(3):88–89, Mar. 1999.

26. P. Maes, Agents that reduce work and information overload, *Commun. ACM*, 37 (7):30–40, 1987.

27. M. Strasser and K. Rothermel, Reliability concepts for mobile agents. *Int. J. Coop. Inf. Sys.*, 7(4):355–382, 1998.

28. R. S. Gray, Agent Tcl: A flexible and secure mobile-agent system, in M. Diekhans and M. Roseman (Eds.), *Proceedings of the 4th Annual Tcl/Tk Workshop (TCL 96)*, Monterey, CA, USENIX Association Berkeley, CA, USA, 1996, pp. 9–23.

29. D. Johansen, R. van Renesse, and F. Schneider, Operating system support for mobile agents, in *Proceedings of the 5th IEEE Workshop on Hot Topics in Operating Systems*, 1995, pp. 42–45.

30. D. Milojicic, W. LaForge, and D. Chauhan, Mobile objects and agents (MOA), in *Proceedings of the 4th USENIX Conf. Object-Oriented Technologies and Systems (COOTS'98)*, USENIX Association Berkeley, CA, USA, 1998.

31. C. Baeumer, M. Breugst, S. Choy, and T. Magedanz, Grasshopper—A universal agent platform based on OMG MASIF and FIPA standards, available: www .grasshopper.de.

32. D. Milojicic et al., MASIF: The OMG mobile agent system interoperability facility, in *Proceedings of the International Workshop on Mobile Agents (MA'98)*, Springer, 1998.

33. T. Finin, R. Fritzson, D. McKay, and R. McEntire, KQML as an agent communication language, in *Proceedings of the 3rd ACM International Conference on Information and Knowledge Management (CIKM'94)*, ACM New York, NY, USA, 1994, pp. 456–463, ISBN:0-89791-674-3.

34. FIPA, ACL message structure specification, Foundation for Intelligent Physical Agents, 2001, available:www.fipa.com.

35. J. Cao, L. Zhang, J. Yang, and S. Das, A reliable mobile agent communication protocol, in *Proceedings 24th International Conference on Distributed Computing Systems*, Hachioji, Tokyo, Japan, Mar. 24–26 2004.

36. S. Mishra and P. Xie, Interagent communication and synchronization support in the daagent mobile agent-based computing system, *IEEE Trans. Parallel Distributed Syst.* 14(3), Mar. 2003, pp: 290–306.

37. S. Fu and C.-Z. Xu, Mobile codes and security, in *Handbook of Information Security*, Wiley, Hoboken, NJ, Hossein Bidgoli, 2005.

38. L. Gong, *Inside Java 2 Platform Security: Architecture, API Design, and Implementation*, Addison-Wesley, Reading, MA, 1999.

39. G. Czajkowski, S. Hahn, G. Skinner, P. Soper, and C. Bryce, A resource management interface for the java platform. *Software: Practice and Experience*, 35(2):123–157, 2005.

12 Performance Evaluation of Mobile Agent Platforms and Comparison with Client–Server Technologies

LUÍS MOURA SILVA

Departamento Engenharia Informática, Universidade de
Coimbra—POLO II, Coimbra, Portugal

12.1 INTRODUCTION

Mobile agents are autonomous programs that can migrate through the machines of the network to accomplish some tasks on behalf of some user. The agents carry their code and their internal data while they are migrating between different machines. According to [1] mobile code can be seen as the merging of two concepts that have been successfully deployed in distributed computing: code-on-demand and remote evaluation. In both cases, the mobility of code is only between a client and a single server. Mobile agents go further than these concepts since the agents can migrate across several machines and provide a more decentralized approach.

According to [2] there is a potential list of advantages for using mobile agent technology, namely, mobile agents reduce the network traffic, give support for disconnected computing, facilitate the process of software upgrading and introduction of new services in the network, and achieve higher scalability and easy integration with legacy systems. During almost a decade there has been intensive research in this paradigm, taken as an assumption that mobile agents can bring some clear advantages against traditional client-server solutions [3].

Tens of papers have been published in the literature and several commercial mobile agent implementations have been presented in the market, including

Mobile Agents in Networking and Distributed Computing, First Edition.
Edited by Jiannong Cao and Sajal K. Das.
© 2012 John Wiley & Sons, Inc. Published 2012 by John Wiley & Sons, Inc.

Aglets from IBM [4], Concordia from Mitsubishi [5], Voyager from ObjectSpace [6], Odyssey from General Magic [7], Jumping Beans from AdAstra [8], Kafka from Fujitsu [9], and Grasshopper from IKV [10].

We developed a Java-based mobile agent platform, called JAMES, in a research project between the University of Coimbra and Siemens [11]. This platform is mainly oriented for the field of telecommunications and included some particular features for better performance, fault-tolerance and resource-control. Our industrial partners (Siemens S.A. and Siemens A.G.) have adopted the JAMES platform to develop some applications of large-scale network management and performance monitoring in telecommunications networks, and the results were quite convincing.

While the list of advantages of using this paradigm seems theoretically quite appealing, it is now a fact that, after all the hype about mobile agents, this model did not take off on a large-scale in the IT industry. The main reason to explain this has to do with the fact that software architects soon understood that they could achieve similar results by using traditional client-server solutions, such as remote invocation and web-technologies. Some people might say mobile agents have been "over-hyped" and it may be a fact.

Today that the technology is more mature and there is no big fuzz about it, software designers may look at this paradigm as a complement solution that can be used in some particular cases, where the decentralized nature of active objects and agents may bring some clear advantages in terms of scalability and autonomous execution.

This chapter presents some performance results collected with two experimental studies: a benchmark study that compares mobile-agent platforms; and a performance comparison between mobile agents and client-server solutions like Java RMI, Corba-IIOP and Java Servlets.

The results that will be presented might be of interest to the reader that is thinking in the future to adopt the mobile agent paradigm to develop some large-scale applications with some specific needs for autonomous execution and versatile software upgrading.

The rest of the chapter is organized as follows: section 12.2 presents a brief overview of the JAMES platform; section 12.3 gives a summary of other commercial platforms; section 12.4 describes the benchmarking comparison between mobile agent platforms and includes some remarks about the performance of targeted platforms; section 12.5 presents a performance comparison between mobile agents and client-server solutions; section 12.6 presents some conclusions about our study.

12.2   BRIEF DESCRIPTION OF JAMES PLATFORM

This study was conducted with the JAMES platform, a mobile agent platform that was developed by a consortium with three partners: the University of Coimbra (Portugal), SIEMENS S.A. (Portugal) and SIEMENS AG

(Germany). This platform was mainly targeted to the field of telecommunication and network management [11].

Our industrial partners used this platform to develop some mobile agent applications that were integrated into commercial products. These applications had something in common: they had to deal with very large amounts of data, distributed over the nodes of telecommunication networks, and required a computing paradigm that would bring software services "on-demand" to do some processing in the sources of data. With this project we learned that this technology, when appropriately used, provides competitive advantages to develop distributed management applications.

The JAMES platform provides the running environment for mobile agents. There is a distinction between the software environment that runs in the manager host and the software that executes in the network elements (NEs): the central host executes the JAMES manager while the nodes in the network run a JAMES agency. The agents are written by application programmers and executed on top of that platform. The JAMES system provides a programming interface that allows the full manipulation of mobile agents. Figure 12.1

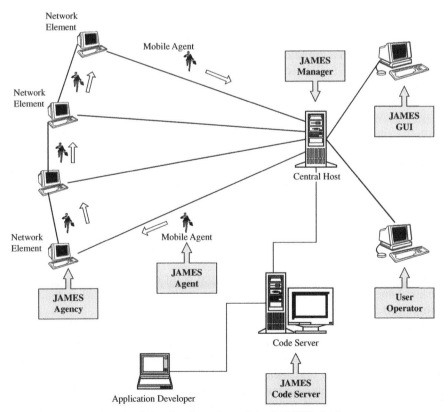

FIGURE 12.1 An Overview of the JAMES platform.

presents a global snapshot of the system with a special description of a possible scenario where the mobile agents will be used.

Every NE runs a Java virtual machine and executes a JAMES agency that enables the execution of the mobile agents. The JAMES agents will migrate through these machines of the network to access some data, execute some tasks and produce reports that will be sent back to the JAMES manager. There is a mechanism of authentication in the JAMES agencies to control the execution of agents and avoid the intrusion of non-official agents. Communication between the different machines is done through stream sockets. A special protocol was developed to transfer the agents across the machines in a robust way and is atomic to the occurrence of failures.

The application developer writes applications based on a set of mobile agents. These applications are written in Java and should use the JAMES API for the control of mobility. After writing an application the programmer should create a JAR with all the classes that make part of the mobile agent. This JAR file is placed in a JAMES code server. This server can be a different machine or in the same machine where the JAMES Manager is executing. In both cases, it maintains a code directory with all the JAR files available and the mapping to the corresponding mobile agents.

The host machine that runs the JAMES manager is responsible for the whole management of the mobile agent system. It provides the interface to the end-user, together with a graphical UI for the remote control and monitoring of the agents. The JAMES GUI is the main tool for management and administration of the platform.

Due to lack of space we will not describe the inner details of the JAMES platform. We have paid special attention to algorithms and techniques to optimize the performance of the applications and we did include some support for fault-tolerance and security. We do not advocate that this platform is better than other ones. Our focus is just on the lessons we have taken from this project which might be important to other researchers that want to conduct further work on this topic of distributed computing.

12.3   BRIEF DESCRIPTION OF OTHER PLATFORMS

In our experimental study we have selected eight different mobile agent platforms: James, Swarm, Odyssey, Grasshopper, Voyager, Concordia, Aglets and Jumping Beans.

There are several reports in the literature about the functionality of some of these platforms. Kiniry and Zimmerman [12] presented a direct comparison between Odyssey, Aglets, and Voyager. In [13] there is another comparison of agent system features that includes Aglets, Voyager, Odyssey, and Kafka. In [14] the authors presented a comprehensive review of three platforms: Aglets, Voyager, and Odyssey. Another evaluation was presented in [15] that included a feature comparison between the following platforms: D'Agents, April, Aglets, Grasshopper, Odyssey, and Voyager.

These reports only focus on the list of features of each platform and present some conclusions about the overall functionality of the platforms. However, no performance results have been reported so far. Next, we will present a short description of each platform.

12.3.1   Aglets SDK

The Aglets Software Developer Kit (ASDK) was developed at the IBM Research Laboratory in Japan [4]. The migration of Aglets is based on a proprietary Agent Transfer Protocol (ATP). The ASDK run-time consists of the Aglets server and a visual agent manager, called Tahiti. There is an additional module of software (called Fiji) that allows the installation of an Aglets server on a HTTP browser. The ASDK provides a modular structure and an easy-to-use API for the programming of Aglets. This platform has extensive support for security and agent communication and provides an excellent package of documentation. In our experiments we used ASDK 1.0.3.

12.3.2   Concordia

Concordia has been developed by Mitsubishi Electric [5]. This platform provides a rich set of features, like support for security, reliable transmission of agents, access to legacy applications, interagent communication, support for disconnected computing, remote administration, and agent debugging. This system also provides good documentation. In our experiments we used version 1.1.2 of Concordia.

12.3.3   Voyager

Voyager is an object request broker with support for mobile objects and autonomous agents. It was developed by ObjectSpace [6]. The agent transport and communication is based on a proprietary ORB on top of TCP-IP. Voyager has a comprehensive set of features, including support for agent communication and agent security. Voyager provides support for Corba and RMI. Due to its dynamic proxy generation, these technologies can be used without the need for stub generators. Thereby, Voyager objects can be used as Corba objects. In our experiments we used version 3.0 beta of Voyager.

12.3.4   Odyssey

Odyssey is a Java-based mobile agent system from General Magic [7]. The platform has a transport-independent API that works with Java RMI, IIOP, and DCOM. Odyssey provides the basic functionality and a small set of features.

Currently, it is not clear if General Magic will continue the efforts in this platform. In our experiments we used version 1.0 beta 2 of Odyssey.

12.3.5 Jumping Beans

Jumping Beans platform is a platform from AdAstra [8]. The main strengths of this platform include the support for security, agent management, easy integration with existing environments, and a small footprint. As far as we know it is also the only platform that claims to support the mobility of Corba objects. However, this platform uses a client-server approach for agent migration: if an agent wants to migrate between two agencies it has to go first to the agent manager. This approach may represent a point of bottleneck in large-scale applications. In our experiments we used version 1.0.4 of Jumping Beans.

12.3.6 Grasshopper

Grasshopper is a mobile agent platform that was distributed commercially by IKV++, a company from Berlin [10]. Grasshopper supports several transport protocols through the use of an internal ORB: a proprietary protocol based on TCP/IP, Java RMI, Corba IIOP, TCP/IP with SSL and RMI with SSL. The platform supports comprehensive support for security, agent communication, and agent persistency. In our experiments we used release 1.2 of the light edition of Grasshopper.

12.3.7 Swarm

The Swarm platform was developed by a research center of Siemens A.G. It is based on version alpha 1.0 of the Mole platform from the University of Stuttgart, Germany. Swarm provides an extensive support for inter-agent communication and agent management. In our experiments we used version 1.0 of Swarm. More details about this platform can be obtained in [16].

In the next section we present the methodology of the benchmarking study, a description of the test environment, the test parameters, and the benchmark application.

12.4 BENCHMARKING STUDY: COMPARING AGENT PLATFORMS

12.4.1 Test Environment

In our experiments we used a dedicated cluster of six machines connected through a 10 Mb/sec switched Ethernet. All the results were taken when the machines were fully dedicated. Every machine has a Pentium II (300 Mhz) processor and 128 Mb of RAM. These machines were running Microsoft Windows NT 4.0 and all the mobile agent platforms have used JDK 1.1.7 with the JIT option.

12.4.2 Application Benchmark

In the experiments presented in this section we have used a simple benchmark application composed by a migratory agent that roams the network to get a report about the current memory usage of each machine. This application has been written in eight different versions for all those platforms.

12.4.3 Test Parameters

In our experiments we changed some of the application and platform parameters, namely the number of agencies, the number of laps performed by the agent, the agent size, and the use of caching and prefetching techniques. This way, we have made tests with one, three and five agencies. The number of itinerary laps performed by the agent has been changed between 1, 10, and 100. The size of additional data that was carried by the mobile agent has been set to none, 100 Kb and 1 Mb. For this application, the size of the jar file was 3.66 Kb and the size of the serialized object with no additional data was around 1 Kb.

12.4.4 Methodology of Benchmarking

All platforms have been tested in the same conditions using the same application, the same test parameters, the same agent itinerary, and the same configuration. Before every set of tests all the machines of the cluster have been rebooted for operating system rejuvenation. The agencies were restarted before each experiment, except for the case where we wanted to measure the effect of code caching. We tried to make all the tests with the agent manager running uninterrupted. Some platforms were not able to survive to some situations of stress testing and the agent manager had to be restarted when it failed. All the experiments were repeated at least four times and the standard deviation was within 5% of the average values.

12.4.5 Experimental Results

The benchmark application was executed in all eight platforms by changing all the test parameters (number of agencies, number of laps, agent data size, caching mode). We measured two main metrics: performance of the application and network traffic. Due to lack of space we will only present the most relevant results, corresponding to 12 experiments.

Experiment 1   In this experiment we measured the execution time of all eight platforms using the following parameters: 1 agency; 1 lap; data size = none; no caching. The results are presented in Figure 12.2. JAMES, Odyssey, and Swarm present the best results for this small-size agent. Jumping Beans is the slower platform: for instance, in this case it executed five times slower than JAMES.

Experiment 2   In this experiment we increased the size of the agent to 100 Kb. The results are presented in Figure 12.3, and as can be seen, they were quite similar: JAMES, Odyssey, and Swarm present the best results while Jumping Beans presented the worst results, being eight times slower than JAMES. When using the caching mechanism, the difference was even higher: Jumping Beans was 48 times slower than JAMES.

Experiment 3   In this third experiment we increased the size of the agent to 1 Mb. The results are presented in Figure 12.4. In this case, it was interesting to observe that Odyssey and JAMES presented the best results. However, the most important result was the fact that two of the platforms crashed in this test: Jumping Beans and Concordia. We tried several runs with different system parameters but apparently these two platforms were not ready for agents with large data sets.

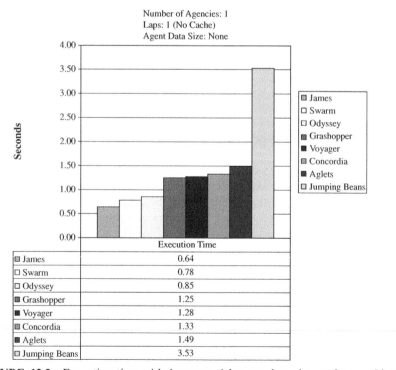

	Execution Time
James	0.64
Swarm	0.78
Odyssey	0.85
Grashopper	1.25
Voyager	1.28
Concordia	1.33
Aglets	1.49
Jumping Beans	3.53

FIGURE 12.2   Execution time with 1 agency, 1 lap, no data size, and no caching.

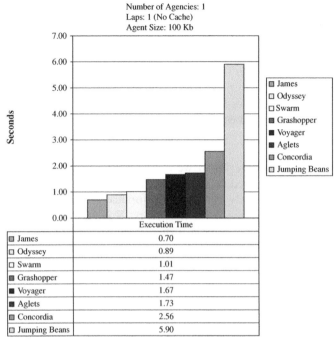

FIGURE 12.3   Execution time with 1 agency, 1 lap, data size = 100 Kb, and no caching.

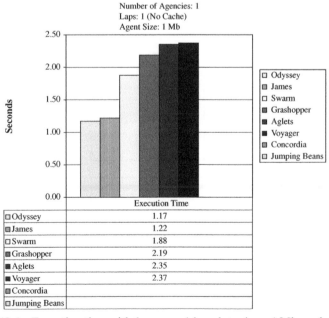

FIGURE 12.4   Execution time with 1 agency, 1 lap, data size = 1 Mb, and no caching.

Experiment 4 In this experiment we used five agencies and we only have results for seven platforms since the evaluation copy we had from Jumping Beans only executed in three platform servers. This test was done without using the caching mechanisms of the platforms and measured the impact of using the code prefetching techniques that we implemented in the JAMES platform. For more details about these mechanisms please refer to [17]. The results are presented in Figure 12.5. As can be seen, the version of JAMES that uses code prefetching achieved the best results: it was two times faster than running without prefetching and it was four times faster than the Aglets SDK.

Experiment 5 In this experiment we executed the benchmark application in five agencies of the dedicated network. We exploited the caching mechanisms of the platforms by previously running the application in those agencies. The results are presented in Figure 12.6: Odyssey, Voyager, JAMES, and Swarm were faster than the other three platforms.

Experiment 6 In this experiment we did not use caching and the size of the agent was increased by 1 Mb. With this agent size the Concordia system always

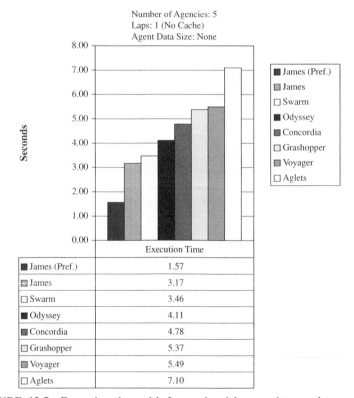

	Execution Time
■ James (Pref.)	1.57
▤ James	3.17
☐ Swarm	3.46
■ Odyssey	4.11
▨ Concordia	4.78
☐ Grashopper	5.37
▨ Voyager	5.49
☐ Aglets	7.10

FIGURE 12.5 Execution time with 5 agencies, 1 lap, no data, and no caching.

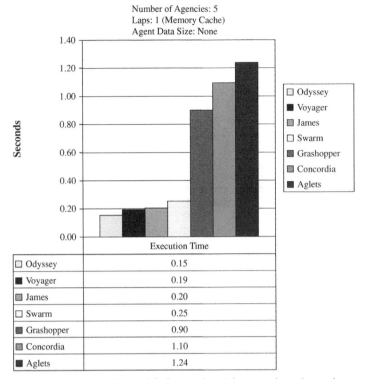

FIGURE 12.6   Execution time with 5 agencies, 1 lap, no data, but using caching.

crashed. The best results were achieved with the JAMES platform and using the code prefetching scheme. As can be seen in Figure 12.7, this version was three times faster than Aglets SDK and the Grasshopper platform.

Experiment 7   This experiment was similar to the previous one, but this time we exploited the use of memory caching by the platforms. When there is caching, the use of code prefetching makes no sense. Once again, the Concordia system was not able to execute the application with a mobile agent of (~) 1 Mb. The results are presented in Figure 12.8 and show that Odyssey and JAMES achieved the best results. In this experiment, Grasshopper was the slowest platform: six times slower than Odyssey and JAMES.

Experiment 8   This experiment departs from the previous ones: this time the agent had to execute 10 laps in the itinerary of five agencies. Increasing the number of laps allowed us to observe the behavior of the caching mechanisms and the way the platforms recycle the communication channels that are used by the mobility subsystem. We started by using a small-size agent. The results are

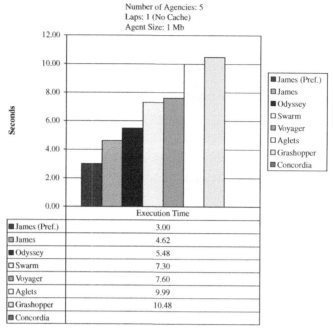

FIGURE 12.7 Execution time with 5 agencies, 1 lap, data size = 1 Mb, and no caching.

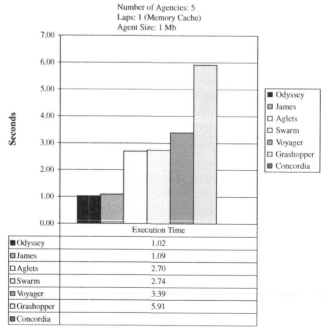

FIGURE 12.8 Execution time with 5 agencies, 1 lap, data size = 1 Mb, but using caching.

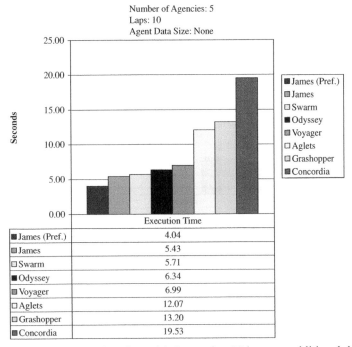

FIGURE 12.9 Execution time with 5 agencies, 10 laps, no additional data.

presented in Figure 12.9. The performance of JAMES was still the best one, although with minimal differences to Swarm, Odyssey, and Voyager. When using the prefetching scheme JAMES was five times faster than Concordia.

Experiment 9 This experiment was similar to the previous one, but this time we increased the agent size by 1 Mb. Once again the Concordia system was not able to execute the application without crashing. JAMES and Odyssey were reasonably faster than the other platforms. Once again, Grasshopper was the slowest platform. The results are shown in Figure 12.10.

Experiment 10 In this final experiment for the execution time we did some stress testing of the platforms by using a mobile agent of about 1 Mb and running it 100 laps over the itinerary of five agencies. The results are presented in Figure 12.11. Two of the platforms, Concordia and Swarm, were not able to execute this agent without crashing. Odyssey and JAMES were the fastest platforms, while Grasshopper was the slowest one. It was five times slower than the Odyssey system.

12.4.5.1 *Measuring the Network Traffic*

Experiment 11 In this experiment we measured the whole traffic in the network that is imposed by the application and the platform protocols. These

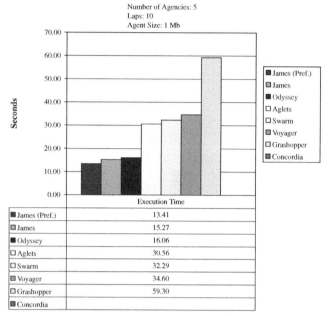

FIGURE 12.10   Execution time with 5 agencies, 10 laps, data size = 1 Mb.

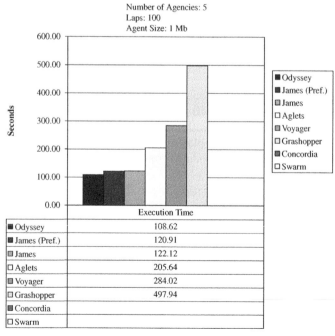

FIGURE 12.11   Execution time with 5 agencies, 100 laps, data size = 1 Mb.

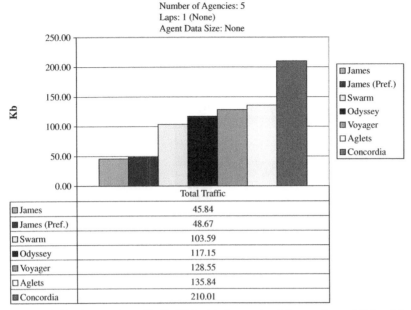

Number of Agencies: 5
Laps: 1 (None)
Agent Data Size: None

	Total Traffic
☐ James	45.84
■ James (Pref.)	48.67
☐ Swarm	103.59
■ Odyssey	117.15
▨ Voyager	128.55
☐ Aglets	135.84
▨ Concordia	210.01

FIGURE 12.12 Network traffic (in Kb) with 5 agencies, 1 lap, and no additional data.

results were collected by using a network sniffer (Sniffer Pro). This metric is useful to evaluate the degree of optimization that was introduced in the mobility subsystem and the network overhead that is introduced by the protocols. Figure 12.12 presents the first results which were taken with five agencies. The agent had no additional data and executed one lap in its itinerary without making use of caching. The network traffic is represented in Kbytes.

As can be seen, the JAMES platform introduces the smallest amount of traffic in the network when compared with the other platforms. This shows some benefits from the optimizations we have in the platform protocols. The version of JAMES that used code prefetching imposed more traffic than the other version due to the additional messages that are necessary to implement that scheme. Concordia was the platform that introduced more traffic in the network.

Experiment 12 In this second experiment we used five agencies, a small-size agent that runs one lap of its itinerary. However, we activated the use of caching in all the platforms to reduce some of the network traffic due to the distribution of code. The results are shown in Figure 12.13. In this situation, JAMES was still the platform that introduced the smallest amount of traffic in the network. Voyager and Swarm presented similar results. Odyssey, Aglets, and Concordia were the platforms that introduced more traffic.

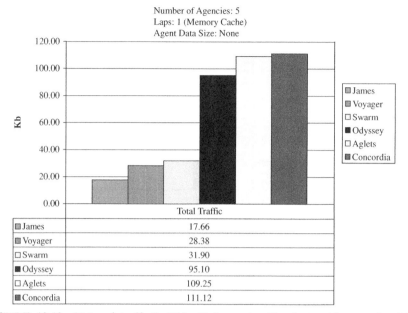

Number of Agencies: 5
Laps: 1 (Memory Cache)
Agent Data Size: None

	Total Traffic
▨ James	17.66
▧ Voyager	28.38
□ Swarm	31.90
■ Odyssey	95.10
□ Aglets	109.25
▩ Concordia	111.12

FIGURE 12.13   Network traffic (in Kb) with 5 agencies, 1 lap, but making use of caching.

12.4.6   Concluding Overview about the Platforms

In this section, we present some discussions about the experience we had with these eight agent platforms.

12.4.6.1   Aglets

Aglets SDK is probably the most famous platform of mobile agents. The results show that it is quite a robust platform and has passed all the tests without crashing. The performance is not so good when compared with other platforms. For instance, the JAMES platform is two or fourteen times faster than Aglets, depending on the test cases. The caching mechanisms seem to be not so efficient. We have done some profiling experiments where we detected there is some garbage left in the memory of the agencies. This memory leak can lead to a deterioration of the performance of the application over time. The network traffic is also a weak point of this platform.

12.4.6.2   Concordia

Concordia is another well-known platform. Unfortunately, the results show that this platform is not very robust in situations of stress testing. We could not run the benchmark with a big size agent (~1 Mb). Although the agency did not hang up the first time the GUI interface crashed every time we tried to create a second agent of that size. The garbage collection within the platform is also not done in an appropriate way and there is a big deterioration in the execution of agents when we perform some consecutive experiments. Performance is another weak point of Concordia, as can be seen by the results presented. This platform is also the one that generated more network traffic.

12.4.6.3 Voyager

Voyager is a commercial platform with hundreds of users. The performance results are not brilliant, and we can say this platform is in the middle of the table. However, there are issues related to lack of robustness of this platform. Sometimes we got memory errors and the platform crashed completely. There were some test cases that could not be done at any time. It is important to notice that this situation happened with big size agents ($\sim$1 Mb) but also with small-size agents. The platform produces a big amount of network traffic when using no cache, but it significantly reduces the traffic when we activated the use of a memory cache.

12.4.6.4 Odyssey

Odyssey is the Java-based successor of Telescript. The results have shown that this platform is very robust: it did not crash in any test we made. The performance is also very good and it presented the best execution times, together with JAMES. The only drawbacks we found was some lack of functionality and the absence of graphical interface for the management of the application and the launching of mobile agents.

12.4.6.5 Jumping Beans

The evaluation copy we had from Jumping Beans only allows execution with three machines. The number of tests we could perform was therefore quite limited. However, those tests were enough to conclude that this platform has a really poor performance. In some cases, it was 40 times slower than the other platforms. The reason for this poor performance is simple: every time a mobile agent wants to migrate from machine A to machine B it has to go first to the agent manager. This manager is a point of bottleneck and the platform is not scalable. The platform is also not very robust in situations of stress testing: for instance, it was not possible to execute the big size agent ($\sim$1 Mb) without giving any memory errors.

12.4.6.6 Grasshopper

This platform has a very user-friendly graphical interface and a comprehensive set of features. In fact, this is the platform that presents the higher functionality. However, the performance of Grasshopper is not very good: it was two to five times slower than the JAMES platform. The robustness is also not as good as we expected: the platform crashed several times for big-size agents ($\sim$1 Mb). Upon the occurrence of a memory error the platform had to be fully restarted. The GUI also has a few bugs that result in crashes of the interface.

12.4.6.7 *Swarm*

This platform presented some problems of stability, although it can have some good performance results. The platform seems to open a channel between all the agencies. If some of these channels are not well established in the beginning of the execution, the agent's migration cannot be done properly and the application hangs up. This situation has happened several times, showing that there are some problems to be solved in this system. The GUI interface is a bit confusing and crashes periodically. The platform also had some problems when using big-size agents: it crashed very often and had to be completely rebooted. The performance was good for the majority of the test cases, being better for small-size agents. The network traffic generated by this platform was also one of the smallest among the other platforms.

12.4.6.8 *JAMES*

The JAMES platform was devised and implemented with performance and robustness in minds. Several mechanisms have been introduced to optimize the migration of mobile agents and it seems those techniques have introduced clear benefits. In most of the test cases, JAMES was the platform with the best level of performance, and it presented a very good level of robustness. The resource control mechanisms have been quite useful to increase the stability of the system and the applications. This platform has been used by Siemens to develop some commercial applications in the field of telecommunications.

While this section presents some benchmark results about a full set of mobile agent platforms the reader might be interested in knowing if this mobile agent paradigm has benefits in terms of performance when compared with traditional client-server solutions.

12.5 COMPARING AGENTS WITH RMI, CORBA AND SERVLETS

In this section we present the results of another experimental study. This time we compared the performance of mobile agents running in the JAMES platform against similar versions of the application that were implemented in Java RMI, Java Servlets, and Corba IIOP.

12.5.1 Test Environment

In these experiments we used the same dedicated cluster of six machines connected through a 10 Mb/sec switched Ethernet. All the results were taken when the machines were fully dedicated. Every machine has a Pentium II (300 Mhz) processor and 128 Mb of RAM. These machines were running Microsoft Windows NT 4.0 and we used JDK 1.1.7 with the JIT option. We used the version 1.0.3 of the JAMES platform. For the CORBA version of the application we used Visibroker (version 3.1). The servlet version was implemented with the JavaWebServer (version 1.1).

12.5.2 Application Benchmark

In this performance study we used a different synthetic application: this time the goal of the application was a simple string search in a set of data files. The data files were located in the servers of the network and the goal was to scan every file to determine the number of occurrences of the target string.

12.5.3 Five Different Versions of the Application

This application was implemented in five different versions: (i) a version with a migratory agent that visits all the nodes in the network to scan the files in the different machines; (ii) a second version with mobile agents, but this time the client created a set of N agents that are sent to the N machines and will conduct the search in parallel; (iii) a third version that was implemented in Java RMI; (iv) a fourth version that was implemented in Corba IIOP (Visibroker ORB); (v) and a fifth version that was implemented in Servlets and communication was done by HTTP.

The versions implemented in RMI, Servlets, and Corba just follow the typical client-server model: the client issues a remote invocation in each server, asking to scan the data file searching the target string. The two other versions made use of the JAMES platform: one made use of a single agent that was sent to the network with a predefined itinerary. In this itinerary the migratory agent should visit one server at a time, scan the local time, store the partial results, and then migrate to the next server. At the end of the itinerary the agent will be sent back to the client requester with the aggregate of the partial results collected. This migratory agent was presented in Figure 12.1.

The second version with agents used parallel and asynchronous execution: when the client application wants to scan the data files in N servers, it creates N cloned agents that will be sent to each server and will work in parallel and bring back the results to the client. This is the popular master-worker model, highly used in parallel computing. Figure 12.14 represents this execution model.

While it would be possible to also execute remote invocations in parallel, when using Java RMI, Servlets, and Corba, we did not use this optimization. It would require a threaded model on the client side and some synchronization mechanisms, not actually used by programmers when they deploy client-server applications in these technologies.

With mobile agents, the creation of several agents to work in parallel is a very straightforward task, and this is why we tried this master-worker version to show some of the potentials of this computing paradigm.

12.5.4 Test Parameters

In our experiments we have taken results with different conditions for the execution: we tried the agent's version with run time downloading of the JAR file the first time it is executed and when the JAR file was already prefetched to

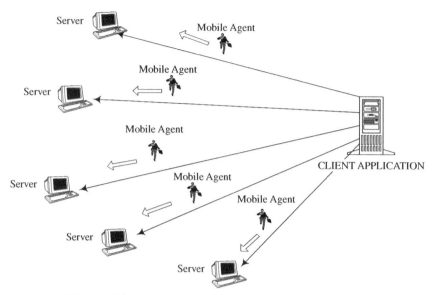

FIGURE 12.14 Master-worker version with mobile agents.

the target servers. We collected results with different servers (1 and 5) and different searches per execution (1 and 5). Finally, we ran the application with different data files in the target servers: 1 Mb, 10 Mb, 100 Mb and 1 Gb.

12.5.5 Methodology of the Performance Study

The five different versions of the application were tested in the same conditions of the network, the same data files, and the same target searches. Before every set of tests all the machines of the cluster were rebooted for operating system rejuvenation. The agencies were restarted before each experiment, except for the case where we wanted to measure the effect of code caching. All the experiments were repeated at least four times and the standard deviation was within 5% of the average values.

12.5.6 Experimental Results

Experiment 1 In the first experiment we measured the total execution time of the five different versions while considering two different scenarios: (a) running with just one server; (b) running with five servers. Results are presented in Figure 12.15.

In this experiment the client only issued one search request and the data file in each server had 1 Mbytes. As can be seen in Figure 12.15, with only one server the Servlet version was the fastest one, which is per se an interesting result. The migratory agent did not pay off the cost of code transfer. When we

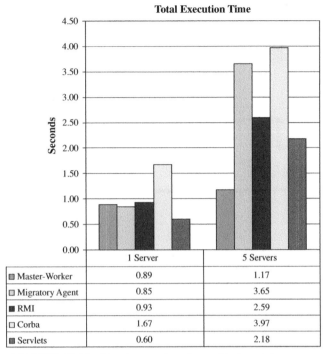

Total Execution Time

	1 Server	5 Servers
▨ Master-Worker	0.89	1.17
☐ Migratory Agent	0.85	3.65
▪ RMI	0.93	2.59
☐ Corba	1.67	3.97
▨ Servlets	0.60	2.18

FIGURE 12.15   Total execution time with 1 and 5 servers.

increased to five servers, the fastest version was the one with mobile agents running in parallel: the master-worker version. This is a quite expected result since the application has work to do in parallel and the usage of concurrent agents brought some clear benefits in terms of performance. The master-worker version with agents achieved a speedup of 3.4 when compared with the Corba version and a speedup of 2.2 when compared with the RMI version. It is also interesting to see that the simple migratory agent took more time to execute than the versions with RMI and Java Servlet.

Experiment 2   In this second experiment the client application issued one and five search requests. Results are presented in Figure 12.16. These results were taken with five servers, and the local data files had 1 Mb. When issuing five different searches to the five servers, the master-worker version with the parallel mobile agents achieved expectably the best results, being 2 times faster than the client-server versions (RMI, Corba and Servlets). Curiously, the single migratory agent got the worst results.

Experiment 3   In the third experiment we changed the size of the local data files: 1 Mbytes, 10 Mbytes, 100 Mbytes and 1 Gbyte. This means we are increasing the computational time in the scanning process of the data files.

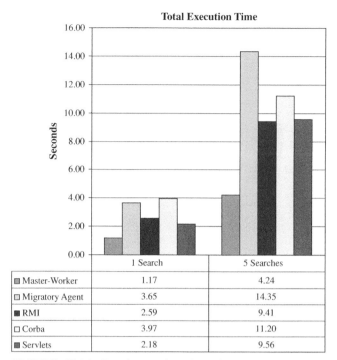

Total Execution Time

	1 Search	5 Searches
■ Master-Worker	1.17	4.24
□ Migratory Agent	3.65	14.35
■ RMI	2.59	9.41
□ Corba	3.97	11.20
■ Servlets	2.18	9.56

FIGURE 12.16   Total execution time with 1 and 5 searches.

The results are presented in Figure 12.17. They were collected with five servers and one search request. When the data files had a size of 1 Gbyte there was almost no perceptible difference between the version with the single migratory agent and the client-server versions. The master-worker version with five running agents did remarkably well: it achieved a speedup of almost five. When the data files got bigger this was the case where the potential for parallel computing can be exploited by sending several agents to the network and exploiting the asynchronous and parallel execution.

12.6   FINAL REMARKS

In this chapter, we presented a rich set of experimental results: in section 12.4 we presented a benchmarking study that compared the performance of eight mobile-agent platforms: James, Swarm, Odyssey, Grasshopper, Voyager, Concordia, Aglets and Jumping Beans. The conclusions of that study might be relevant to the reader who is looking for a platform and wants to get more insight about the differences in performance.

In section 12.5 we presented a performance study comparing mobile agents with traditional client-server solutions. It was clear that in those experiments,

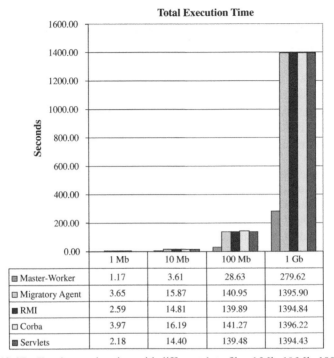

Total Execution Time

	1 Mb	10 Mb	100 Mb	1 Gb
▨ Master-Worker	1.17	3.61	28.63	279.62
▢ Migratory Agent	3.65	15.87	140.95	1395.90
■ RMI	2.59	14.81	139.89	1394.84
▢ Corba	3.97	16.19	141.27	1396.22
▨ Servlets	2.18	14.40	139.48	1394.43

FIGURE 12.17 Total execution time with different data-files: 1 Mb, 10 Mb, 100 Mb, 1 Gb.

the single migratory agent has almost no advantage against solid technologies like Java RMI, Java Servlets, and Corba IIOP.

We were only able to observe some gains in performance when the application had some work to do in parallel. In this case, the client created a set of agents that were sent to the network, do the work in parallel, and bring back the outcome. The results are encouraging and we were able to obtain almost linear speedups. While this study presents some results taken some time ago, the relative conclusions might still be valid. It remains to be seen how the adoption of mobile-agents will be in production code.

REFERENCES

1. V. Pham and A. Karmouch, Mobile software agents: An overview, *IEEE Commun. Mag.*, July 1998, pp. 26–37.

2. C. G. Harrison, Mobile agents: Are they a good idea?, Technical Report, IBM T. J. Watson Research Center, RC 19887, Mar. 1995.

3. D. Lange and M. Oshima, Seven good reasons for mobile agents, *Commun. ACM*, 42(3):88–89, Mar. 1999.

4. IBM Aglets Workbench, available: http://www.trl.ibm.co.jp/aglets/.

5. Concordia Platform, available: http://www.meitca.com/HLS/Projects/Concordia/.

6. Voyager System, available: http://www.objectspace.com/products/voyager/.

7. General Magic Odyssey, available: http://www.genmagic.com/technology/odyssey.html.

8. Jumping Beans Platform, available: http://www.JumpingBeans.com.

9. Kafka, available: http://www.fujitsu.co.jp/.

10. Grasshopper Platform, available: http://www.ikv.de/products/grasshopper.

11. L. M. Silva, G. Soares, P. Martins, V. Batista, C. Renato, L. Almeida, and N. Stohr, JAMES: A platform of mobile agents for the management of telecommunication networks, in S. Albayrak (Ed.), *Proceedings of Intelligent Agents for Telecommunication Applications: 3rd International Workshop, IATA'99*, Lecture Notes in Artificial Intelligence, Springer, Stockholm, Sweden, Aug. 1999.

12. J. Kiniry and D. Zimmerman, A hands-on look at Java mobile agents, *IEEE Internet Comput.*, July–Aug. 1997, pp. 21–30.

13. T. Ugai, and M. Bursell, Comparison of autonomous mobile agent technologies, Internal Report, FollowMe Project, APM, Cambridge, Oct. 1997.

14. M. Corkery, A review of state of the art in mobile agent systems, Technical Report, Dept. Computer Science, National University of Ireland, Maynooth, Ireland, 1998.

15. A. Guther and M. Zell, Platform enhancement requirements, Internal Report, Project MIAMI (ACTS Program AC338), 1998, available: http://www.fokus.gmd.de/research/cc/ima/Miami.

16. E. Kovacs, K. Rohrle, and M. Reich, Integrating mobile agents into the mobile middleware, in K. Rothermel and F. Hohl (Eds.), *Proceedings of the 2nd International Workshop, MA'98*, Lecture Notes in Computer Science, Stuttgart, Germany, Sept. 1998, pp. 124–135.

17. G. Soares, and L. M. Silva, Optimizing the migration of mobile agents, in A. Karmouch and R. Impley (Eds.), *Proceedings of the 1st International Workshop on Mobile Agents for Telecommunication Applications (MATA'99)*, World Scientific Publishing Ltd, Ottawa, Canada, Oct. 1999.

INDEX

Mobile Agents in Networking and Distributed Computing, First Edition.
Edited by Jiannong Cao and Sajal K. Das.
© 2012 John Wiley & Sons, Inc. Published 2012 by John Wiley & Sons, Inc.

Printed and bound by CPI Group (UK) Ltd, Croydon, CR0 4YY

27/10/2024

14580330-0002